Statistical Methods for Meta-Analysis

Statistical Methods for Meta-Analysis

Larry V. Hedges

Department of Education
University of Chicago
Chicago, Illinois

Ingram Olkin

Department of Statistics
and School of Education
Stanford University
Stanford, California

ACADEMIC PRESS, INC.
Harcourt Brace Jovanovich, Publishers
San Diego New York Berkeley Boston
London Sydney Tokyo Toronto

ACADEMIC PRESS, INC.
San Diego, California 92101

United Kingdom Edition published by
ACADEMIC PRESS LIMITED
24-28 Oval Road, London NW1 7DX

LIBRARY OF CONGRESS CATALOGING IN PUBLICATION DATA

Hedges, Larry V.
Statistical methods for meta-analysis.

Includes index.
1. Social sciences—Statistical methods. I. Olkin,
Ingram. II. Title. III. Title: Meta-analysis.
HA29.H425 1985 300'.72 84-12469
ISBN 0-12-336380-2 (alk. paper)
ISBN 0-12-336381-0 (paper)

PRINTED IN THE UNITED STATES OF AMERICA

89 90 91 92 93 9 8 7 6 5 4 3

To the memory of Alfred and Aillene Hedges

To Julia Ann Olkin

Contents

Preface xv

1. Introduction

A. The Use of Statistical Procedures for Combining the Results of Research Studies in the Social Sciences 2
A.1 The Misuse of Statistical Significance in Reviewing Research 3
A.2 Statistical Procedures for Combining Estimates of Effect Magnitude 6
B. Failings in Conventional Statistical Methodology in Research Synthesis 7
B.1 Goals of Statistical Procedures in Research Synthesis 7
B.2 Conventional Analyses for Effect Size Data 9
B.3 Conceptual Problems with Conventional Analyses 10
B.4 Statistical Problems with Conventional Analyses 11
C. Statistics for Research Synthesis 12

2. Data Sets

A. Cognitive Gender Differences 16
B. Sex Differences in Conformity 21
C. The Effects of Open Education 23

D. The Relationship between Teacher Indirectness and Student Achievement 26

3. **Tests of Statistical Significance of Combined Results**

A. Preliminaries and Notation 28
B. General Results on Tests of Significance of Combined Results 31
C. Combined Test Procedures 33
C.1 Methods Based on the Uniform Distribution 34
C.2 The Inverse Chi-Square Method 37
C.3 The Inverse Normal Method 39
C.4 The Logit Method 40
C.5 Other Procedures for Combining Tests 42
C.6 Summary of Combined Test Procedures 43
D. The Uses and Interpretation of Combined Test Procedures in Research Synthesis 45
E. Technical Commentary 46

4. **Vote-Counting Methods**

A. The Inadequacy of Conventional Vote-Counting Methodology 48
B. Counting Estimators of Continuous Parameters 52
C. Confidence Intervals for Parameters Based on Vote Counts 53
C.1 Use of Normal Theory Approach 54
C.2 Use of Chi-Square Theory 54
D. Choosing a Critical Value 56
E. Estimating an Effect Size 56
F. Estimating a Correlation 63
G. Limitations of Vote-Counting Estimators 67
H. Vote-Counting Methods for Unequal Sample Sizes 69
H.1 The Large Sample Variance of the Maximum Likelihood Estimator 70
H.2 Estimating Effect Size 71

5. **Estimation of a Single Effect Size: Parametric and Nonparametric Methods**

A. Estimation of Effect Size from a Single Experiment 76
A.1 Interpreting Effect Sizes 76
A.2 An Estimator of Effect Size Based on the Standard Mean Difference 78

A.3 An Unbiased Estimator of Effect Size 81
A.4 The Maximum Likelihood Estimator of Effect Size 81
A.5 Shrunken Estimators of Effect Size 82
A.6 Comparing Parametric Estimators of Effect Size 82
B. Distribution Theory and Confidence Intervals for Effect Sizes 85
B.1 The Asymptotic Distribution of Estimators of Effect Size 85
B.2 Confidence Intervals for Effect Sizes Based on Transformations 88
B.3 Exact Confidence Intervals for Effect Sizes 91
C. Robust and Nonparametric Estimation of Effect Size 92
C.1 Estimates of Effect Size that are Robust Against Outliers 93
C.2 Nonparametric Estimators of Effect Size 93
C.3 Estimators Based on Differences of Control versus Treatment Proportions 95
C.4 Estimators Based on Gain Scores in the Experimental Group Relative to the Control Group 97
C.5 Nonparametric Estimators Involving Only Posttest Scores 98
C.6 Relationships between Estimators 99
D. Other Measures of Effect Magnitude 100
D.1 The Correlation Coefficient and Correlation Ratio 101
D.2 The Intraclass Correlation Coefficient 102
D.3 The Omega-Squared Index 103
D.4 Problems with Variance-Accounted-For Measures 103
E. Technical Commentary 104

6. Parametric Estimation of Effect Size From a Series of Experiments

A. Model and Notation 108
B. Weighted Linear Combinations of Estimates 109
B.1 Estimating Weights 110
B.2 Efficiency of Weighted Estimators 113
B.3 The Accuracy of the Large Sample Approximation to the Distribution of Weighted Estimators of Effect Size 114
C. Other Methods of Estimation of Effect Size from a Series of Experiments 117
C.1 The Maximum Likelihood Estimator of Effect Size from a Series of Experiments 118

C.2 Estimators of Effect Size Based on Transformed Estimates 119
D. Testing for Homogeneity of Effect Sizes 122
D.1 Small Sample Significance Levels for the Homogeneity Test Statistic 124
D.2 Other Procedures for Testing Homogeneity of Effect Sizes 124
E. Computation of Homogeneity Test Statistics 127
F. Estimation of Effect Size for Small Sample Sizes 128
F.1 Estimation of Effect Size from a Linear Combination of Estimates 129
F.2 Modified Maximum Likelihood Estimation of Effect Size 131
G. The Effects of Measurement Error and Invalidity 131
G.1 The Effects of Measurement Error 132
G.2 The Effect of Validity of Response Measures 138

7. Fitting Parametric Fixed Effect Models to Effect Sizes: Categorical Models

A. An Analogue to the Analysis of Variance for Effect Sizes 149
B. Model and Notation 149
C. Some Tests of Homogeneity 153
C.1 Testing Whether All Studies Share a Common Effect Size 153
C.2 Testing Homogeneity of Effect Sizes Across Classes 154
C.3 Testing Homogeneity within Classes 155
C.4 An Analogy to the Analysis of Variance 156
C.5 Small Sample Accuracy of the Asymptotic Distributions of the Test Statistics 156
D. Fitting Effect Size Models to a Series of Studies 157
E. Comparisons among Classes 159
E.1 Simultaneous Tests for Many Comparisons 160
F. Computational Formulas for Weighted Means and Homogeneity Statistics 163

8. Fitting Parametric Fixed Effect Models to Effect Sizes: General Linear Models

A. Model and Notation 168
B. A Weighted Least Squares Estimator of Regression Coefficients 169

C. Testing Model Specification 172
D. Computation of Estimates and Test Statistics 173
E. The Accuracy of Large Sample Approximations 174
F. Other Methods of Estimating Regression Coefficients 183
F.1 Maximum Likelihood Estimators of Regression Coefficients 183
F.2 Estimators Based on Transformations of Sample Effect Sizes 184
G. Technical Commentary 187

9. Random Effects Models for Effect Sizes

A. Model and Notation 191
B. The Variance of Estimates of Effect Size 193
C. Estimating the Effect Size Variance Component 193
D. The Variance of the Effect Size Variance Component 195
E. Testing That the Effect Size Variance Component Is Zero 197
F. Estimating the Mean Effect Size 198
G. Empirical Bayes Estimation for Random Effects Models 200

10. Multivariate Models for Effect Sizes

A. Model and Notation 206
B. The Multivariate Distribution of Effect Sizes 209
C. Estimating a Common Effect Size from a Vector of Correlated Estimates 210
C.1 Testing Homogeneity of Correlated Effect Sizes 210
C.2 Estimation of Effect Size From Correlated Estimates 212
D. Estimating a Common Effect Size and Testing for Homogeneity of Effect Sizes 213
D.1 An Estimator of Effect Size 213
D.2 Testing Homogeneity of Effect Sizes 215
E. Estimating a Vector of Effect Sizes 216
F. Testing Homogeneity of Vectors of Effect Sizes 219
G. Is Pooling of Correlated Estimators Necessary? 221

11. Combining Estimates of Correlation Coefficients

A. Estimating a Correlation from a Single Study 224
A.1 Point Estimation of a Correlation from a Single Study 224
A.2 Approximations to the Distribution of the Sample Correlation Coefficient 226

A.3 Exact Confidence Intervals for Correlations 228
B. Effects of Measurement Error 228
C. Estimating a Common Correlation from Several Studies 229
C.1 Weighted Estimators of a Common Correlation 230
C.2 The Maximum Likelihood Estimator of a Common Correlation 232
D. Testing Homogeneity of Correlations across Studies 234
D.1 A Test of Homogeneity Based on Fisher's z-Transform 235
D.2 The Likelihood Ratio Test of Homogeneity of Correlations 236
E. Fitting General Linear Models to Correlations 237
E.1 Model and Notation 237
E.2 Estimating Regression Coefficients 238
E.3 Testing Model Specification 240
E.4 Computation of Estimates and Test Statistics 241
F. Random Effects Models for Correlations 242
F.1 Model and Notation 242
F.2 Testing That the Variance of Population Correlations is Zero 243
F.3 An Unbiased Estimate of the Correlation Variance Component 244

12. Diagnostic Procedures for Research Synthesis Models

A. How Many Observations Should Be Set Aside 249
B. Diagnostic Procedures for Homogeneous Effect Size Models 251
B.1 Graphic Method for Identifying Outliers 251
B.2 The Use of Residuals to Locate Outliers 253
B.3 The Use of Homogeneity Statistics to Locate Outliers 256
C. Diagnostic Procedures for Categorical Models 257
D. Diagnostic Procedures for General Linear Models 257
D.1 The Use of Residuals to Locate Outliers in General Linear Models 258
D.2 Changes in Regression Coefficients 260
D.3 The Use of Model Specification Statistics to Locate Outliers 261
E. Diagnostic Procedures for Combining Estimates of Correlation Coefficients 262
F. Technical Commentary 262

13. Clustering Estimates of Effect Magnitude

A. Theory for Clustering Unit Normal Random Variables 266
A.1 Disjoint Clustering 267
A.2 Overlapping Clustering 269
B. Clustering Correlation Coefficients 271
C. Clustering Effect Sizes 273
D. The Effect of Unequal Sample Sizes 276
D.1 The Effect on the Disjoint Clustering Procedure 277
D.2 The Effect on the Overlapping Clustering Procedure 278
D.3 An Alternative Method for Handling Unequal Sample Sizes 279
E. Relative Merits of the Clustering Procedures 281
F. Computation of Tables 282
F.1 The Significance of Gaps 282
F.2 Significance Values for the Overlapping Clustering Procedure 283

14. Estimation of Effect Size When Not All Study Outcomes Are Observed

A. The Existence of Sampling Bias in Observed Effect Size Estimates 286
B. Consequences of Observing Only Significant Effect Sizes 287
B.1 Model and Notation 288
B.2 The Distribution of the Observed Effect Size 288
C. Estimation of Effect Size from a Single Study When Only Significant Results Are Observed 290
C.1 Estimation of Effect Size 291
C.2 The Distribution of the Maximum Likelihood Estimator 292
D. Estimation of Effect Size from a Series of Independent Experiments When Only Significant Results are Observed 297
D.1 Estimation of Effect Size Using Counting Procedures 298
D.2 The Maximum Likelihood Estimator of Effect Size 301
D.3 Weighted Estimators of Effect Size 302
D.4 Applications of the Methods That Assume Censoring 303
E. Other Methods for Estimating Effect Sizes When Not All Study Outcomes Are Observed 304
E.1 Assessing the Number of Studies (with Null Results) Needed to Overturn a Conclusion 305
E.2 Random Effects Models When Not All Study Outcomes Are Observed 306

F. Technical Commentary 307

15. Meta-Analysis in the Physical and Biological Sciences

A. A Multiplicative Model 314
B. Estimating Displacement and Potency Factor 318
C. A Multiplicative Model with Scaling 321
D. An Additive Model for Interlaboratory Differences 322
E. An Additive Model With a Control 323

Appendix

A. Table of the Standard Normal Cumulative Distribution Function 328
B. Percentiles of Chi-Square Distributions 330
C. Values of $z = \frac{1}{2}\log[(1+r)/(1-r)]$ and $r = (e^{2z}-1)/(e^{2z}+1)$ and Graph of Fisher's z-Transformation 333
D. Values of $\sqrt{2}\sinh^{-1} x$ and $\sinh(x/\sqrt{2})$ 335
E. Confidence Intervals of the Parameter p of the Binomial Distribution 337
F. Nomographs for Exact Confidence Intervals for Effect Size δ When $2 \leqslant n \leqslant 10$ 341
G. Confidence Intervals for the Correlation Coefficient for Different Sample Sizes 342

References 347

Index 361

Preface

Methodology for combining findings from repeated research studies has a long history. Early examples of combining evidence are found in replicated astronomical and physical measurements. Agricultural experiments particularly lend themselves to replication and led to the development of statistical techniques for merging results.

In recent years a plethora of meta-analyses have emerged in social science research. The need to arrive at policy decisions affecting social institutions fostered the momentum toward summarizing research. But, as with most methodologies, abuse frequently accompanies use. Two central aspects of meta-analysis were quickly recognized. One involved methods for collecting the body of information to be summarized. This pinpointed a variety of problems, pitfalls, and questions. For example, what steps should be taken to guarantee objectivity? Should some studies be omitted because of inadequacies in design or execution?

The second aspect of meta-analysis assumes as a starting point that we have available a set of reasonably well-designed studies that address the same question using similar outcome measures and focuses on the methodology needed for summarizing the data. Because classical statistics primarily addresses the analysis of single experiments, new formulations, models, and methods are required.

The main purpose of this book is to address the statistical issues for integrating independent studies. There exist a number of papers and books that discuss the mechanics of collecting, coding, and preparing data for a meta-analysis, and we do not deal with these.

It is not unusual in the early development of a field for terms to be used that later may be less than adequate or more restrictive than need be. In particular, the term *effect size* has been used to refer to standardized mean differences. In the beginning this usage was very natural in that the particular studies of interest did indeed involve differences between means. However, with more elaborate experimentation and more diverse applications, differences between treatments may depend not only on means but also on variances, medians, correlations, order statistics, distances, etc. Thus it would behoove us to now use the term *effect size* to refer to any such indices. But because this would be contrary to much of the existing literature, we do not, somewhat regrettably, do so. Instead we introduce the term *effect magnitude* to refer to measures in general.

The problem is further compounded. For large samples, quantities such as variances, medians, correlations, etc., will frequently have a normal distribution in which some of the population parameters will indeed be means. Consequently, although we may begin with statistics that are not means, we often end up with statistics that are effect sizes in the original sense.

Because this book concerns methodology, the content necessarily is statistical, and at times mathematical. In order to make the material accessible to a wider audience, we have not provided proofs in the text. Where proofs are given, they are placed as commentary at the end of a chapter. These can be omitted at the discretion of the reader.

We make a number of technical statements such as ''The statistic Q has a chi-square distribution with degrees of freedom,'' or ''The statistics Q_1 and Q_2 are independent.'' Each of these statements warrants a proof or a reference. However, for the sake of simplicity, readability, and accessibility to the materials, we have taken the liberty of omitting many proofs and references. However, as a compromise, we include a few proofs and references for statements that might be considered typical.

At times, mathematical expressions are needed. When possible we provide tables and graphs to make these expressions simple to use.

In our writing we have in mind a prototypical reader who is familiar with basic statistics at an applied level. This normally means the completion of a one-year sequence in statistics at a noncalculus level, and includes the ideas of statistical inference, regression and correlation, and analysis of variance. Concepts such as distribution (cumulative distribution function), expected value, bias, variance, and mean-squared error are generally defined in standard introductory statistics textbooks and are not defined in this book.

Occasionally, we use more advanced statistical concepts, such as consistency, efficiency, invariance, or asymptotic distributions. Because these concepts are used infrequently, we do not give formal definitions and on occasion give only a brief explanation or no explanation at all. Elementary expositions of these concepts can be found in more advanced statistics books. Our main reason for not discussing these concepts is that they are not essential for an understanding of the principles. A lengthy explanation would destroy readability. On the other hand, these comments can be useful for those readers familiar with the concepts.

Throughout the book we describe computational procedures whenever required. Many computations can be completed on a hand calculator, whereas some require the use of a standard statistical package such as SAS, SPSS, or BMD. Readers with experience using a statistical package or who conduct analyses such as multiple regression or analysis of variance should be able to carry out the analyses described with the aid of a statistical package.

Because of the inclusion of so many tables, a commentary on interpolation may be in order. For any two-way table (see figure), the simplest method of interpolation is linear in each direction; and in general, linear interpolation will suffice for most practical purposes. For more accurate interpolation, the values can be plotted horizontally or vertically. Thus, for example, vertical plots give interpolated values between c_1 and d_1 and between c_2 and d_2, denoted by crosses. Similarly, horizontal plots given interpolated values between c_1 and c_2 and between d_1 and d_2, denoted by dots. Subsequent linear interpolation in the other direction will provide a more accurate result than two-way linear interpolation. Of course, plotting a complete row or column of interpolated values gives still more accurate results, but this may be more effort than is warranted in practice.

A comment is in order concerning the calculations presented in the examples. Individual terms are presented after rounding, whereas totals are computed without roundoff. Consequently, discrepancies in the last decimal may exist in some of the computations.

There is an inherent difficulty in trying to use a single letter to denote a particular characteristic. For example, if we denote a population mean by μ, then how shall we denote the mean of the means of an experiment and control group? If we label the two means as μ^E and μ^C and $\mu = \frac{1}{2}(\mu^E + \mu^C)$, then we have denoted by μ two different types of means. Such inconsistencies are inherent in the subject and occur throughout the book, so it is important to make quite explicit the underlying context and thereby remove potential confusion. Because

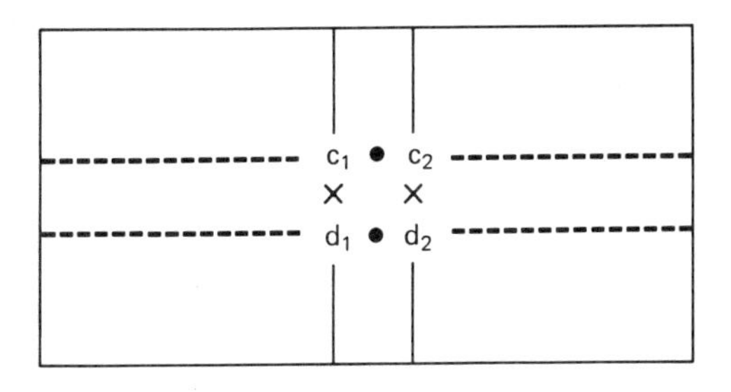

of this, we at times belabor the structure of the data and the populations from which they are drawn.

Several symbols and notations are used throughout. Population parameters are generally denoted by Greek letters, e.g., δ, σ, ρ, and their estimates are denoted by the corresponding Latin letters d, s, r, or by hats, $\hat{\delta}$, $\hat{\sigma}$, $\hat{\rho}$. The expectation (mean) and variance of an estimator $\hat{\delta}$, say, are denoted $E(\hat{\delta})$ and either $\mathrm{Var}(\hat{\delta})$ or $\sigma^2(\hat{\delta})$.

When several estimators are provided for a single parameter δ, say, we may need more symbols than $\hat{\delta}$ and d. In that case we use other symbols, such as, $\tilde{\delta}$, or other letters, such as g, $\tilde{g}$.

Normally we use the symbol $\hat{\delta}$ to denote the maximum likelihood estimator of δ. An important fact about maximum likelihood estimation is that the maximum likelihood estimator of a monotone function of a parameter $g(\delta)$ is $g(\hat{\delta})$. This provides a simple mechanism to estimate transformed parameters. Unfortunately, this phenomenon does not hold for unbiased estimators, that is, if $\hat{\delta}$ is an unbiased estimator of δ, $g(\hat{\delta})$ is not an unbiased estimator of $g(\delta)$.

The expression $\hat{\delta} \sim N(\delta, \tau^2)$ means that the distribution of the estimator is normal with mean δ and variance τ^2. By the same token, $(\hat{\delta} - \delta)/\tau \sim N(0,1)$ generally means that the distribution of the expression $(\hat{\delta} - \delta)/\tau$ is standard normal.

In order to carry out a statistical test or to obtain a confidence interval, we have to know the distribution of the statistic being used. In standard examples this distribution is usually known. However, when the models become complicated the distributions required are generally unknown. One way of dealing with this problem is to simulate the distribution using a computer. This can still be difficult, and at the very least expensive. Another alternative is to use large sample theory. It turns out that large sample results lead to normal distributions, which can be handled directly. The deficiency with large sample approximations is that they might be poor for small samples. We do use large sample results throughout, and have carried out some simulations that show that the results remain valid for relatively small samples.

When we write $\hat{\tau} \sim N(\tau, \sigma^2_\infty(\tau))$ we mean that the large sample distribution of $\hat{\tau}$ is approximately normal with a mean of τ and an asymptotic variance $\sigma^2_\infty(\tau)$. This variance may be a constant independent of any parameters, or it may be a function of parameters. In the latter case, an estimate of $\sigma^2_\infty(\tau)$ is denoted by $\hat{\sigma}^2_\infty(\hat{\tau})$. Although it would be more accurate to distinguish between an asymptotic variance σ^2_∞ and an exact variance σ^2, the notation is more cumbersome than is warranted. For simplicity, we omit the infinity symbol. However, the context will generally clarify whether the variance is exact or approximate.

Analysis of variance dot notation as in $d_{i.}$, $d_{.j}$ or $d_{..}$ is quite standard. Because the dot is not always typographically clear, we use a plus sign instead—d_{i+}, d_{+j}, d_{++}.

We write $a \approx b$ to mean that a and b are "approximately" equal. We use the symbol $a \equiv b$ to signify a definition.

Independent multiple opinions and replications of experiments are but two examples of corroborative evidence which are currently in vogue, and which we believe will increase in frequency during the next decade. We hope that the present development of methodology will provide some guidelines for the rigorous interpretation of data from independent sources.

Acknowledgments

It is a pleasure to express our appreciation to Dr. Betsy Becker (Michigan State University), Mr. Brad Hanson (Stanford University), Dr. Nan Laird (Harvard University), Ms. Therese Pigott (University of Chicago), Dr. Marie Tisak (University of California, Berkeley), and Dr. Margaret Uguroglu (DeVry, Inc.) for reading the manuscript and providing us with numerous comments and suggestions. The authors are particularly grateful to Dr. Betsy Becker for carrying out the simulation studies and for the computations in many of the tables. Therese Pigott (University of Chicago) checked the computations in all of our examples. Although it would be comforting to be able to blame these readers for errors, honesty compels us to say that any shortcomings are solely ours.

Unfortunately, the first draft of this book was not typed onto a word processor, thereby necessitating many retypings. For this task we thank Julie Less, Irene Miura, Jerri Rudnick and Tonda West who managed to maintain a calm exterior at all times.

We are grateful to the Spencer Foundation and to the National Science Foundation for their continued support of our research. Their generous support has been essential in making this work possible.

The authors wish to acknowledge, with thanks, the following associations and societies for kindly granting us permission to use published materials:

The American Psychological Association for Tables 1, 2, 3, and Fig. 1 of Chapter 4; Tables 4, 5, 6 and Figure 3 of Chapter 5; Tables 3, 4, and 7 of Chapter

6; Tables 1, 2, 3, 4, 5, and 6 of Chapter 13; and Table 4 of Chapter 14.

The American Educational Research Association for Table 2 of Chapter 5, Tables 1, 2 and 3 of Chapter 8, and Table 1, 2 and 3 of Chapter 14.

The Institute of Mathematical Statistics for Table 1 of Chapter 11.

The Biometrika Trust for Appendix E and Appendix G.

CHAPTER 1

Introduction

Replication of experimental results has long been a central feature of scientific inquiry, and it raises questions concerning how to combine the results obtained. In the early part of this century modern statistical methods were constructed for the analysis of individual agricultural experiments, and shortly thereafter statistical methods for combining the results of such experiments were developed.

Two distinctly different directions have been taken for combining evidence from different studies in agriculture almost from the very beginning of statistical analysis in that area. One approach relies on testing for statistical significance of combined results across studies, and the other relies on estimating treatment effects across studies. Both methods date from as early as the 1930s (and perhaps earlier) and continue to generate interest among the statistical research community to the present day.

Testing for the statistical significance of combined data from agricultural experiments is perhaps the older of the two traditions. One of the first proposals for a test of the statistical significance of combined results (now called testing the minimum p or Tippett method) was given by L. H. C. Tippett in 1931. Soon afterwards, R. A. Fisher (1932) proposed a method for combining statistical significance, or p-values, across studies. Karl Pearson (1933) independently derived the same method shortly thereafter, and the method variously called Fisher's method or Pearson's method was established. Research on tests of the significance of combined results has flourished since that time, and now well over 100 papers in the statistical literature have been devoted to such tests.

Tests of the significance of combined results are sometimes called *omnibus* or *nonparametric tests* because these tests do not depend on the type of data or the statistical distribution of those data. Instead, tests of the statistical significance of combined results rely only on the fact that p-values are uniformly distributed between zero and unity. Omnibus tests make use of this fact. Although omnibus tests have a strong appeal in that they can be applied universally, they suffer from an inability to provide estimates of the magnitude of the effects being considered. Thus, omnibus tests do not tell the experimenter *how much* of an effect a treatment has.

In order to determine the magnitude of the effect of an agricultural treatment, a second approach was developed which involved combining numerical estimates of treatment effects. One of the early papers on the subject (Cochran, 1937) appeared a few years after the first papers on omnibus procedures. Additional work in this tradition appeared thereafter (e.g., Yates & Cochran, 1938; Cochran, 1943). Even the earliest writers on the question of combining numerical estimates of treatment effects recognized some of the substantive and methodological problems in research synthesis that we face today. For example, it was recognized that not all studies provide equally good data, and that the estimates of experimental error (e.g., the quality of the data) reported with each study are not to be trusted completely.

A. THE USE OF STATISTICAL PROCEDURES FOR COMBINING THE RESULTS OF RESEARCH STUDIES IN THE SOCIAL SCIENCES

Procedures for testing the statistical significance of combined results are readily applicable to the problem of combining the results of studies in the social sciences. Combined significance tests have been advocated periodically in the social sciences for at least 30 years (e.g., Jones & Fiske, 1953; Winer, 1971). The use of these omnibus test procedures remained infrequent until quite recently, however. Statistical significance was often used in less formal (and often misleading) procedures in research reviews.

A.1 THE MISUSE OF STATISTICAL SIGNIFICANCE IN REVIEWING RESEARCH

Researchers in the social sciences often use statistical significance to help interpret the results of individual research studies. It seems quite intuitive to use the outcomes of significance tests in each study to assess the average effect and the consistency of effects across studies. In this section we examine ways in which reviewers have used the outcomes of significance tests to make this assessment. One of the procedures for using the results of significance tests to assess whether average effects are nonzero is shown to be misleading. A related procedure that is frequently used to assess consistency of research results is also faulty, and the combined effect of these two procedures can lead to overly pessimistic conclusions about the consistency and existence of treatment effects.

A.1.a The Misuse of Statistical Significance to Assess Whether Effects Are Nonzero

Research reviewers must assess whether empirical data in a series of studies support the conclusion that an experimental effect or relationship between variables is nonzero. Reviewers start with a series of replicated studies. Each study is believed to provide evidence about the existence of an effect or a relationship. In order to assess the overall effect by using the outcomes of significance tests in the individual experiments, a reviewer might, in essence, let each study "cast a vote" for or against the existence of the hypothetical effect. Intuitively, if a large proportion of studies obtain statistically significant results, then this should be evidence that the effect is nonzero. Conversely, if few studies find statistically significant results, then the combined evidence for a nonzero effect would seem to be weak.

In spite of the intuitive appeal of such vote-counting procedures, it can be shown that they can be faulty in drawing inferences about treatment effects. Studies of the properties of vote-counting procedures have shown that such procedures can be strongly biased toward the conclusion that the treatment has no effect. Moreover, this bias is not reduced as the amount of evidence (number of studies) increases. (See Section A of Chapter 4.)

A.1.b Assessing the Consistency of Effects across Studies

Reviewers must also draw conclusions about the consistency of research results or effect magnitudes across replicated studies. Outcomes of the significance tests in the individual studies have been used to assess the consistency of effect magnitudes by sorting the studies into categories according to the outcomes of the statistical significance tests. Effect magnitudes of the studies are considered to be consistent if the outcomes of the significance tests are largely consistent. Unfortunately the logic of this procedure is flawed, and conclusions about consistency of effect magnitudes derived from it can be incorrect.

The fallacy in using the outcomes of significance tests to determine consistency is identical to a common fallacy about when replication attempts do or do not succeed. This fallacy is best illustrated by an example. Suppose that two identical studies test for a relationship between two variables. Then there are four possible categorized outcomes of those two studies as illustrated in Table 1.

Most reviewers would regard configurations of results 1 and 2 (two significant or two nonsignificant results) as evidence of consistency of research results. They would similarly regard configurations of results 3 and 4 (one significant and one nonsignificant result) as a "failure to replicate." Neither conclusion is necessarily correct. It is entirely possible for two relationships (e.g., correlations or effect sizes) to be both significant and significantly different from one another. Similarly, relationships may be both statistically insignificant and significantly different from one another. Finally, the fact that the relationship is significant in one study but is not significant in a second study does not guarantee that the relationships evidenced in the two studies are significantly different from one another. There is no easy way to tell whether the results of studies are consistent with one another from the outcomes of their individual significance tests alone. A more detailed discussion of this fallacy in research reviewing is given by Humphreys (1980).

TABLE 1

Possible Configurations of Outcomes of Two Studies

	Outcome	
Configuration	Study 1	Study 2
1	Significant	Significant
2	Not significant	Not significant
3	Significant	Not significant
4	Not significant	Significant

A.1.c The Combined Effect of Erroneous Procedures for Evaluating Existence and Consistency of Effects

The combined effect of the two fallacies in the use of statistical significance testing is to lead reviewers to very pessimistic conclusions about research results. The small to moderate sample sizes and effect sizes, δ, usually found in educational research lead to a situation in which many if not most studies will fail to reject the null hypothesis at the $\alpha = 0.05$ significance level. In Fig. 1 the expected proportion of significant results (significant two-sample t-statistics) is plotted as a function of sample size (assuming that all studies have the same sample size) for three effect sizes. This figure shows the expected proportion of significant results for $\delta = 0.20$, 0.50, and 0.80. These three values constitute small, medium, and large effect sizes. The curves show that for sample sizes less than 100 and a "small" effect size, the expected proportion of significant results never exceeds 0.20. Even when $\delta = 0.50$, the expected proportion of significant results exceeds 0.50 only when the sample size nears 70.

One conclusion is that if most effects are small to medium and sample sizes are moderate, then there is likely to be a predominance of nonsignificant results in the studies examined by a research reviewer. Use of the vote-counting method to decide whether the treatment effect is nonzero can lead to the conclusion that the average effect of the treatment is not different from zero. Moreover, the outcomes of most of the significance tests are consistent (that is, not significant), which could lead to the conclusion that the results of the studies *consistently suggest no effect.* Alternatively, a reviewer

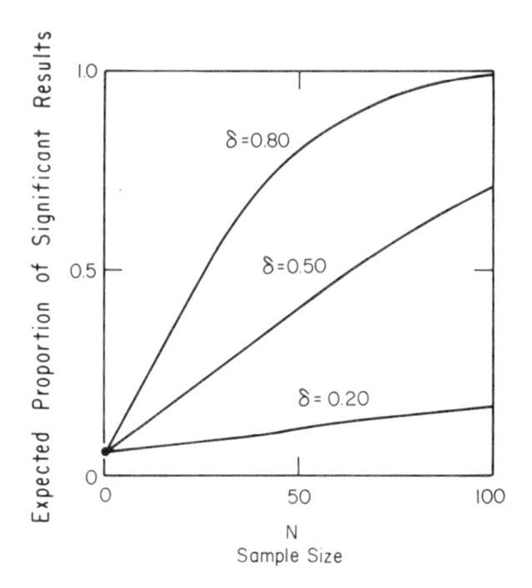

Fig. 1 The expected proportion of significant results as a function of sample size $N = n^E + n^C$ and effect size δ.

might regard the few significant outcomes as evidence that the effect of the treatment, while negligible overall, has an inconsistent effect. Thus the combined effect of the two erroneous procedures is to suggest an unwarranted pessimism about the magnitude and possibly the consistency of the treatment effects under consideration.

A.2 STATISTICAL PROCEDURES FOR COMBINING ESTIMATES OF EFFECT MAGNITUDE

The tradition of combining estimates from different studies was not much used in educational or psychological research until Glass's proposals in 1976. It would be incorrect to say that it was never used, particularly among reviews of studies that produced correlation coefficients as the descriptive statistics. For example, an outstanding and influential example of combining evidence by aggregating correlation coefficients is provided by Bloom's (1964) *Stability and Change in Human Characteristics.*

Why were these methods not generally used in educational research? A major stumbling block is that the methods used for combining estimates in agriculture (unlike combined significance tests) are not generally applicable for educational data. The reason is that in agriculture all of the studies measure the same variable on the same scale of measurement. The yield of a field of wheat can be measured (with the usual normal fluctuations) regardless of who does the measuring. The same cannot be said of academic achievement or self-concept or attitude toward school, because these outcomes of research in education and psychology are typically psychological constructs that have no natural scale of measurement. Thus different research studies may repeat the same experiment and measure the same construct but use instruments that yield numbers on completely different scales. Although it has been argued (e.g., Campbell, 1969) that the ability to measure the same construct in different ways has important theoretical advantages, it complicates the process of combining statistical evidence from different studies.

Glass deserves considerable credit for recognizing a solution to the problem of how to combine the results of studies using different scales of measurement. The solution proposed by Glass (1976) was to estimate an index of effect magnitude that did not depend on the arbitrary scaling of the dependent variable. The value of the index of effect magnitude would therefore not depend on which of a variety of linearly equatable measurement devices (psychological tests) was used in a study. Consequently if a series of studies used different, but approximately linearly equatable, measures of an outcome variable, the indices of effect magnitude would be comparable across studies. Statistical methods could then be used to study the variation across studies of these "scale-free" indices of effect magnitude.

Two scale-free indices of effect magnitude were suggested by Glass. One index is Cohen's (1969) standardized mean difference, which we will call the *effect size*. The effect size is now used extensively for expressing and for combining the results of studies that assess the effectiveness of an experimental treatment. A second scale-free index of effect magnitude is the product–moment correlation coefficient. The correlation coefficient is most often used to summarize the relationship between two continuous variables.

In meta-analysis, an estimate of the index of effect magnitude is obtained for each study. Individual effect magnitude estimates can then be averaged to obtain an overall estimate of effect magnitude. Other statistical analyses can also be performed to study the variation of effect magnitudes across studies. Until recently, conventional statistical methods, such as multiple regression analysis and the analysis of variance, have been used to analyze effect magnitude data in meta-analysis (Glass, McGaw, & Smith, 1981). The use of conventional statistical procedures for the analysis of effect sizes or correlations initially seemed to be an innocuous extension of statistical methods to a new situation. Recent research on statistical methods for meta-analysis has demonstrated that the use of statistical procedures such as the analysis of variance or regression analysis cannot be justified on either statistical or conceptual grounds.

B. FAILINGS IN CONVENTIONAL STATISTICAL METHODOLOGY IN RESEARCH SYNTHESIS

In order to discuss the shortcomings of conventional statistical procedures in meta-analysis, we begin by exploring the goals of statistical techniques in research synthesis under optimal circumstances. Our exploration of goals and conventional analyses is conducted in terms of effect sizes because the corresponding argument for analyses based on correlations is completely analogous. We consider the application of conventional statistical procedures in meta-analysis and show that, even under the best circumstances, these methods fail to attain one or more of the goals of research synthesis. Next we show that the inherent assumptions of conventional statistical methods are seriously violated when these methods are used to analyze sample effect magnitudes.

B.1 GOALS OF STATISTICAL PROCEDURES IN RESEARCH SYNTHESIS

Before discussing specific statistical procedures for quantitative synthesis of research, it is useful to consider what can be expected of a statistical analysis in the best possible situation. This process may help us in view seemingly novel statistical analyses in more familiar ways.

Perhaps the simplest research synthesis is one in which the raw data from several experiments are available and can be pooled directly. For example, suppose that we have a series of k, two-group experiments, each of which is designed to investigate the effect of a treatment using an experimental/control group design. Assume that each study measures the normally distributed outcome variable using the same instrument and the same sampling plan so that the within-group population variances of the outcome scores are identical. For convenience we can arbitrarily fix the common within-group variances to be unity, although this is not essential.

The situation defined above is one in which the raw data from the individuals in all the studies are directly comparable. Consequently, the outcome scores of the individuals could be combined and analyzed in one global statistical analysis. This is a format familiar to most social scientists, who would probably use the data from all individuals in a $2 \times k$ (two treatments and k studies) analysis of variance. In our idealized case, the assumptions of the analysis of variance will be met.

What does one learn from the analysis of variance? There are three omnibus F-tests in textbooks. The F-test for the main effect of studies tests whether the average value of the outcome variable (averaged over both experimental and control groups) differs across studies. This test is not particularly interesting. The other two F-tests are more fruitful. The F-test for the treatment factor tests whether the treatment group performs better than the control group *on the average across all k experiments.* The F-test for the treatment-by-studies interaction tests whether the treatment effect is consistent across studies. The interpretation of the statistical analysis rests largely on these latter two tests. A large treatment effect with a negligible interaction is interpreted to mean that the treatment produces a large consistent effect across studies. If the interaction is *not* negligible, then interpretations become more complicated. An interaction suggests that the treatment effect is larger in some studies than in others. Any blanket statements about the main effects must be qualified by the fact that treatment effects vary significantly across studies.

A significant interaction signifies that one should begin to look for reasons why the treatment effect varies across studies. Variations across studies in treatment, experimental procedure, conditions of measurement, or sample composition might enter into an explanation of variations in treatment effect. If a suitable explanatory variable were found, it should be included in the statistical analysis as a (blocking) factor. The new statistical analysis would reveal (by an appropriate F-test) whether the new factor accounted for a significant amount of variation in the treatment effects and whether variations in the treatment effect across studies within levels of the new factor remained substantial. That is, we can test whether a proposed explanatory factor succeeds in removing or "explaining" the variations in treatment effects across studies. This test is conceptually analogous to the original test for the treatment-by-studies interaction.

Thus in the best possible case, where data from all studies can be combined directly, the statistical analysis has several features:

(1) The average treatment effect can be estimated (and tested) across studies.
(2) The consistency of treatment effects can also be tested (via the treatment-by-studies interactions).
(3) The effect of explanatory variables that define differences among studies can be tested.
(4) The significance of variation in treatment effects across levels of the explanatory variables can be tested to determine if all variations in treatment effects are essentially explained.

In evaluating statistical methods for research synthesis, it is useful to determine which features of the "best case" analysis are available in any proposed analysis. New statistical methods have been developed that provide the advantages of the best-case analysis for any meta-analysis. These methods permit the research synthesizer to answer essentially the same questions that can be answered when the raw data from all studies can be directly combined. Conventional statistical procedures in meta-analysis fail to answer one or more of the questions of interest. Moreover, the use of some conventional analyses for effect size data frequently involves serious violations of the assumptions of these techniques. Thus conventional statistical procedures in meta-analysis are problematic for both statistical and conceptual reasons. We now turn to the specific problems of conventional statistical procedures in meta-analysis.

B.2 CONVENTIONAL ANALYSES FOR EFFECT SIZE DATA

Conventional analyses in research synthesis have been greatly influenced by the pioneering work of Glass. He suggested combining the results of studies by first calculating an estimate of effect size g, which is the standardized difference between the experimental and control group means:

$$g = (\overline{Y}^{E} - \overline{Y}^{C})/s.$$

The estimates of effect size from different studies are standardized, so they are, in effect, on the same scale. Consequently the research synthesizer can combine these estimates across studies, or treat the effect sizes as raw data for statistical analyses (analysis of variance or multiple linear regression) that relate characteristics of studies to treatment effects (Glass, McGaw, & Smith, 1981).

B.3 CONCEPTUAL PROBLEMS WITH CONVENTIONAL ANALYSES

Now compare conventional effect size analyses with the best-case analysis in which all of the raw data can be directly combined. In our idealized best case, the treatment effect (mean difference) corresponds to the effect size for the study. In conventional analysis the effect sizes can be averaged to obtain an estimate of the average treatment effect. Similarly, the effect of any particular explanatory variable can be tested by using that variable as a blocking factor in an analysis of variance (or as a predictor in a regression analysis) that uses the effect size as the dependent variable. Thus the conventional effect size analysis has two of the features of the best-case analysis.

However, the conventional analysis lacks two important features of the best case analysis. First, it is impossible to directly test the consistency of effect sizes across studies in the conventional analysis. That is, there is no analogue to the test for treatment-by-study interactions. The conventional analysis for testing systematic variation among k effect sizes has $k - 1$ degrees of freedom for systematic variation among effect sizes and one degree of freedom for the grand mean, so that there are no degrees of freedom left over for estimation of the error or nonsystematic variation. Consequently it is impossible, in the conventional framework, to construct a test to determine whether the systematic variation in k effect sizes is larger than the nonsystematic variation exhibited by those effect sizes.

Note that it is possible, in the conventional analysis, to construct a test for differences among the average effect sizes of two or more *groups of studies*, as long as at least one of the groups contains two or more effect sizes. The multiple effect sizes within the group(s) serve as replicates from which an estimate of nonsystematic variance is obtained. Then the test is constructed by comparing "systematic" variance among group mean effect sizes to the "nonsystematic" variance of effect sizes within groups. However, such a test is conceptually and statistically perilous. How does the investigator know that the effect sizes exhibit only nonsystematic variability within the groups? If the investigator chooses the wrong groups, considerable *systematic variance* may be pooled into the estimate of the error variance. This was the essence of Presby's (1978) criticism of Glass's meta-analysis of psychotherapy outcome studies. She argued that Glass's analysis of differences among types of psychotherapy was flawed because he used overly broad categories of therapy—categories that included considerable systematic variation. The effect of including systematic variation in the estimates of error terms decreases the sensitivity of the statistical test for systematic variation. The conceptual problem in the conventional analysis is that the amount of variation among observed effect sizes that is systematic is unknown.

Precisely the same problem plagues an attempt to construct a test for the variation in effect sizes that remains after employing an explanatory variable. If the investigator tries to "explain" variation in effect sizes by grouping studies with similar characteristics (or using a linear predictor), there is no way to assess whether the remaining variation among the effect sizes is systematic or random.

B.4 STATISTICAL PROBLEMS WITH CONVENTIONAL ANALYSES

The analysis of effect sizes or correlation coefficients by using conventional statistical methods is also problematic for purely statistical reasons. Conventional statistical procedures (t-tests, the analysis of variance, multiple regression analysis) rely on parametric assumptions about the data. All of these procedures require that the nonsystematic variance associated with every observation be the same (the so-called homoscedasticity assumption). That is, if we think of each observation as composed of a systematic part and an error part, then the errors for all observations must be equally variable. In the analysis of variance we check that within-cell variances are reasonably similar in value for all cells in the design. In regression analysis we check this assumption by determining whether the residual variance about the regression line is reasonably constant for all values of the predictor variable.

In the case of estimates of effect magnitude (either correlation coefficients or effect sizes), the nonsystematic variance of an observation can be calculated analytically. In fact, the nonsystematic variance of estimates of effect size is inversely proportional to the sample size of the study on which the estimate is based. Therefore if studies have different sample sizes, which is usually the case, effect size estimates will have different error variances. If the sample sizes of the studies vary over a wide range, so will the error variances. In many meta-analyses it is not unusual for the range of sample sizes to be on the order of 50 to 1. In such a case the error variances are substantially heterogeneous.

The effects of heterogeneity of variance on analysis of variance F-tests have been studied extensively (see, for example, Glass, Peckham, & Sanders, 1972). Furthermore, heterogeneous variances have only small effects on the validity of the F-tests in a conventional analysis of variance. However, the situation in research synthesis is usually quite different from that in which robustness of F-tests is usually studied. Studies of the effects of heterogeneity of variance in ANOVA usually give a different variance to one or more *groups* in the design. Thus every observation in the same group has the same variance and there are at most two to three different variances in the entire experiment. In the case of research synthesis the heterogeneity is usually more pronounced. Every observation (study) may have a different variance. Moreover, the range of variances studied in connection with the robustness of F-tests is usually rather limited, often less than 5 to 1. The studies that examine the effects of very wide ranges of variances and groups of unequal size find that the F-test is not necessarily robust to substantial heterogeneity of variance. For example, Glass, Peckham, and Sanders (1972) note that when the ratio of variances is 5 to 1 and the sample sizes are unequal, then the actual significance level of the F-test can be six times as large as the nominal significance level, say, 0.30 instead of 0.05.

Thus the violation of the homogeneity of variance assumption in the analysis of variance and in regression analysis is severe in research synthesis. Moreover, the type of violation of the assumptions has not been extensively studied. There is little reason to believe that the usual robustness of the F-test will somehow prevail. The statistical problem of violation of the assumptions of conventional statistical procedures, and the potential problem of bias due to pooling of systematic variation into estimates of error variance, raise severe questions about the validity of conventional statistical procedures in meta-analysis. There does not appear to be any rigorously defensible argument for the use of conventional t-tests, analysis of variance, or regression analysis to analyze effect sizes or correlations.

C. STATISTICS FOR RESEARCH SYNTHESIS

This book is about statistical methods for combining the results of studies in the social and behavioral sciences. Much of the recent impetus for work in this area stems from the early work of Glass and the growing body of literature in this subject. In 1976 Glass distinguished types of statistical analyses in the social sciences. The original analysis of a set of data is termed *primary analysis*. Statistical texts and methodological training in the social sciences concentrate almost exclusively on primary statistical analysis. *Secondary analysis* is a reanalysis of data that have already been collected by another investigator. Some secondary analyses are conducted to reaffirm answers to questions raised in the primary analysis, whereas other secondary analyses attempt to answer new questions. The methods, potential, and limitations of secondary analysis have been discussed by Cook (1974).

Meta-analysis is the rubric used to describe quantitative methods for combining evidence across studies. Because meta-analysis usually relies on "data" in the form of summary statistics derived from the primary analyses of studies, it is truly an analysis of the results of statistical analyses. We attempt to provide a set of statistical procedures designed specifically for meta-analysis. These procedures exploit the properties of effect magnitudes to avoid the difficulties of conventional statistical procedures in meta-analysis.

Chapter 3 consists of a review of omnibus procedures for testing the statistical significance of combined results and a discussion of the properties and limitations of such procedures. Some simple estimators of effect magnitude based on so-called vote-counting methods are given in Chapter 4. In Chapter 5 we present several estimators of effect size and develop sampling theory for those estimators. Chapter 6 develops procedures for combining several estimates of effect size and for testing whether a series of studies share a common effect size. Analogues to the analysis of variance and multiple regression analysis for effect sizes are developed in Chapters 7 and 8. An analogue to random effects analysis of variance for effect sizes which treats the population effect sizes as random variables is given in Chapter 9. Chapter 10 is an exploration of the properties of correlated effect size estimates. Statistical procedures for analyses involving correlations are given in Chapter 11. Procedures for detecting outliers in research synthesis are discussed in Chapter 12, and Chapter 13 is a discussion of clustering procedures for effect sizes and correlations. Chapter 14 demonstrates the effects of censoring of effect size estimates corresponding to nonsignificant mean differences and provides some estimation procedures under this form of nonrandom sampling. The final chapter deals with models in the physical sciences which have characteristics quite different from those of models in the social sciences. The Appendices contain tables and nomographs used throughout the text. Some of these have been constructed in order to make the methods more readily available.

Many meta-analyses on a diverse range of topics have already been published. Because some of these may be of interest from both a substantive and a methodological point of view, we include a selected bibliography of recent meta-analytic studies.

This book does not discuss some of the issues that are crucially important in the *practice* of research synthesis. For example, we do not discuss the extraction of estimates of effect size or correlation coefficients from the data reported in individual studies. Such methods are given in Glass, McGaw, & Smith, (1981), and range from simple algebraic manipulations to elaborate estimation procedures with extensive assumptions. These methods must often be used to obtain effect magnitude estimates when all studies do not report summary statistics in the same way. The calculation of effect size estimates that are likely to be comparable across studies is often a time-consuming task involving a great deal of good judgment.

Similarly, we do not discuss the problem of how to develop hypotheses about explanatory models that account for the variation in effect sizes across studies. This is one of the most difficult and creative tasks in research synthesis and parallels the activities and conceptualization of primary research (see Jackson, 1980 and Cooper, 1982). The development of explanatory models invariably involves a combination of substantive and methodological insight about the particular domain of research under review. Such insight is probably best obtained by a thorough grounding in the theory and methodology of research in the domain of research that is under review. Some useful generalizations are discussed in Light and Smith (1971), Pillemer and Light (1980), Glass, McGaw, & Smith, (1981), Light and Pillemer (1982), and Light (1983). One general observation arising from our experience is that variations in the outcomes of well-controlled studies are considerably easier to model than are variations in the outcomes of poorly controlled studies. We believe this is a consequence of the fact that biases resulting from poor controls perturb measures of effect magnitude.

We do not attempt to provide guidelines to differentiate studies of high quality from those of low quality. This is a substantive issue that must be addressed in each research synthesis. We believe that no statistical procedure can perform the magic of extracting valid and reliable conclusions from data of poor quality. Any research synthesis is only as strong as the data on which it is based. Poor statistical procedures, however, can lead to incorrect conclusions from high-quality data. Both high-quality studies and correct statistical procedures are necessary to assure that a review is not misleading. In this vein we note that many of the hundreds of research syntheses conducted to date use statistical procedures that are of questionable validity or are demonstrably incorrect. The conclusions of these meta-analyses may indeed be correct, but the statistical reasoning in support of these conclusions is not.

CHAPTER 2

Data Sets

In this chapter we introduce several prototypical data sets that are used throughout this book to illustrate the various statistical methods presented. The data sets were chosen to represent a range of areas in education and psychology. Each data set has been previously analyzed and published, and some of the data sets have been reanalyzed by different investigators.

The data sets contain typical information that is available to the research reviewer. That is, the outcome data consist of a single summary statistic used as an index of effect magnitude, usually a standardized mean difference or a correlation coefficient. In particular, the original sets of observations are not provided, because for these data sets, and indeed in virtually all other examples of research synthesis, the original observations are not available.

In addition to the index of effect magnitude, the sample size and characteristics of the experimental conditions in the studies are given. The selection of these characteristics in the studies is not entirely haphazard. Because the object of research synthesis is to determine how broadly a result may generalize, reviewers usually use as independent variables several characteristics that they believe to be related to the experimental outcome. The characteristics that we have selected were found to be useful theoretically or empirically by previous reviewers of the research from which these data sets are derived.

Remark: Additional parts of the data sets are included in order to provide the reader with data to use as an exercise in carrying out some of the statistical procedures.

We also provide in the Bibliography a selected list of meta-analyses carried out in recent years.

A. COGNITIVE GENDER DIFFERENCES

The first set of studies deals with sex differences in cognitive abilities as collected by Hyde (1981). The effect sizes were calculated from studies reported in the review by Maccoby and Jacklin (1974). Although Hyde's article includes studies on sex differences in four cognitive abilities (quantitative, verbal, visual–spatial, and field articulation), we use only the effect size estimates derived from studies of quantitative ability. This data set has attracted considerable attention and has been reanalyzed at least twice, by Rosenthal and Rubin (1982a) and by Becker and Hedges (1984). The data for other cognitive abilities are included for the reader to use.

Although Hyde's data table reports the directions of sex differences for 16 studies of quantitative ability, effect size estimates could be calculated for only seven of the studies because the data necessary for effect size calculations were not reported in the other studies. The data reported by Hyde for each study include characteristics of studies such as the year of publication, the age of subject, the total sample size N (the sum of male and female sample sizes), and a description of the sample. These characteristics are reported in Table 1.

The sampling plans in the studies reported by Hyde differ considerably. Some of the studies used national representative samples, whereas others used more select samples, such as Harvard undergraduates. Because the effect size depends on the standard deviations of the scores in the samples, Becker and Hedges (1984) introduced an index of sample selectivity in their analysis of these data.

Selectivity of the sample was coded trichotomously, where a value of 1 represents relatively unselected samples, including nationwide samples, school children not described as being in special programs or advanced courses, and most adult groups. Samples regarded as somewhat selective are assigned a selectivity value of 2. These include samples of college students, school children enrolled in advanced or special courses (e.g., physics), and the like. A selectivity value of 3 represents highly select samples such as students from prestigeous colleges. All selectivity ratings were coded by two individuals and the intercoder agreement was 97 percent.

TABLE 1

Studies of Gender Differences in Quantitative Ability

Study authorship	Year of publication	Age of subject	Total sample size N	Standardized mean difference g	Unbiased standardized mean difference d	Sample description	Sample selectivity code
Bieri, Bradburn, & Galinsky	1958	18–21	76	0.72	0.71	Radcliffe women, Harvard men	3
Droege	1967	18	6,167	0.06	0.06	High-school students	1
Very	1967	18–21	355	0.59	0.59	College students	2
Walberg	1969	16–17	1,050	0.43	0.43	High-school physics students	1
Jacobson, Berger, & Milham	1970	18–21	136	0.27	0.27	College students	2
Backman	1972	17	2,925	0.89	0.89	Project TALENT sample, high-school students	1
American College Testing Program	1976–1977	18	45,222	0.35	0.35	First-year college students	2

TABLE 2

Studies of Gender Differences in Verbal Ability

Study authorship	Year of publication	Age of subject	Total sample size N	Standardized mean difference g	Unbiased standardized mean difference d	Sample description	Sample selectivity code
Bayley & Oden	1955	29	1,102	−0.25	−0.25	Gifted adults and their spouses	3
Bieri, Bradburn, & Galinsky	1958	18–21	76	0.19	0.19	Radcliffe women, Harvard men	3
Gates	1961	13	1,657	0.22	0.22	School children	1
Mendelsohn & Griswold	1966	18–21	223	0.16	0.16	U.C. Berkeley students	2
Droege	1967	18	6,167	0.22	0.22	Representative sample of high-school students	1
Very	1967	18–21	355	0.41	0.41	Penn State students	2
Laughlin, Branch, & Johnson	1969	18–21	528	0.03	0.03	College students	1
Walberg	1969	16–17	2,074	0.33	0.33	High-school physics students	2
Backman	1972	17	2,925	1.40	1.40	Project TALENT sample, high-school students	1
Blum, Fosshage, & Jarvik	1972	64	54	0.58	0.57	Longitudinal development sample	1
Matarazzo	1972	16–64	1,700	0.12	0.12	WAIS 1955 standardization sample	1
American College Testing Program	1976–1977	18	45,222	0.26	0.26	First-year college students	2

TABLE 3

Studies of Gender Differences in Visual–Spatial Ability

Study authorship	Year of publication	Age of subject	Total sample size N	Standardized mean difference g	Unbiased standardized mean difference d	Sample description	Sample selectivity code
Stafford	1961	14–17	128	0.60	0.60	High-school students	1
Droege	1967	18	6167	0.41	0.41	High-school seniors	1
Very	1967	18–21	355	0.52	0.52	College students	2
Backman	1972	17	2925	0.83	0.83	Project TALENT sample, high-school students	1
Nash	1975	11–12	105	0.04	0.04	Sixth and ninth grades, New York public schools	1
		14–15	102	0.48	0.48		
Sherman	1978	15–16	1233	0.31	0.31	High-school mathematics students	1

TABLE 4
Studies of Gender Differences in Field Articulation Ability

Study authorship	Year of publication	Age of subject	Total sample size N	Standardized mean difference g	Unbiased standardized mean difference d	Sample description	Sample selectivity code
Gruen	1955	19–25	60	0.77	0.76	Trained dancers	1
Gross	1959	18–25	140	1.16	1.15	College students	2
Fiebert	1967	12	30	0.49	0.48	Deaf subjects	1
		15	30	0.30	0.29	Deaf subjects	1
		18	30	0.67	0.65	Deaf subjects	1
Schwartz & Karp	1967	17	46	0.85	0.84	Paid volunteers	1
		30–39	40	0.71	0.70	Paid volunteers	1
		58–80	34	0.51	0.50	Paid volunteers	1
Willoughby	1967	18–21	76	0.18	0.18	College students	2
Oltman	1968	18–21	163	0.17	0.17	College students	2
Bogo, Winget, & Gleser	1970	18–21	97	0.78	0.77	College students	2
Morf & Howitt	1970	18–27	44	0.27	0.27	College students	2
Morf, Kavanaugh & McConville	1971	18–24	78	0.40	0.40	College students	2
Blum, Fosshage, & Jarvik	1972	64	43	−0.46	0.45	Aging twins	2

Hyde's effect size data for studies of sex differences in verbal ability, visual–spatial ability, and field articulation ability are listed in Tables 2, 3, and 4, respectively. The reader may find these data useful for practicing the procedures described in this book.

B. SEX DIFFERENCES IN CONFORMITY

Are females more conforming than males? This question has received considerable attention in social psychology. There have been several attempts to synthesize the results reported in this literature (Eagly, 1978; Cooper, 1979; Eagly & Carli, 1981; Becker, 1983). With the exception of Eagly (1978), these reviews relied on quantitative methods to combine effect size estimates extracted from the studies. The most comprehensive collection of studies was that reported and analyzed by Eagly and Carli (1981), and subsequently reanalyzed by Becker (1983). Both of these papers report analyses of three data sets. Each data set consists of effect sizes extracted from conformity studies using a particular experimental paradigm. For expository purposes we examine the effect size data for only one of the three data sets, called "other conformity" studies by Eagly and Carli. Because these studies use an experimental paradigm involving a nonexistent norm group, we call these studies "fictitious norm group" studies.

Fictitious norm group studies examine the effect of knowledge of other people's responses on an individual's response. Typically, an experimental subject is presented with an opportunity to respond to a question of opinion. Before responding, the individual is shown some "data" on the responses of other individuals. The "data" are manipulated by the experimenters, and the "other individuals" are the fictitious norm group. For example, the subject might be asked for an opinion on a work of art and told that 75 percent of Harvard undergraduates liked the artwork "a great deal."

Several methodological characteristics of fictitious norm group studies were examined by Eagly and Carli, and by Becker (1983) including the number and sex of agents in the fictitious norm group, the sex of the experimenters, and the age of the subjects. Methodological variables such as the number of items on the instrument used to measure the outcome and the type of outcome measure used were also considered by Becker (1983).

The effect size estimates calculated by Eagly and Carli, and in some cases recalculated by Becker, are presented in Table 5. The reference for each study, the total sample size (sum of male and female sample sizes), the number of items on the outcome measure used, the percentage of male authors, and a description of the sample are also given.

TABLE 5

Studies of Sex Differences in Conformity Using the Fictitious Norm Group Paradigm

Study authorship	Year of publication	Total sample size N	Standardized mean difference g	Unbiased standardized mean difference d	No. of items	Percentage of male authors	Sample description
King	1959	254	0.35	0.35	38	100	Precollege students
Wyer	1966	80	0.37	0.37	5	100	Precollege students
Wyer	1967	125	−0.06	−0.06	5	100	College students
Sampson & Hancock	1967	191	−0.30	−0.30	2	50	Precollege students
Sistrunk	1971	64	0.70	0.69	30	100	College students
Sistrunk & McDavid, study II	1971	90	0.40	0.40	45	100	College students
Sistrunk & McDavid study IV	1971	60	0.48	0.47	45	100	College students
Sistrunk	1972	20	0.85	0.81	45	100	College students
Feldman-Summers, Montano, Kasprzyk, & Wagner, study I	1977	141	−0.33	−0.33	2	25	College students
Feldman-Summers, Montano, Kasprzyk, & Wagner, study II	1977	119	0.07	0.07	2	25	College students

C. THE EFFECTS OF OPEN EDUCATION

A very large data set derived from studies of the effectiveness of open education programs was assembled by Hedges, Giaconia, and Gage (1981). Their analysis included nearly 200 studies that examined the effects of open education on student outcomes by comparing students in experimental open classroom schools with students from traditional schools. The studies used a variety of different outcome variables that were classified into 16 different dependent variable clusters for the statistical analyses:

achievement motivation
adjustment
anxiety
attitude toward school
attitude toward teacher
cooperativeness
creativity
curiosity
general mental ability
independence and self-reliance
language achievement
locus of control
mathematics achievement
reading achievement
science and social studies achievement
self-concept

We use only a few of these data sets. In particular, we use a data set corresponding to randomized experiments and other well-controlled studies examining the effect of open education on attitude toward school. Data for 11 well-controlled studies of the effect of open education on attitude toward school are given in Table 6.

We also use a data set derived from seven of the studies of the effects of open education on independence and self-reliance. These seven studies were selected, in part, because they had the same sample size $n^E = n^C = 30$. The results of these studies, another data set derived from randomized experiments and other well-controlled studies of the effects of open education on self-concept, and studies of the effects of open education on creativity are summarized in Tables 7, 8, and 9.

TABLE 6

Studies of the Effects of Open Education on Attitude toward School

Study	Method of defining openness	Grade level	Sample size: Open education group, n^E	Sample size: Traditional school group, n^C	Standardized mean difference g	Unbiased standardized mean difference d
1	Space	4–6	40	40	−0.256	−0.254
2	Space	4–6	40	40	0.264	0.261
3	Observation	4–6	90	90	−0.043	−0.043
4	Observation	4–6	40	40	0.655	0.649
5	Judgment	K–3	79	49	0.506	0.503
6	Judgment	K–3	84	45	0.461	0.458
7	Judgment	K–3	78	55	0.580	0.577
8	Judgment	4–6	131	138	0.158	0.158
9	Judgment	4–6	38	110	0.591	0.588
10	Judgment	4–6	38	93	0.394	0.392
11	Judgment	4–6	20	23	−0.056	−0.055

TABLE 7

Studies of the Effects of Open Education on Student Independence and Self-Reliance

Study	Sample size $n^E = n^C$	Standardized mean difference g	Unbiased standardized mean difference d
1	30	0.708	0.699
2	30	0.092	0.091
3	30	−0.059	−0.058
4	30	−0.080	−0.079
5	30	−0.238	−0.235
6	30	−0.500	−0.494
7	30	−0.595	−0.587

TABLE 8

Studies of the Effects of Open Education on Student Self-Concept

Study	Grade level	Sample size: Open education group, n^E	Sample size: Traditional school group, n^C	Standardized mean difference g	Unbiased standardized mean difference d
1	4–6	100	180	0.100	0.100
2	4–6	131	138	−0.162	−0.162
3	4–6	40	40	−0.091	−0.090
4	4–6	40	40	−0.049	−0.049
5	4–6	97	47	−0.046	−0.046
6	K–3	28	61	−0.010	−0.010
7	K–3	60	55	−0.434	−0.431
8	4–6	72	102	−0.262	−0.261
9	4–6	87	45	0.135	0.134
10	K–3	80	49	0.019	0.019
11	K–3	79	55	0.176	0.175
12	4–6	40	109	0.056	0.056
13	4–6	36	93	0.045	0.045
14	K–3	9	18	0.106	0.103
15	K–3	14	16	0.124	0.121
16	4–6	21	22	−0.491	−0.482
17	4–6	133	124	0.291	0.290
18	K–3	83	45	0.344	0.342

TABLE 9

Studies of the Effects of Open versus Traditional Education on Student Creativity

Study	Grade level	Sample size $n^E = n^C$	Standardized mean difference g	Unbiased standardized mean difference d
1	6	90	−0.583	−0.581
2	5	40	0.535	0.530
3	3	36	0.779	0.771
4	3	20	1.052	1.031
5	2	22	0.563	0.553
6	4	10	0.308	0.295
7	8	10	0.081	0.078
8	1	10	0.598	0.573
9	3	39	−0.178	−0.176
10	5	50	−0.234	−0.232

D. THE RELATIONSHIP BETWEEN TEACHER INDIRECTNESS AND STUDENT ACHIEVEMENT

Studies of the relationship of teaching variables and student achievement were reviewed by Gage (1978). Several of the studies that reported correlations between an observational measure of teacher indirectness and student achievement are reported in Table 10. In these data the sample size is the number of teachers. The correlation coefficient reflects the relationship between a score on teacher indirectness derived from an observational measure and mean class achievement.

TABLE 10

Studies of the Relationship between an Observational Measure of Teacher Indirectness and Student Achievement

Study	Grade level	No. of teachers	Correlation coefficient r
1	2	15	−0.073
2	4	16	0.308
3	7	15	0.481
4	8	16	0.428
5	10–12	15	0.180
6	12	17	0.290
7	4	15	0.400

CHAPTER 3

Tests of Statistical Significance of Combined Results

Interestingly, tests for the statistical significance of combined results were possibly the first statistical procedures developed for quantitative research synthesis. The earliest reference to a statistical procedure for combining significance tests appears to be in a book by L. H. C. Tippett published in 1931. Shortly thereafter, the problem of combining significance tests attracted the attention of R. A. Fisher (1932) and Karl Pearson (1933). It has been a source of considerable research in the years since this early period. The problem of combining significance tests continues to be of fundamental interest in statistics, and the statistical literature on the subject continues to grow.

In spite of the burgeoning statistical literature on combining significance tests, these techniques have not been used much in the social sciences until recently. Tippett's procedure was explained and generalized by Wilkinson (1951), and Fisher's method was similarly explained by Jones and Fiske (1953). Several procedures for testing the significance of combined results were reviewed by Mosteller and Bush (1954), and some new procedures for combining significance tests in psychology were proposed by Edgington (1972a,b). The appearance of the earlier expositions for combining significance tests in the social science literature did not increase their use by social scientists. However, an expository article by Rosenthal (1978) did help to popularize some of the ideas and procedures. This paper describes several alternative methods for testing the significance of combined results. It provides detailed descriptions of the test procedures and numerical examples of the use of each procedure. Rosenthal also compares properties of the combined significance test procedures.

This chapter is a review of so-called omnibus statistical methods for testing the statistical significance of combined results. The procedures are called *omnibus* or *nonparametric* because they do not depend on the form of the underlying data, but only on the exact significance levels commonly called p-values. A key point is that observed p-values derived from continuous test statistics have a uniform distribution under the null hypothesis *regardless of the test statistics* or distribution from which they arise. Therefore, combined significance tests that depend on the underlying data only through p-values are nonparametric in the sense that they do not depend on the parametric form of the data.

The nonparametric nature of combined significance tests gives them great flexibility in applications. Such tests can be used to combine any independent tests of hypotheses, even though the individual tests examine somewhat different hypotheses. For example, we might use combined significance tests to summarize the results of 10 studies each of which examined the effect of a treatment on a different outcome variable. Such a procedure would test whether the treatment produced a superior outcome on *any* of the dimensions investigated. These procedures can also be used in research synthesis to combine the results of studies that test the same conceptual hypothesis by different methods.

Omnibus tests of statistical significance can almost always be applied to data collected for the synthesis of social science research. However, they do not always provide a test of the hypothesis of interest to the research reviewer. Such tests do not, for example, support inferences about the average magnitude of effects or about consistency of results across studies (see Section D).

A. PRELIMINARIES AND NOTATION

Consider a collection of k independent studies characterized by parameters $\theta_1, \ldots, \theta_k$, such as means, mean differences, or correlations. Assume further that the ith study produces a test statistic T_i to be used to test the null hypothesis

$$H_{0i}: \qquad \theta_i = 0, \quad i = 1, \ldots, k,$$

where large values of the test statistic lead to rejection of the null hypothesis. The hypotheses $H_{01}, \ldots, H_{0k}$ need not have the same substantive meaning, and similarly, the statistics $T_1, \ldots, T_k$ need not be of related form. The omnibus null hypothesis H_0 is that none of the effects is significant, that is, that all the θ's are zero:

$$H_0: \qquad \theta_1 = \theta_2 = \cdots = \theta_k = 0.$$

Note that the composite hypothesis H_0 holds only if each of the subhypotheses $H_{01}, \ldots, H_{0k}$ holds.

The one-tailed p-value for the ith study is

$$p_i = \text{Prob}\{T_i \geq t_{i0}\}, \tag{1}$$

where t_{i0} is the value of the statistic actually obtained (the sample realization of T_i) in the ith study. Here the distributions of the k statistics $T_1, \ldots, T_k$ need not be the same. If H_{0i} is true, then p_i is uniformly distributed in the interval [0, 1].

Remark: The p-values $p_1, \ldots, p_k$ defined in (1) require one-tailed p-values. When two-tailed p-values are used a modification of the method is often required. The direction of the tail to be used must be selected a priori and should not depend on the data.

The problem of selecting a test for H_0 is complicated by the fact that there are many different ways in which the omnibus null hypothesis H_0 can be false. For example, H_0 is false if all the θ's but one are zero, e.g.,

$$\theta_1 = \cdots = \theta_{k-1} = 0, \qquad \theta_k > 0.$$

But it is also false if none of the θ's is zero, e.g., $\theta_1 = \cdots = \theta_k > 0$. However, a test that is sensitive to the departure from zero of one deviant θ may not be sensitive to departures from zero of all the parameters. In general, we cannot expect one test to be sensitive to all possible alternatives, so that the appropriateness of a particular combined significance test depends strongly on the alternative of interest.

The way in which alternative hypotheses can vary is illustrated in two dimensions in Fig. 1, where regions in the (θ_1, θ_2) plane are plotted. The null hypothesis H_0 corresponds to the origin (0, 0). Region A is a region where both θ_1 and θ_2 are close to zero. Region B is a region where one θ is close to zero and the other θ is not close to zero. Regions C and D correspond to cases where both θ_1 and θ_2 are far from zero. Although regions A–D constitute alternative hypotheses, we would expect a test to perform differently for one alternative than for another.

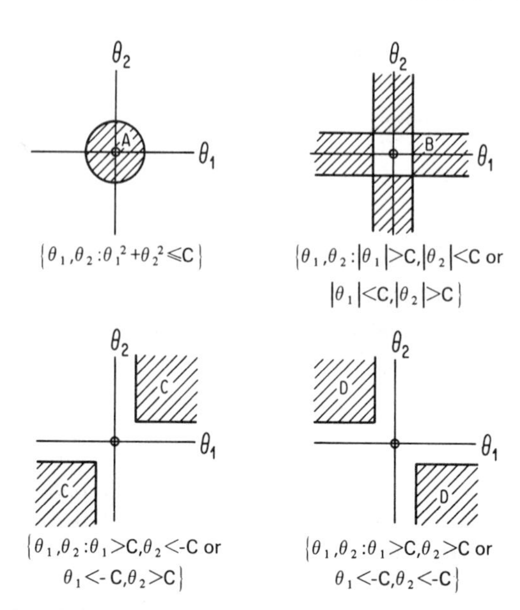

Fig. 1 Examples of alternative hypotheses in two-dimensional parameter spaces.

It is useful to distinguish between several general alternative hypotheses. In the first situation, all of the deviations from the null hypothesis occur in one direction that is known in advance. This alternative hypothesis is given by

$$H_1: \qquad \theta_i \geq 0, \quad i = 1, \ldots k, \text{ and at least one } \theta_i > 0.$$

Hypothesis H_1 is relevant as in the case of F-statistics in the analysis of variance or chi-square statistics for goodness of fit.

A second general alternative hypothesis states that differences are consistent, but may be in either direction. This alternative hypothesis is given by

$$H_2: \qquad \theta_i \leq 0 \text{ or } \theta_i \geq 0, \quad i = 1, \ldots, k, \text{ and at least one } \theta_i \neq 0,$$

as in the case of a t-statistic or a correlation coefficient.

Thus, several studies might examine the effect of a treatment on the *same* outcome construct, so that we would expect a consistent effect; but we might not know whether that effect was positive or negative.

A third general alternative hypothesis is when nonzero effects can occur in either direction. This hypothesis is given by

$$H_3: \qquad \text{at least one } \theta_i \neq 0, \quad i = 1, \ldots, k.$$

This alternative hypothesis is relevant when combining studies using statistics that reject for large or small values, and in which effects in different studies need not all have the same sign. For example, a series of studies might provide data on the effect of a treatment on several *different* outcomes. The treatment could enhance performance on some outcomes but retard performance on other outcomes.

B. GENERAL RESULTS ON TESTS OF SIGNIFICANCE OF COMBINED RESULTS*

In some cases there may exist tests of hypotheses that have good power for a particular alternative. It is less common for a test to have good power regardless of the alternative. Of course, when this happens we would select such a test.

But the existence of a best test is not the norm. Rather, it is more usual for a test not to be unequivocally optimal, but perhaps to be optimal only within a narrower framework. Because of this we find tests described as uniformly most powerful unbiased, invariant, similar, etc. Even some standard procedures are not unequivocally optimal.

In the present instance there is no best test for testing for the significance of combined results against any of the three general alternatives. Therefore, the selection of a combined significance test requires a more thorough analysis of competing tests. There are several general analytic results that guide the choice among the various candidate procedures. Each of these general results is designed to uncover those test procedures for combining results that are optimal according to some criterion. None of the criteria encompasses all the desired properties of a combined test procedure, and procedures that are optimal according to one criterion may not be optimal according to another criterion.

One principle that leads to general analytic results is *admissibility* of a combined test procedure. A combined test procedure is said to be admissible if it provides *a* (not necessarily *the only*) most powerful test against *some* alternative hypothesis for combining *some* collection of tests. More colloquially, a test that is admissible is best for something. (Here the term *best* includes the case where another test is equally as good.) It may therefore seem reasonable to narrow the collection of all combined test procedures to those that are admissible. Unfortunately, the class of admissible combined test procedures is still quite large.

Although admissibility is a comforting principle, the practical usefulness of admissibility is problematic for other reasons. Suppose we have a candidate test that we know is inadmissible. Then we know that there exists a test that

* Starred sections contain more advanced mathematical concepts and may be omitted upon a first reading.

is better than our candidate. But we may not be able to find the test that is better for some specific alternative—even though we know it exists. Because of this, we may want to consider inadmissible tests except when we can find a specific test that is superior.

A second principle that is appealing is called *monotonicity*. The property of monotonicity requires that if a combined test procedure rejects the null hypothesis for one set of p-values, it must also reject the hypothesis for any set of componentwise smaller p-values. More explicitly, a combined test procedure is called *monotone* if whenever the set of significance levels $p_1, \ldots, p_k$ leads to rejection of H_0, and if we have a set of significance levels $p_1^*, \ldots, p_k^*$ such that

$$p_1^* \leq p_1, \ldots, p_k^* \leq p_k,$$

then the set of significance levels $p_1^*, \ldots, p_k^*$ also leads to rejection of H_0.

A key result that connects the principles of monotonicity and admissibility is that every monotone combined test procedure is admissible and therefore optimal for some testing situation (Birnbaum, 1954). Under certain conditions, the class of admissible tests can be narrowed further and can be characterized as in the following fact.

Fact: If the distribution of each of the test statistics $T_1, \ldots, T_k$ is in the one-parameter exponential family (a relatively broad family), then a combined test procedure is admissible if and only if the acceptance region of the combined test is *convex*.

Thus, the determination of admissibility is reduced to the determination of convexity. But convexity is a simple property in that it is generally readily determined whether a region is convex or not. For completeness of exposition we review some of the essential ideas involving convexity.

A region is convex if for any two points in the region the line connecting the two points is in the region. Convexity is illustrated in two dimensions in Fig. 2. Regions A–C in Fig. 2 are convex, whereas regions D–F are not convex.

The above fact is useful in showing that many test procedures (e.g., Pearson's method and Wilkinson's extensions of Tippett's method described in the next section) are inadmissible for combining test statistics whose distributions are members of the one-parameter exponential family. Unfortunately, many of the test statistics of concern to social scientists (the t-test, the correlation coefficient, the F-test) have distributions that are *not* members of the one-parameter exponential family. Therefore, Birnbaum's second result may be irrelevant to many testing situations faced by social scientists.

Another principle leading to general analytic results concerns the behavior of the test statistic of the combined test procedure as the sample size in each

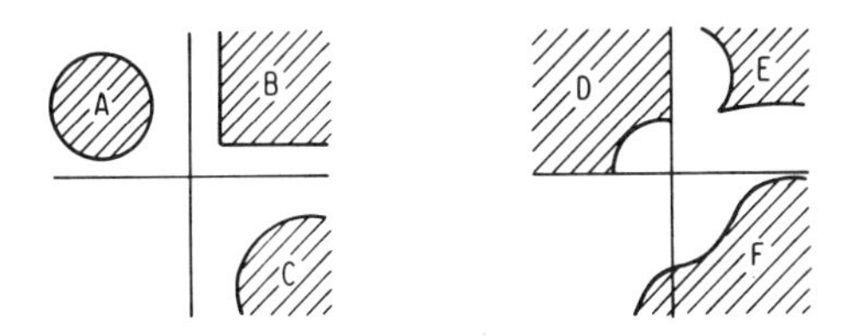

Fig. 2 Illustrations of two-dimensional regions that are convex and not convex.

study becomes large. This is the criterion of *efficiency*. There are numerous definitions of efficiency, in part because it is usually not easy to compute a particular measure of efficiency. Some measures of efficiency involve comparisons of sample sizes needed to obtain a given power level, others involve comparisons of power at the same sample size, and still others compare significance levels. One particular measure is called Bahadur efficiency (Bahadur, 1965, 1967). We call a test B-optimal (in the sense of Bahadur efficiency) if it gives the "most significant" results when the sample size is large. That is, the test will have the smallest possible significance level for large samples.

This definition of efficiency is based on the rate that the significance level tends to zero. The significance level of a test that is B-optimal tends to zero at least as fast as that of any other test as the sample size gets larger. However, several competing tests can be B-optimal, so that this notion does not always serve to single out a particular test. (For a discussion of conditions leading to B-optimality see Littell & Folks, 1971, 1973.)

Both admissibility and asymptotic efficiency have been used to study combined test procedures. Neither principle gives a complete picture of the performance of any statistical test. Moreover, these criteria sometimes disagree about the optimality of combined test procedures. The strength of these criteria is that they are intuitively meaningful and lead to general analytic results on the performance of combined test procedures.

C. COMBINED TEST PROCEDURES

In this section, we discuss the most widely used tests of significance for combining results. Each of these methods satisfies the monotonicity principle and is therefore optimal for some testing situation. These methods can be classified into two general categories depending on the nature of the combined test statistic. Some procedures use selected ordered values of $p_1, \ldots, p_k$ as the test statistic, whereas others use linear combinations of monotone functions of the p_i. We begin with procedures involving order statistics of $p_1, \ldots, p_k$.

Remark: Note that we assume that the statistical tests to be combined have a continuous test statistic, which in turn leads to a p-value that is uniformly distributed under the null hypothesis. Test statistics with discrete distributions (e.g., test statistics based on discrete data) will not yield p-values with uniform distributions. In fact, discrete test statistics lead to p-values with discrete distributions. Consequently, the combination procedures described in this chapter have to be modified by incorporating "corrections for continuity" before they are used with discrete data. Such corrections are described by Wallis (1942) and Lancaster (1949). An alternative approach to the problem of combining discrete p-values is to make the p-values into continuous variates by adding to them an appropriate uniform random variable. The latter approach to combining discrete p-values is discussed in Pearson (1950).

C.1 METHODS BASED ON THE UNIFORM DISTRIBUTION

The first test of the significance of combined results was proposed by Tippett (1931), who pointed out that if $p_1, \ldots, p_k$ are independent p-values (from continuous test statistics), then each has a uniform distribution under H_0. Therefore, if $p_{[1]}$ is the minimum of $p_1, \ldots, p_k$, a test of H_0 at significance level α is obtained by comparing $p_{[1]}$ with $1 - (1 - \alpha)^{1/k}$, so that the test procedure is to

$$\text{reject } H_0 \text{ if } p_{[1]} < 1 - (1 - \alpha)^{1/k}. \tag{2}$$

Remark: The test (2) is a monotone procedure with convex acceptance region; therefore, Tippett's procedure is admissible for tests using statistics that are members of the exponential family.

Example

The results of 10 studies of sex differences in conformity that used the fictitious norm group paradigm are given in Table 1. (See Section B of Chapter 2.) The table also includes the two-sample t-statistic and the corresponding p-value.

Note that in this case alternative hypothesis H_1 is relevant, since sex differences in conformity, if they exist, are expected to reflect a higher degree of conformity for females than for males. Such a difference is reflected in a large t-statistic and a small one-tailed p-value.

The combined null hypothesis is that none of the studies exhibits a sex difference. Note that the smallest p-value, that associated with study 1, is

TABLE 1

Data on 10 Studies of Sex Differences in Conformity Using the Fictitious Norm Group Paradigm

Study	Sample size		Effect size	Student's	Significance level			
	Control n^C	Experimental n^E	d	t	p	$-2 \log p$	$\Phi^{-1}(p)$	$\log[p/(1-p)]$
1	118	136	0.35	2.78	0.0029	11.682	−2.758	−5.838
2	40	40	0.37	1.65	0.0510	5.952	−1.635	−2.923
3	61	64	−0.06	−0.33	0.6310	0.921	0.335	0.537
4	77	114	−0.30	−2.03	0.9783	0.044	2.020	3.809
5	32	32	0.70	2.80	0.0034	11.367	−2.706	−5.680
6	45	45	0.40	1.90	0.0305	6.978	−1.873	−3.458
7	30	30	0.48	1.86	0.0341	6.760	−1.824	−3.345
8	10	10	0.85	1.90	0.0367	6.608	−1.790	−3.266
9	70	71	−0.33	−1.96	0.9740	0.053	1.942	3.622
10	60	59	0.07	0.38	0.3517	2.090	−0.381	−0.612

$p_{[1]} = 0.0029$. Comparing this value with $0.005 = 1 - (1 - 0.05)^{1/10}$, we see that $p_{[1]}$ is smaller than the critical value associated with the Tippett method. Therefore, we reject the hypothesis of no sex difference in any study and conclude that a sex difference is exhibited in at least one study.

A generalization of Tippett's procedure is due to Wilkinson (1951) and uses not just the smallest but the rth smallest p-value $p_{[r]}$ as a test statistic. Let

$$p_{[1]} \leq p_{[2]} \leq \cdots \leq p_{[k]}$$

be the ordered p-values (order statistics) obtained from k independent studies.

We can think of a test using $p_{[r]}$ as a comparison of $p_{[r]}$ with a critical value of $p_{r,\alpha}$. Alternatively, $p_{[r]} < p_{r,\alpha}$ implies that at least r of the p-values are less than $p_{r,\alpha}$. Therefore, one might use a critical value $p_{r,\alpha}$ for $p_{[r]}$ or use a critical number $m_{r,\alpha}$ of p-values that are smaller than a fixed level α. Wilkinson described his test procedure in terms of the number of p-values that are significant, i.e., smaller than α. He gave tables of the probability of obtaining m or more significant results at the $\alpha = 0.05$ and $\alpha = 0.01$ levels (that is, m or more p-values less than 0.05 or 0.01) for $k \leq 25$. Nomographs extending Wilkinson's tables to $k = 100$ for $\alpha = 0.05$ and to $k = 500$ for $\alpha = 0.01$ are given in Sakoda, Cohen, and Beall (1954).

Because $p_{[r]}$ has a beta distribution with parameters r and $k - r + 1$, tables of the incomplete beta function can be used to obtain significance levels of $p_{[r]}$ directly. Table 2 is derived from the beta distribution and gives the $\alpha = 0.05$ and $\alpha = 0.01$ critical values of $p_{[r]}$ for $r = 2, 3, 4, 5$ and $k = 5$ to 50.

When $r > 1$, this test does not have a convex acceptance region and is therefore inadmissible for combining test statistics that are members of the one-parameter exponential family. However, the statistics $p_{[2]}, p_{[3]}, \ldots, p_{[k-1]}$ have the advantage that they do not depend on the one or more most extreme observations. Thus although these statistics may not yield the most sensitive test of H_0 when the data are "clean," they will be relatively insensitive to outlying studies whose results may deviate greatly from the rest.

Example

Using the data in Table 1 as in the previous example, we can test the combined null hypothesis using the second smallest p-value. The test consists in comparing the second smallest p-value $p_{[2]}$ (that of study 5) with the critical value for $r = 2, k = 10$ obtained from Table 2. Comparing $p_{[2]} = 0.0034$ with 0.037, at the 5-percent level of significance we reject the null hypothesis that no sex difference is exhibited in any study.

TABLE 2
Critical Values for the rth Smallest p-Value

	$\alpha = 0.05$				$\alpha = 0.01$			
k	$r = 2$	3	4	5	$r = 2$	3	4	5
5	0.076	0.189	0.343	0.549	0.033	0.106	0.222	0.398
6	0.063	0.153	0.271	0.418	0.027	0.085	0.173	0.294
7	0.053	0.129	0.225	0.341	0.023	0.071	0.142	0.236
8	0.046	0.111	0.193	0.289	0.020	0.060	0.121	0.198
9	0.041	0.098	0.169	0.251	0.017	0.053	0.105	0.171
10	0.037	0.087	0.150	0.222	0.016	0.048	0.093	0.150
11	0.033	0.079	0.135	0.200	0.014	0.043	0.084	0.134
12	0.030	0.072	0.123	0.181	0.013	0.040	0.076	0.121
13	0.028	0.066	0.113	0.166	0.012	0.036	0.069	0.111
14	0.026	0.061	0.104	0.153	0.011	0.033	0.064	0.102
15	0.024	0.057	0.097	0.142	0.010	0.031	0.059	0.094
20	0.018	0.042	0.071	0.104	0.008	0.023	0.044	0.069
25	0.014	0.034	0.057	0.082	0.006	0.018	0.034	0.054
30	0.012	0.028	0.047	0.068	0.005	0.015	0.028	0.045
35	0.010	0.024	0.040	0.058	0.004	0.013	0.024	0.038
40	0.009	0.021	0.035	0.051	0.004	0.011	0.021	0.033
45	0.008	0.018	0.031	0.045	0.003	0.010	0.019	0.029
50	0.007	0.017	0.028	0.040	0.003	0.009	0.017	0.026

C.2 THE INVERSE CHI-SQUARE METHOD

Perhaps the most widely used combination procedure is that of Fisher (1932). Given k independent studies and the p-values $p_1, \ldots, p_k$, Fisher's procedure uses the product $p_1 p_2 \cdots p_k$ to combine the p-values. He made use of a connection between the uniform distribution and the chi-square distribution, namely, that if U has a uniform distribution, then $-2 \log U$ has a chi-square distribution with two degrees of freedom. Consequently, when H_{0i} is true, $-2 \log p_i$ has a chi-square distribution with two degrees of freedom. Because the sum of independent chi-square variables has a chi-square distribution, we have the very simple and elegant fact that if H_0 is true, then

$$-2 \log(p_1 p_2 \cdots p_k) = -2 \log p_1 - \cdots -2 \log p_k$$

has a chi-square distribution with $2k$ degrees of freedom. Because of this fact, no special tables are needed for the Fisher method. The test procedure becomes

$$\text{reject } H_0 \text{ if } P = -2 \sum_{i=1}^{k} \log p_i \geq C, \tag{3}$$

where the critical value C is obtained from the upper tail of the chi-square distribution with $2k$ degrees of freedom.

Example

Returning to the studies on sex differences in conformity using the fictitious norm group paradigm, we test the combined null hypothesis using the inverse chi-square method. Table 1 gives $-2 \log p_i$ for each study. The total of these values is

$$-2 \sum_{i=1}^{10} \log p_i = 52.45.$$

Comparing 52.45 with 31.4, the 95-percent critical value of a chi-square with $20 = (2)(10)$ degrees of freedom, we see that the combined null hypothesis is rejected at the 5-percent level.

A variation of the Fisher method proposed by Good (1955) involves the use of weights for the individual p-values. Suppose $p_1, \ldots, p_k$ are the obtained p-values from k studies. Let $v_1, \ldots, v_k$ be nonnegative weights that are selected a priori. The weighted Fisher method combines the p-values via the product

$$P_w = p_1^{v_1} p_2^{v_2} \cdots p_k^{v_k}. \tag{4}$$

Of course, when $v_1 = \cdots = v_k = 1$, we obtain Fisher's method.

The motivation for this procedure is to permit the investigator the flexibility to assign greater weight to the more sensitive studies. The choice of weights is usually a difficult problem that needs to be addressed afresh in each new situation. In a few instances procedures for choosing optimal weights have been devised. One case that has been examined is the combination of F-tests in incomplete block designs (Zelen, 1957). Optimal weights and an analysis of the power of the weighted Fisher method in this situation are given in Zelen and Joel (1959) and Pape (1972).

One major difficulty in applying the weighted Fisher method is that the distribution of the product $p_1^{v_1} \cdots p_k^{v_k}$ is rather complicated. The exact distribution was obtained by Robbins (1948) and Good (1955) when $v_1, \ldots, v_k$ are distinct. The cumulative distribution function of the product P_w is given by

$$\text{Prob}\{P_w \leq q\} = \frac{q^{1/v_1}}{a_1} + \cdots + \frac{q^{1/v_k}}{a_k}, \tag{5}$$

where

$$a_i = (v_i - v_1)(v_i - v_2) \cdots (v_i - v_{i-1})(v_i - v_{i+1}) \cdots (v_i - v_k)/v_i^{k-1}.$$

Formula (5) is valid only if $v_1, \ldots, v_k$ are distinct, and the calculations required are subject to computational inaccuracies if any pair of the v_i are nearly equal. No general expression for the distribution of P_w is currently available if the weights are not distinct.

Note that, under H_0, $-2 \log P_w$ is distributed as a weighted sum of chi-square variables. This distribution has a long history (see, e.g., Johnson & Kotz, 1970, Volume 3, Chapter 29) but simple representations are available only in special cases.

C.3 THE INVERSE NORMAL METHOD

Another procedure for combining p-values is the inverse normal method proposed independently by Stouffer, Suchman, DeVinney, Star and Williams (1949) and by Lipták (1958). This procedure involves transforming each p_i-value to the corresponding normal score, and then "averaging." More specifically, define Z_i by $p_i = \Phi(Z_i)$, where $\Phi(x)$ is the standard normal cumulative distribution function. When H_0 is true, the statistic

$$Z = \frac{Z_1 + \cdots + Z_k}{\sqrt{k}} = \frac{\Phi^{-1}(p_1) + \cdots + \Phi^{-1}(p_k)}{\sqrt{k}} \tag{6}$$

has the standard normal distribution. Hence we reject H_0 whenever Z exceeds the appropriate critical value of the standard normal distribution.

Example

Returning again to our data on sex differences in conformity, we use the inverse normal method to test the combined hypothesis of no sex difference in any study. Table 1 gives $\Phi^{-1}(p_i)$ for each study. The sum of the entries in this column is $\sum \Phi^{-1}(p_i) = -8.67$, so that

$$Z = \frac{\Phi^{-1}(p_1) + \cdots + \Phi^{-1}(p_{10})}{\sqrt{10}} = \frac{-8.67}{\sqrt{10}} - 2.74.$$

Comparing -2.74 with -1.96, the 95-percent two-tailed critical value of the standard normal distribution, we see that the combined null hypothesis is rejected at the 5-percent level of significance.

A modification of the inverse normal method involves the use of non-negative weights $v_1, \ldots, v_k$ and was suggested by Mosteller and Bush (1954). The weighted inverse normal procedure is based on the test statistic

$$Z_w = \frac{v_1 Z_1 + \cdots + v_k Z_k}{\sqrt{v_1^2 + \cdots + v_k^2}} = \frac{v_1 \Phi^{-1}(p_1) + \cdots + v_k \Phi^{-1}(p_k)}{\sqrt{v_1^2 + \cdots + v_k^2}}. \tag{7}$$

When H_0 is true, Z_w has the standard normal distribution, and we reject H_0 when Z_w exceeds the appropriate critical value of the standard normal distribution. An important advantage of this weighted procedure is that no special calculations are needed to obtain critical values of the weighted test statistic. However, an optimal choice of the weights has not been obtained.

Remark: Both the unweighted and weighted inverse normal procedures have convex acceptance regions. Therefore, both inverse normal procedures are admissible combination procedures for tests whose statistics are members of the one-parameter exponential family. The inverse normal procedures are also symmetric in the sense that p-values near zero are accumulated in the same way as p-values near unity. Therefore the inverse normal procedure is suitable for combining results when the direction of deviations from H_0 is not known in advance.

C.4 THE LOGIT METHOD

Yet another method for combining k independent p-values $p_1, \ldots, p_k$ was suggested by George (1977) and investigated by Mudholkar and George (1979). Transform each p-value into a logit, $\log[p/(1-p)]$, and then combine the logits via the statistic

$$L = \log \frac{p_1}{1-p_1} + \cdots + \log \frac{p_k}{1-p_k}. \tag{8}$$

The exact distribution of L is not simple, but when H_0 is true, George and Mudholkar (1977) show that the distribution of L (except for a constant) can be closely approximated by Student's t-distribution with $5k + 4$ degrees of freedom. Therefore, the test procedure using the logit statistic is

$$\text{reject } H_0 \text{ if } L^* = |L| \sqrt{(0.3)(5k+4)/k(5k+2)} > C, \tag{9}$$

where the critical value C is obtained from the t-distribution with $5k + 4$ degrees of freedom. [The term 0.3 in (9) is more accurately given by $3/\pi^2$.]

For large values of k, $\sqrt{3(5k+4)/\pi^2(5k+2)} \cong 0.55$, so that

$$L^* \cong (0.55/\sqrt{k})L.$$

Example

We use the logit method to test the combined null hypothesis in the data on sex differences in conformity given in Section B of Chapter 2. Table 1 gives values of $\log[p/(1-p)]$ for each study. The sum of these values is

$$L = \sum_{i=1}^{10} \log \frac{p_i}{1-p_i} = -17.154,$$

so that the test statistic L^* is

$$L^* = \sqrt{(0.3)(50 + 4)/10(50 + 2)}\,|L| = 0.1765\,|L| = 3.03.$$

Since L^* exceeds 2.01, the 95-percent critical value of Student's t-distribution with $50 + 4 = 54$ degrees of freedom, we reject the combined null hypothesis of no sex difference in any study.

Remark: The logit procedure is monotone, but the acceptance region of the logit procedure is not convex. Therefore, it is not an admissibie procedure for combining tests using statistics from the one-parameter exponential family. The logit procedure, like the Fisher method, is asymptotically B-optimal, however. That is, as the total sample size of the k studies tends to infinity, no other combined test procedure achieves a lower significance level with the same sample size. Note that the logit method is an example of a test that is optimal by one criterion (asymptotic B-efficiency) but suboptimal by the admissibility criterion for combining p-values based on statistics from the exponential family. Simulation studies by Mudholkar and George (1979) of the power of the logit procedure suggest that it may be nearly optimal for a variety of situations.

The logit test also responds symmetrically to p-values near zero and near unity. Therefore, the logit test can also be used in situations where the direction of deviation from H_0 is not known in advance, that is, when alternative H_3 (Section A) obtains.

A modification of the logit method permits the investigator to assign differential weights to all studies in the combined test statistic. If $v_1, \ldots, v_k$ are nonnegative weights, then the weighted logit combination procedure combines $p_1, \ldots, p_k$ by

$$L_w = v_1 \log \frac{p_i}{1 - p_1} + \cdots + v_k \log \frac{p_k}{1 - p_k}. \tag{10}$$

The statistic L_w (except for a constant) has a distribution that can be approximated by Student's t-distribution. More specifically,

$$L_w^* = L_w/\sqrt{c_w}$$

has a t-distribution (approximately) with m degrees of freedom, where

$$c_w = 3m/(m - 2)\pi^2(v_1^2 + \cdots + v_k^2),$$
$$m = 4 + 5(v_1^2 + \cdots + v_k^2)^2/(v_1^4 + \cdots + v_k^4).$$

Thus, the test of H_0 using L_w consists in comparing L_w^* with the appropriate critical value of the t-distribution with m degrees of freedom.

C.5 OTHER PROCEDURES FOR COMBINING TESTS

Several other functions for combining p-values have been proposed. In 1933 Karl Pearson suggested combining p-values via the product

$$(1 - p_1)(1 - p_2)\cdots(1 - p_k).$$

Other functions of the statistics $p_i^* = \text{Min}\{p_i, 1 - p_i\}$, $i = 1, \ldots, k$, were suggested by David (1934) for the combination of two-sided test statistics, which treat large and small values of the p_i symmetrically. Neither of these procedures has a convex acceptance region, so these procedures are not admissible for combining test statistics from the one-parameter exponential family. These proposals have not received a great deal of examination in the literature. The results of simulation studies by Mudholkar and George (1979) suggest that Fisher's method is markedly more powerful than Pearson's method in most situations.

A quite different combined test procedure was proposed by Edgington (1972a,b). Edgington proposed combining p-values by taking the sum

$$S = p_1 + \cdots + p_k, \tag{11}$$

and gave a tedious but straightforward method for obtaining significance levels for S. A large sample approximation to the significance levels of S is given in Edgington (1972b). Although it is a monotone combination procedure and therefore is admissible, Edgington's method is generally thought to be a poor procedure since one large p-value can overwhelm many small values that compose the statistic S. However, there have been almost no numerical investigations of this procedure.

A general method for obtaining admissible combination procedures was proposed by Lipták (1958). He argued that independent p-values $p_1, \ldots, p_k$ could be combined by first transforming each p_i to an inverse probability distribution function, not necessarily the normal distribution. More specifically, for any cumulative distribution function F, let $Z_i = F^{-1}(p_i)$, and let $v_1, \ldots, v_k$ be nonnegative weights. Now combine the tests by the statistic

$$T = v_1 Z_1 + \cdots + v_k Z_k = v_1 F^{-1}(p_1) + \cdots + v_k F^{-1}(p_k). \tag{12}$$

This method is quite general. In fact, the Fisher procedure and the normal procedure are both examples of (12). In the first case F is the chi-square distribution with two degrees of freedom, and in the second case F is the normal distribution. Any other distribution whose convolution is easy to calculate (e.g., any chi-square distribution) could also be used. In fact, Berk and Cohen (1979) have shown that any unweighted procedure using inverse chi-square cumulative distribution functions [i.e., $F^{-1}(p)$ has a chi-square distribution with k degrees of freedom] is asymptotically B-optimal.

Lancaster (1961) suggested using different distributions $F_1, \ldots, F_k$ to provide a weighted statistic whose null distribution is easy to calculate. He suggested the statistic

$$T = F_1^{-1}(p_1) + \cdots + F_k^{-1}(p_k), \tag{13}$$

where now $F_1, \ldots, F_k$ are chi-square cumulative distribution functions with $a_1, \ldots, a_k$ degrees of freedom, respectively. When H_0 is true, T is distributed as a chi-square variate with $a = a_1 + \cdots + a_k$ degrees of freedom. The effect of making some a_i larger than others is to give more weight to some of the p-values. However, there has been relatively little investigation of such methods. One result is that "weighting" by using unequal a_i yields a combined test procedure that is asymptotically B-optimal (Berk & Cohen, 1979).

Still another alternative for the combination of independent test statistics is the direct combination of individual test statistics. For example, if the test statistics $T_1, \ldots, T_k$ from the original studies have chi-square distributions with $a_1, \ldots, a_k$ degrees of freedom, respectively, then

$$T = T_1 + \cdots + T_k \tag{14}$$

has a chi-square distribution with $a = a_1 + \cdots + a_k$ degrees of freedom. The use of (14) is identical to Fisher's method when $a_1 = a_2 = \cdots = a_k = 2$ and has been found to have about the same power as Fisher's method in many situations (Bhattacharya, 1961; Koziol & Perlman, 1978). If the individual test statistics are normally distributed with known variance, then a direct combination of test statistics is also feasible (Schaafsma, 1968).

The direct combination of individual test statistics is less attractive when the convolution of the statistics does not have a simple form. For example, the sum of independent Student t-variates is known (Walker & Saw, 1978), but the distribution is not simple. One alternative (Winer, 1971) is to treat a t-variate with m_i degrees of freedom as if it were normally distributed with variance $m_i/(m_i - 2)$, and treat the sum of t-variates as a sum of normal variates. This approximation will obviously be satisfactory only when m is large. For a further discussion of combining independent t-tests, see Oosterhoff (1969).

C.6 SUMMARY OF COMBINED TEST PROCEDURES

Several combined test procedures were reviewed in Section C. General analytic results about these procedures were also described. The nature and interrelationship of these procedures are illustrated in Table 3, which gives a description of each procedure, a summary of the analytic results relevant to that procedure, and the reference distribution for the test statistic. It seems that Fisher's test is perhaps the best one to use if there is no indication of particular alternatives.

TABLE 3

Summary of Combined Test Procedures

Procedure	Test statistic	Critical value distribution	Admissible for exponential family	Asymptotically *B*-optimal
Uniform (Tippett)	$p_{[1]} = \text{Min}\{p_1, \ldots, p_k\}$	$p_{[1]} < 1 - (1 - \alpha)^{1/k}$	Yes	No
Uniform (Wilkinson)	$p_{[r]} = r$th smallest value	$\text{Beta}(r, k - r + 1)$	No	No
Inverse chi-square (Fisher)	$P = -2 \sum_{i=1}^{k} \log p_i$	Chi-square with $2k$ degrees of freedom	Yes	Yes
Inverse normal	$Z = \frac{1}{\sqrt{k}} \sum_{i=1}^{k} \Phi^{-1}(p_i)$	Standard normal	Yes	No
Logit	$L = \sum_{i=1}^{k} \log \frac{p_i}{1 - p_i}$, $L^* = \sqrt{cL}$, $c = 3(5k + 4)/\pi^2 k(5k + 2)$	Student's t with $5k + 4$ degrees of freedom	No	Yes

D. THE USES AND INTERPRETATION OF COMBINED TEST PROCEDURES IN RESEARCH SYNTHESIS

Combined test procedures were developed to combine the results of significance tests from different research studies. It may therefore seem peculiar to discuss the uses of these techniques in quantitative research synthesis in the social sciences. Suppose we have a series of k independent studies. If each study tests a null hypothesis about the effect of (the same) treatment, an abvious application of combined test procedures is to combine the k independent tests of the treatment effect. Some researchers (e.g., Rosenthal, 1978; Cooper, 1979) advocate the use of omnibus combined test procedures to combine tests of treatment effects.

In spite of the intuitive appeal of using combined test procedures to combine tests of treatment effects, there frequently are problems in the interpretation of results of such a test of the significance of combined results (see, e.g., Wallis, 1942, or Adcock, 1960). Just what can be concluded from the results of an omnibus test of the significance of combined results? Recall that the null hypothesis of the combined test procedure is

$$H_0: \qquad \theta_1 = \theta_2 = \cdots = \theta_k = 0;$$

that is, H_0 states that the treatment effect is zero in *every* study. If we reject H_0 using a combined test procedure, we may safely conclude that H_0 is false. However, H_0 is false if *at least one* of $\theta_1, \ldots, \theta_k$ is different from zero. Therefore, H_0 could be false when $\theta_1 > 0$ and $\theta_2 = \cdots = \theta_k = 0$. It is doubtful if a researcher would regard such a situation as persuasive evidence of the efficacy of a treatment.

The difficulty in the interpretation of omnibus tests of the significance of combined results stems from the nonparametric nature of the tests. Rejection of the combined null hypothesis allows the investigator to conclude only that the omnibus null hypothesis is false. Errors of interpretation usually involve attempts to attend a parametric interpretation to the rejection of H_0. For example, an investigator might incorrectly conclude that because H_0 is rejected, the treatment effects are greater than zero (Adcock, 1960). Alternatively, an investigator might incorrectly conclude that the treatment showed a consistent effect across studies (i.e., $\theta_1 = \cdots = \theta_k$). Neither parametric interpretation of the rejection of H_0 is warranted.

The purpose of quantitative research synthesis in the social sciences is usually to make rather general yet precise conclusions from the data. Frequently this means that the investigator wishes to draw conclusions about the magnitude, direction, and consistency of experimental effects. Omnibus tests of the significance of combined results are poorly suited for this purpose.

On the other hand, techniques based on combination of effect sizes do support inferences about direction, magnitude, and consistency of effects. Therefore, statistical analyses based on effect sizes are preferable for most research synthesis problems in the social sciences.

An important application of omnibus test procedures is to combine the results of dissimilar studies to screen for any effect. For example, combined test procedures can be used to test whether a treatment has an effect *on any* of a series of different outcome variables. Alternatively, combined test procedures can be used to combine the results of effect size analyses based on different outcome variables. Combined test procedures can even be used to combine the results of related analyses computed using different parameters such as correlation coefficients or effect sizes.

E. TECHNICAL COMMENTARY

Section C.1: To show that Tippett's procedure is monotone, note that $p_1^* \leq p_1, \ldots, p_k^* \leq p_k$ implies that $p_{[1]}^* \leq p_{[1]}$. Consequently, if

$$p_{[1]} \leq 1 - (1 - \alpha)^{1/k},$$

then $p_{[1]}^* \leq 1 - (1 - \alpha)^{1/k}$, and hence the procedure is monotone.

To show that the acceptance region is convex, suppose that $(p_1', \ldots, p_k')$ and $(p_1'', \ldots, p_k'')$ are in the acceptance region:

$$p_{[1]}' > 1 - (1 - \alpha)^{1/k}, \qquad p_{[1]}'' > 1 - (1 - \alpha)^{1/k}.$$

Let $\bar{p}_i = (p_i' + p_i'')/2$, $i = 1, \ldots, k$. Then

$$\begin{aligned} \bar{p}_{[1]} &= \text{Min}\{\bar{p}_1, \ldots, \bar{p}_k\} \geq \tfrac{1}{2}\,\text{Min}\{p_1', \ldots, p_k'\} + \tfrac{1}{2}\,\text{Min}\{p_1'', \ldots, p_k''\} \\ &= \tfrac{1}{2}(p_{[1]}' + p_{[1]}'') > 1 - (1 - \alpha)^{1/k}, \end{aligned}$$

so that $(\bar{p}_1, \ldots, \bar{p}_k)$ is also in the acceptance region, which means that the acceptance region is convex.

CHAPTER 4

Vote-Counting Methods

In Chapter 3 we discussed distribution-free or nonparametric methods based on p-values for testing the statistical significance of combined results. Although these omnibus tests can be used to summarize social science data when very little information is available on each study, the extreme nonparametric nature of these tests limits their usefulness. In a sense, these tests ignore too much of the parametric structure to be useful in obtaining detailed conclusions about indices of effect magnitude such as correlation coefficients or standardized mean differences.

Vote-counting methods also require very little information from each study. We typically need to know the sign of a mean difference or correlation or whether a hypothesis test yields a statistically significant result. Vote-counting methods are partially parametric in the sense that they permit inferences about scale-invariant indices of effect size.

Note that the vote-counting methods described in this chapter differ from the vote-counting or "box-score" techniques that are sometimes used in research reviews (see Light & Smith, 1971). The conventional vote-counting or box-score review attempts to draw inferences about the existence of treatment effects by sorting studies into categories according to the outcome of tests of hypotheses reported in the studies. The inference procedures used in conventional vote-counting procedures are inherently flawed and are likely to be misleading (see Section A). In this chapter we discuss statistical methods for the analysis of vote-count data that provide explicit estimates of effect magnitude parameters such as the correlation coefficient or standardized mean difference.

A. THE INADEQUACY OF CONVENTIONAL VOTE-COUNTING METHODOLOGY

Conventional vote-counting or box-score methodology uses the outcome of the test of significance in a series of replicated studies to draw conclusions about the magnitude of the treatment effect. Intuitively, if a large proportion of studies obtain statistically significant results, then this could be evidence that the effect is not zero. Conversely, if few studies find statistically significant results, then the combined evidence for a nonzero effect would seem to be weak.

Light and Smith (1971) first called attention to the logic described above and coined the term *vote counting* to describe the procedure. In their formulation a number of studies compare a group that receives some experimental treatment with one that receives no treatment. In the vote-counting method, the available studies are sorted into three categories: those that yield significant results in which the mean difference is positive, those that yield significant results in which the mean difference is negative, and those that yield nonsignificant results.

> If a plurality of studies falls into any of these three categories, with fewer falling into the other two, the modal category is declared the winner. This modal categorization is then assumed to give the best estimate of the true relationship between the independent and dependent variables. (Light & Smith, 1971, p. 433)

In particular, the effect (assumed to be the same for all studies) is greater than zero if a plurality, that is, more than one-third of the studies, shows statistically significant results in the positive direction. The use of other criteria, such as a majority rather than a plurality or a four-fifths majority, is also common among research reviewers. Is one fraction preferable to another? And, if so, which fraction is best? The arguments of this section suggest that the use of any such fraction can be a poor practice.

Suppose that k "identical" studies are to be integrated and that in each study a statistic T is calculated. For example, in studies of two groups, T might correspond to Student's t, whereas in correlational studies, T might be a sample correlation coefficient. Assume that the population effect magnitude that determines the distribution of T is the same for all studies. This assumption corresponds to assuming that the standardized mean difference (in the case of two-group studies) or the population correlation coefficient (in the case of correlational studies) is the same for all studies to be integrated. If these assumptions are met, then we can calculate the probability that any given study yields a positive significant result. In general, this probability depends on the sample size n for the study and the population effect magnitude δ.

The probability of a positive significant result is

$$p \equiv \text{Prob}\{\text{significant result} \mid \delta, n\} = \int_{C_\alpha}^{\infty} f(t; \delta, n)\, dt,$$

where $f(t; \delta, n)$ is the probability density function of the statistic T in samples of size n with effect magnitudes δ, and C_α is the critical value of the statistic T.

The process of accumulating positive significant results is viewed as a series of trials, in which each trial is a success (a positive significant result) or a failure (a nonsignificant or negative significant result). Furthermore, each study has an equal probability p of yielding a significant result. This probability is simply the power of the statistical test in each study. It is well known that the number of successes arising from a series of such trials has the binomial distribution with parameter p.

In vote-counting procedures we decide that an effect δ is greater than zero if the proportion of studies with positive significant results exceeds one-third (or more generally the cutoff value C_0). In small samples the binomial distribution is used to calculate the probability that the proportion of a cluster of k studies yielding positive significant results exceeds a preset criterion C_0. If X denotes the number of successes, then

$$\begin{aligned}\text{Prob}\{\text{proportion of successes} > C_0\} &= \text{Prob}\left\{\frac{X}{k} > C_0\right\} \\ &= \sum_{i=[C_0 k]+1}^{k} \binom{k}{i} p^i (1-p)^{k-i},\end{aligned}$$

where $[a]$ is the greatest integer less than or equal to a, and $0 \leq C_0 \leq 1$.

This procedure permits us to evaluate the probability that vote counting using the usual criterion of $C_0 = \frac{1}{3}$ detects an effect for various sample sizes n and effect sizes δ. For example, suppose that the statistical tests used are two-tailed t-tests ($\alpha = 0.05$) and that each study in the cluster has an identical two-group design, with n subjects per group. Then the effect size is

$$\delta = (\mu^{\mathrm{E}} - \mu^{\mathrm{C}})/\sigma,$$

where μ^E and μ^C are the population means for the experimental and control groups, respectively, and σ is the common population standard deviation for each group. Table 1 provides a tabulation of the probability that a vote count with $C_0 = \frac{1}{3}$ fails to detect an effect for various sample sizes, effect sizes, and numbers of studies to be integrated. When effect sizes are moderate to small ($\delta \le 0.5$), standard vote counting frequently fails to detect the effects. More important, the situation does not always improve as the number k of studies integrated increases. For example, when $\delta \le 0.3$, the probability that a standard vote count detects an effect decreases as k increases from 10 to 50.

The surprising fact that the vote-counting method may tend to make the wrong decision more often as the amount of evidence (number of studies) increases was explained by Hedges and Olkin (1980). On an intuitive level, the hypothesis test in each study has a chance of yielding an incorrect decision. In this case the effect is nonzero and the incorrect decision is failing to detect a real (nonzero) effect: a type II error. Because the possible errors are of the same type (failure to detect the nonzero effect), they *do not* cancel one another.

TABLE 1

Probability That a Vote Count with $C_0 = \frac{1}{3}$ Fails to Detect an Effect for Various Sample Sizes per Group, n, Effect Sizes δ, and Total Numbers of Studies, k[a]

		Effect size $\delta \equiv (\mu^E - \mu^C)/\sigma$							
k	n	0.1	0.2	0.3	0.4	0.5	0.6	0.7	0.8
10	10	1.000	0.999	0.998	0.994	0.985	0.968	0.935	0.880
10	20	1.000	0.998	0.990	0.966	0.906	0.987	0.606	0.395
10	30	0.999	0.995	0.975	0.906	0.947	0.502	0.252	0.089
10	40	0.999	0.991	0.950	0.813	0.547	0.254	0.073	0.012
10	50	0.999	0.986	0.914	0.694	0.358	0.105	0.016	0.001
20	10	1.000	1.000	1.000	0.999	0.997	0.989	0.966	0.914
20	20	1.000	1.000	0.998	0.988	0.941	0.800	0.545	0.265
20	30	1.000	1.000	0.993	0.941	0.747	0.400	0.118	0.016
20	40	1.000	0.999	0.978	0.834	0.463	0.119	0.011	0.000
20	50	1.000	0.997	0.948	0.672	0.222	0.023	0.001	0.000
50	10	1.000	1.000	1.000	1.000	1.000	1.000	0.998	0.986
50	20	1.000	1.000	1.000	1.000	0.994	0.915	0.589	0.174
50	30	1.000	1.000	1.000	0.994	0.862	0.363	0.035	0.000
50	40	1.000	1.000	0.999	0.942	0.461	0.036	0.000	0.000
50	50	1.000	1.000	0.995	0.773	0.124	0.001	0.000	0.000

[a] Each of the k replicated studies has common sample size n. A two-tail t-test is used to test mean differences at the 0.05 level of significance. An effect is detected if the proportion of positive significant results exceeds one-third.

In fact, as the number of studies becomes large the proportion of studies yielding significant results is approximately the average power of the test. Indeed, if the average power p of the statistical tests is smaller than the cutoff criterion C_0, then the probability that the vote count makes the correct decision tends to zero as the number k of studies increases (Hedges & Olkin, 1980). Thus if $p < C_0$ the power of the vote count as a decision procedure tends to zero as k becomes large. Moreover, for any population effect size δ, there is some average sample size for which the power of the vote count will tend to zero.

Clearly, if the combinations of effect sizes and sample sizes for which vote counting fails (has power tending to zero) are not typical of social science research, then the theoretical failure of the procedure might not have practical consequences. Unfortunately vote counting can fail for sample sizes and effect sizes that most commonly appear in educational and psychological research. Suppose that two-tailed t-tests are used in a sequence of studies each of which has n subjects per group and an effect size $\delta = (\mu^E - \mu^C)/\sigma$ as defined previously. Figure 1 is a graph of the relation between total sample size in each study and the effect size δ below which the power of a vote count tends to zero for various values of C_0. When the total sample size for each study ($N = 2n$) is 20, δ must be greater than 0.5 for the power of the vote count to be greater than zero for large k with $C_0 = \frac{1}{3}$. (A *medium* effect size is defined by Cohen, 1969, to be $\delta = 0.5$.) Hence, for studies with samples of size 20 and a medium-sized effect, the power of the standard vote count approaches zero. With a sample size of 50 in each study, the power of the standard vote count ($C_0 = \frac{1}{3}$) tends to zero for δ less than 0.36. Even when each study has a sample size of 100, the power tends to zero if δ is less than 0.26 with $C_0 = \frac{1}{3}$.

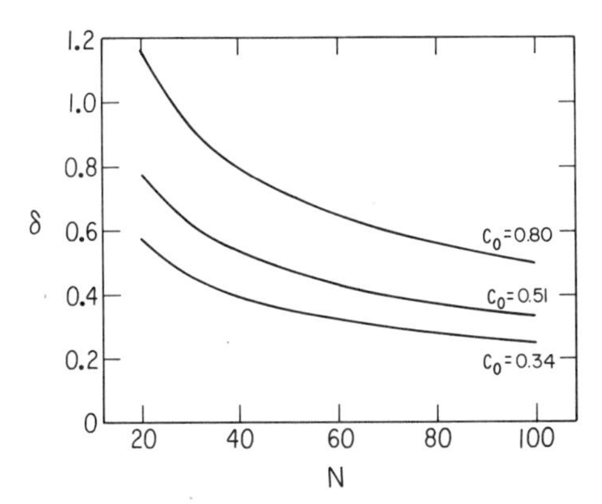

Fig. 1 Effect size $\delta = (\mu^E - \mu^C)/\sigma$ below which the power of a vote count tends to zero for a large number of studies with common sample size $n^E + n^C = N$.

Cohen (1969) defines a *small* effect size to be $\delta = 0.20$ and argues that many interesting educational and psychological effects are of that magnitude. Thus vote counts that use the plurality criterion ($C_0 = \frac{1}{3}$) have to detect small effects with power that tends to zero, even when each study has a sample size as large as 100. This suggests that the plurality criterion is inadequate for the range of effect sizes and sample sizes most common in educational research.

Remark: For simplicity of exposition, the arguments presented are developed using a model of equal sample and effect sizes for each study. However, a similar result on the large sample properties of vote counts holds in more general situations. The p for each study is, in effect, the power of the statistical test for each study. If, for each study, the power is bounded away from zero, then the large sample power of the vote counting method tends to zero whenever the average p is less than C_0. Thus if this mild regularity condition of nonvanishing powers is satisfied and if the average power is less than C_0, then the power of the vote-counting method tends to zero as k becomes large.

B. COUNTING ESTIMATORS OF CONTINUOUS PARAMETERS

Consider a collection of k independent studies designed to estimate a common parameter θ. Each study produces an estimator T of the parameter θ yielding estimates $T_1, \ldots, T_k$ of the respective parameters $\theta_1, \ldots, \theta_k$. If there is no effect in the ith study, then the parameter θ_i will be zero. For the ith experiment (study) we test the hypothesis

$$H_{0i}: \qquad \theta_i = 0$$

by the procedure

$$\text{reject } H_{0i} \text{ if } T_i > C,$$

where C is a critical value obtained from the distribution of T_i.

Remark: Here we have treated the hypothesis test as one-sided. The case of two-sided tests can be treated in this formulation by using the statistic T_i^2 as an estimator of θ_i^2 or using T_i^2 to test the hypothesis that $\theta_i^2 = 0$. Vote-counting estimators derived from the T_i^2 can then be developed to estimate θ_i^2. Note that it will not, in general, be possible to use a two-tailed test statistic T_i^2 to estimate θ if θ can take both positive and negative values.

The essential feature of vote-counting methods is that the values of the estimators (or test statistics) $T_1, \ldots, T_k$ are *not* observed. Instead we observe only the number U of experiments for which the null hypothesis H_0 was rejected, i.e., where $T_i > C$. The problem is to obtain an estimate and confidence interval for the common parameter value θ using only a count of these votes.

The data obtained are the number U of positive results in a series of k independent trials, where each trial results in a success (null hypothesis is rejected) or a failure (null hypothesis is not rejected). The sample proportion U/k of positive results is the maximum likelihood estimate of the probability $p_C(\theta)$ of success. If for a fixed value of C the power function $p_C(\theta)$ of a T is a strictly monotone function of θ, then we can obtain the maximum likelihood estimator $\hat{\theta}$ of θ from the maximum likelihood estimator of $p_C(\theta)$. This is accomplished by solving for $\hat{\theta}$ in the equation

$$p_C(\hat{\theta}) = U/k. \tag{1}$$

Unfortunately, the function $p_C(\theta)$ is usually quite complicated, so that simple solutions may be unavailable. For certain problems tables of $p_C(\theta)$ are available, and these permit one to readily solve Eq. (1). Tables for the effect size and correlation are given in Sections E and F.

Because the power function $p_C(\theta)$ is a monotone function of θ, confidence intervals for $p_C(\theta)$ can be translated to confidence intervals for θ. Thus we first need to discuss methods for determining confidence intervals for $p_C(\theta)$.

Confidence Intervals from alternative vote-counting methods are available for obtaining a confidence interval for a proportion, each of which generates a procedure for obtaining a confidence interval for the parameter θ.

C. CONFIDENCE INTERVALS FOR PARAMETERS BASED ON VOTE COUNTS

There are several methods for obtaining confidence intervals for $p_C(\theta)$. For example, Clopper and Pearson (1934) have provided nomographs for obtaining exact confidence intervals for $p_C(\theta)$ based on the sample size k and the sample proportion U/k. These nomographs (and extensions thereof) are reproduced in Appendix E for $\alpha = 0.01, 0.05, 0.10$, and 0.20. To find a confidence interval for $p_C(\theta)$, find the obtained value of U/k on the abscissa, then move vertically until the bands with the appropriate sample size k are intersected. The ordinate values of the intersections are the confidence limits for $p_C(\theta)$ given U/k. For example, if $k = 10$ and $U/k = 0.070$, the 95-percent confidence interval for $p_C(\theta)$ from Appendix E is (0.35, 0.97). Approximations to the distribution of U/k are also available; these can be used to obtain very accurate confidence intervals for $p_C(\theta)$ (see, e.g., Molenaar, 1970). We do not deal with these methods, but instead use the simpler asymptotic theory for the distribution of U/k to obtain confidence intervals for $p_C(\theta)$.

Two principal methods for obtaining confidence intervals for $p_C(\theta)$ are based on the large sample normal approximation to the binomial distribution. For large k, the sample proportion $\hat{p} = U/k$ is approximately normally distributed with a mean $p \equiv p_C(\theta)$ and variance $p(1-p)/k$. Consequently, for large k,

$$z = \sqrt{k}(\hat{p} - p)/\sqrt{p(1-p)} \tag{2}$$

has a standard normal distribution.

C.1 USE OF NORMAL THEORY

The normal approximation (2) can be used to obtain a confidence interval for p by estimating the variance $p(1-p)/k$ by $\hat{p}(1-\hat{p})k$ and then using this estimated variance to obtain an approximate confidence interval. [Actually, any consistent estimator of $\hat{p}$ will serve to estimate the variance $\hat{p}(1-\hat{p})/k$.] Consequently a $100(1-\alpha)$-percent confidence interval (p_L, p_U) for p is given by

$$p_L = \hat{p} - C_{\alpha/2}\sqrt{\frac{\hat{p}(1-\hat{p})}{k}}, \qquad p_U = \hat{p} + C_{\alpha/2}\sqrt{\frac{\hat{p}(1-\hat{p})}{k}}, \tag{3}$$

where C_α is the two-tailed critical value of the standard normal distribution. Using a table of $p_C(\theta)$ as a function of θ, we can find that value of θ which solves the equation $p_C(\theta) = a$, whenever a is between zero and unity. The confidence interval (p_L, p_U) for p can be translated to a confidence interval (θ_L, θ_U) for θ by solving

$$p_C(\theta_L) = p_L, \qquad p_C(\theta_U) = p_U.$$

C.2 USE OF CHI-SQUARE THEORY

A second approach for obtaining a confidence interval for p is to note that the square of a standard normal variate has a chi-square distribution with one degree of freedom, so that from (2), for large k,

$$z^2 = k(\hat{p} - p)^2/p(1-p) \tag{4}$$

has a chi-square distribution with one degree of freedom.

The chi-square distribution can be used to obtain a confidence interval for p by setting

$$k(\hat{p} - p)^2/p(1 - p) = C_\alpha, \tag{5}$$

where C_α is the upper critical value of the chi-square distribution with one degree of freedom. Using (5), we need to find the two values of p that solve this equation. It is informative to plot the function

$$g(p) = (\hat{p} - p)^2/p(1 - p).$$

Plots for $\hat{p} = 0.6$ and $\hat{p} = 0.7$ are given in Fig. 2.

The value of C_α is usually between 1.28 and 2.32 (for α between 0.20 and 0.01). Since k is larger than 5, the largest set of values for C_α/k is between 0.32 and 1.08. We enlarge the scale in this portion of the range of $g(p)$ to enable easier reading of the graph.

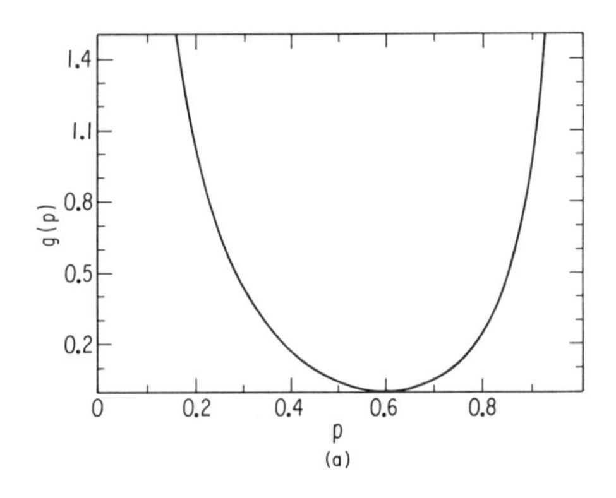

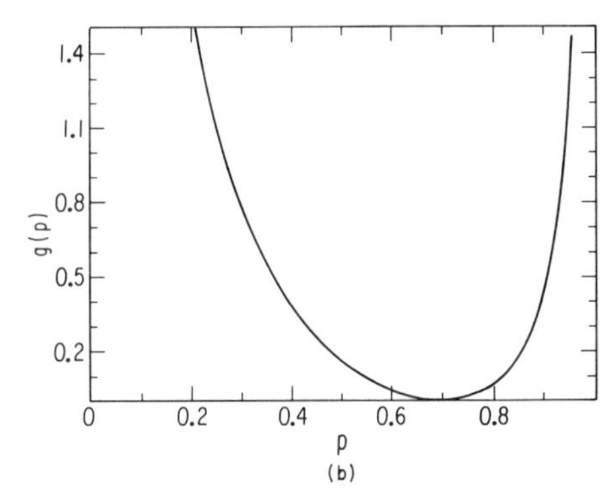

Fig. 2 Graph of $g(p) = (\hat{p} - p)^2/p(1 - p)$ for (*a*) $\hat{p} = 0.6$ and (*b*) $\hat{p} = 0.7$.

We can also use the quadratic equation to find the two points $\tilde{p}_L$ and $\tilde{p}_U$ where $g(\tilde{p}) = C_\alpha/k$. These two points constitute a $100(1 - \alpha)$-percent confidence interval for p. Analytically, we obtain

$$\tilde{p}_L = \frac{2\hat{p} + b - \sqrt{b^2 + 4b\hat{p}(1 - \hat{p})}}{2(1 + b)}, \qquad \tilde{p}_U = \frac{2\hat{p} + b + \sqrt{b^2 + 4b\hat{p}(1 - \hat{p})}}{2(1 + b)}, \tag{6}$$

where $b = C_\alpha/k$. Using a table of $p_C(\theta)$ as a function of θ and solving $p_C(\tilde{\theta}_L) = \tilde{p}_L$ and $p_C(\tilde{\theta}_U) = \tilde{p}_U$ yields the corresponding $100(1 - \alpha)$-percent confidence interval $(\tilde{\theta}_L, \tilde{\theta}_U)$ for θ.

These two methods generally give different confidence intervals, although they give similar answers when k is large. The chi-square procedure does not involve substituting of $\hat{p}$ for p in the expression for the variance of $\hat{p}$. Although one might expect this method to yield more accurate confidence intervals, this is not always the case.

D. CHOOSING A CRITICAL VALUE

The vote-counting method involves counting the number of studies in which the test statistic T_i exceeds an arbitrary critical value C, and is designed for use when we do not have access to the actual values of the statistics $T_1, \ldots, T_k$. Since we are not usually in a position to choose any critical value C, we often have to use a value of C that is a conventional critical value that yields a particular significance level α. In this case, vote counting corresponds to counting the number of studies that have statistically significant results at the 100α-percent significance level. This use of C corresponding to a conventional significance level such as $\alpha = 0.05$ is frequently a good choice because studies often report statistical significance of results even if the studies do not include other summary statistics.

Another useful, if unconventional, critical value is that corresponding to $\alpha = 0.50$. In this case, vote counting corresponds to counting the proportion of positive mean differences or positive sample correlation coefficients obtained in k studies. The use of a critical value of $C = 0$ corresponding to a significance level $\alpha = 0.50$ is often useful when the reports of studies give the direction of the effect but do not give other information about the magnitude or statistical significance of the effect.

E. ESTIMATING AN EFFECT SIZE

Consider a collection of k studies in which each study consists of two groups, an experimental (E) group and a control (C) group. The function that relates the effect size parameter δ to the probability that $T_i > C$ must be the same across studies to use the method developed in this section. Because this probability is a function of n_i^E and n_i^C in each study, we require that the sample sizes $n_i^E = n_i^C = n$ be the same for all k studies. A discussion of ways for dealing with unequal sample sizes is given in Section H.

Denote the score of the jth individual in the experimental and control groups of the ith study by Y_{ij}^E and Y_{ij}^C, respectively. We assume that these studies are independent and that each of the Y_{ij}^E and Y_{ij}^C is normally distributed. Schematically the data are as follows:

	Data	
Study	Experimental	Control
1	$Y_{11}^E, \ldots, Y_{1n}^E$	$Y_{11}^C, \ldots, Y_{1n}^C$
$\vdots$	$\vdots$	$\vdots$
k	$Y_{k1}^E, \ldots, Y_{kn}^E$	$Y_{k1}^C, \ldots, Y_{kn}^C$

The means and variances are

	Experimental		Control	
Study	Mean	Variance	Mean	Variance
1	μ_1^E	σ_1^2	μ_1^C	σ_1^2
$\vdots$	$\vdots$	$\vdots$	$\vdots$	$\vdots$
k	μ_k^E	σ_k^2	μ_k^C	σ_k^2

More succinctly,

$$Y_{ij}^E \sim \mathcal{N}(\mu_i^E, \sigma_i^2), \qquad Y_{ij}^C \sim \mathcal{N}(\mu_i^C, \sigma_i^2), \qquad i = 1, \ldots, k, \quad j = 1, \ldots, n$$

Define the effect size parameter δ_i for the ith experiment by

$$\delta_i = \frac{\mu_i^E - \mu_i^C}{\sigma_i}, \qquad i = 1, \ldots, k. \tag{7}$$

It is important to note that the effect size parameter δ_i is not a function of the arbitrary scale of measurement in any particular experiment.

Assume that the effect sizes δ_i are constant across the studies, that is, that $\delta_1 = \delta_2 = \cdots = \delta_k = \delta$. The (unobserved) estimate of the parameter δ_i is the sample standardized mean difference (Glass's effect size) g_i, given by

$$g_i = \frac{\bar{Y}_i^E - \bar{Y}_i^C}{S_i}, \qquad i = 1, \ldots, k, \tag{8}$$

TABLE 2

Probability That the Sample Mean of the Experimental Group Exceeds the Sample Mean of the Control Group When the Common Size per Group Is n and the Population Standardized Mean Difference Is δ

	Standardized effect size											
n	0.00	0.01	0.02	0.03	0.04	0.05	0.06	0.07	0.08	0.09	0.10	0.1
1	0.500	0.503	0.506	0.508	0.511	0.514	0.517	0.520	0.523	0.525	0.528	0.528
2	0.500	0.504	0.508	0.512	0.516	0.520	0.524	0.528	0.532	0.536	0.540	0.540
3	0.500	0.505	0.510	0.515	0.520	0.524	0.529	0.534	0.539	0.544	0.549	0.549
4	0.500	0.506	0.511	0.517	0.523	0.528	0.534	0.539	0.545	0.551	0.556	0.556
5	0.500	0.506	0.513	0.519	0.525	0.532	0.538	0.544	0.550	0.557	0.563	0.563
6	0.500	0.507	0.514	0.521	0.528	0.534	0.541	0.548	0.555	0.562	0.569	0.569
7	0.500	0.508	0.515	0.522	0.530	0.537	0.545	0.552	0.560	0.567	0.574	0.574
8	0.500	0.508	0.516	0.524	0.532	0.540	0.548	0.556	0.564	0.571	0.579	0.579
9	0.500	0.508	0.517	0.525	0.534	0.542	0.551	0.559	0.567	0.576	0.584	0.584
10	0.500	0.509	0.518	0.527	0.536	0.544	0.553	0.562	0.571	0.580	0.588	0.588
11	0.500	0.509	0.519	0.528	0.537	0.547	0.556	0.565	0.574	0.584	0.593	0.593
12	0.500	0.510	0.520	0.529	0.539	0.549	0.558	0.568	0.578	0.587	0.597	0.597
13	0.500	0.510	0.520	0.530	0.541	0.551	0.561	0.571	0.581	0.591	0.601	0.601
14	0.500	0.511	0.521	0.532	0.542	0.553	0.563	0.574	0.584	0.594	0.604	0.604
15	0.500	0.511	0.522	0.533	0.544	0.554	0.565	0.576	0.587	0.597	0.608	0.608
16	0.500	0.511	0.523	0.534	0.545	0.556	0.567	0.578	0.590	0.600	0.611	0.611
17	0.500	0.512	0.523	0.535	0.546	0.558	0.569	0.581	0.592	0.604	0.615	0.615
18	0.500	0.512	0.524	0.536	0.548	0.560	0.571	0.583	0.595	0.606	0.618	0.618
19	0.500	0.512	0.525	0.537	0.549	0.561	0.573	0.585	0.597	0.609	0.621	0.621
20	0.500	0.513	0.525	0.538	0.550	0.563	0.575	0.588	0.600	0.612	0.624	0.624
21	0.500	0.513	0.526	0.539	0.552	0.564	0.577	0.590	0.602	0.615	0.627	0.627
22	0.500	0.513	0.526	0.540	0.553	0.566	0.579	0.592	0.605	0.617	0.630	0.630
23	0.500	0.514	0.527	0.540	0.554	0.567	0.581	0.594	0.607	0.620	0.633	0.633
24	0.500	0.514	0.528	0.541	0.555	0.569	0.582	0.596	0.609	0.622	0.636	0.636
25	0.500	0.514	0.528	0.542	0.556	0.570	0.584	0.598	0.611	0.625	0.638	0.638
50	0.500	0.520	0.540	0.560	0.579	0.599	0.618	0.637	0.655	0.674	0.692	0.692
100	0.500	0.528	0.556	0.584	0.611	0.638	0.664	0.690	0.714	0.738	0.760	0.760
200	0.500	0.540	0.579	0.618	0.655	0.692	0.726	0.758	0.788	0.816	0.841	0.841
400	0.500	0.556	0.611	0.664	0.714	0.760	0.802	0.839	0.871	0.898	0.921	0.921

TABLE 2 (*Continued*)

Standardized effect size												
0.2	0.3	0.4	0.5	0.6	0.7	0.8	0.9	1.0	1.1	1.2	1.3	1.4
0.556	0.584	0.611	0.638	0.664	0.690	0.714	0.738	0.760	0.782	0.802	0.821	0.839
0.579	0.618	0.655	0.692	0.726	0.758	0.788	0.816	0.841	0.864	0.885	0.903	0.920
0.597	0.643	0.688	0.730	0.769	0.804	0.836	0.865	0.890	0.911	0.929	0.944	0.957
0.611	0.664	0.714	0.760	0.802	0.839	0.871	0.898	0.921	0.940	0.955	0.967	0.976
0.624	0.682	0.736	0.785	0.829	0.866	0.897	0.923	0.943	0.959	0.971	0.980	0.986
0.636	0.698	0.756	0.807	0.851	0.887	0.917	0.940	0.958	0.972	0.981	0.988	0.992
0.646	0.713	0.773	0.825	0.869	0.905	0.933	0.954	0.969	0.980	0.988	0.992	0.996
0.655	0.726	0.788	0.841	0.885	0.919	0.945	0.964	0.977	0.986	0.992	0.995	0.997
0.664	0.738	0.802	0.856	0.898	0.931	0.955	0.972	0.983	0.990	0.994	0.997	0.998
0.673	0.749	0.814	0.868	0.910	0.941	0.963	0.978	0.987	0.993	0.996	0.998	0.999
0.680	0.759	0.826	0.880	0.920	0.950	0.970	0.983	0.990	0.995	0.998	0.999	1.000
0.688	0.769	0.836	0.890	0.929	0.957	0.975	0.986	0.993	0.996	0.998	0.999	1.000
0.695	0.778	0.846	0.899	0.937	0.963	0.979	0.989	0.995	0.998	0.999	1.000	1.000
0.702	0.786	0.855	0.907	0.944	0.968	0.983	0.991	0.996	0.998	0.999	1.000	1.000
0.708	0.794	0.863	0.914	0.950	0.972	0.986	0.993	0.997	0.999	1.000	1.000	1.000
0.714	0.802	0.871	0.921	0.955	0.976	0.988	0.994	0.998	0.999	1.000	1.000	1.000
0.720	0.809	0.878	0.928	0.960	0.979	0.990	0.996	0.998	0.999	1.000	1.000	1.000
0.726	0.816	0.885	0.933	0.964	0.982	0.992	0.996	0.999	1.000	1.000	1.000	1.000
0.731	0.822	0.891	0.938	0.968	0.984	0.993	0.997	0.999	1.000	1.000	1.000	1.000
0.736	0.829	0.897	0.943	0.971	0.987	0.994	0.998	0.999	1.000	1.000	1.000	1.000
0.742	0.834	0.902	0.947	0.974	0.988	0.995	0.998	0.999	1.000	1.000	1.000	1.000
0.746	0.840	0.908	0.951	0.977	0.990	0.996	0.999	1.000	1.000	1.000	1.000	1.000
0.751	0.846	0.912	0.955	0.979	0.991	0.997	0.999	1.000	1.000	1.000	1.000	1.000
0.756	0.851	0.917	0.958	0.981	0.992	0.997	0.999	1.000	1.000	1.000	1.000	1.000
0.760	0.856	0.921	0.962	0.983	0.993	0.998	0.999	1.000	1.000	1.000	1.000	1.000
0.841	0.933	0.977	0.994	0.999	1.000	1.000	1.000	1.000	1.000	1.000	1.000	1.000
0.921	0.983	0.998	1.000	1.000	1.000	1.000	1.000	1.000	1.000	1.000	1.000	1.000
0.977	0.999	1.000	1.000	1.000	1.000	1.000	1.000	1.000	1.000	1.000	1.000	1.000
0.998	1.000	1.000	1.000	1.000	1.000	1.000	1.000	1.000	1.000	1.000	1.000	1.000

TABLE 3

Probability That a Two-Sample t-Test Is Significant at the 5-Percent Level for Common Sample Size n per Group with Population Standardized Mean Difference Is δ

	Standardized effect size											
n	0.00	0.01	0.02	0.03	0.04	0.05	0.06	0.07	0.08	0.09	0.10	0.1
2	0.050	0.051	0.052	0.052	0.053	0.054	0.055	0.056	0.056	0.057	0.058	0.058
3	0.050	0.051	0.052	0.053	0.054	0.056	0.057	0.058	0.059	0.060	0.062	0.062
4	0.050	0.051	0.053	0.054	0.055	0.057	0.058	0.060	0.061	0.063	0.064	0.064
5	0.050	0.052	0.053	0.055	0.056	0.058	0.060	0.061	0.063	0.065	0.067	0.067
6	0.050	0.052	0.053	0.055	0.057	0.059	0.061	0.063	0.065	0.067	0.069	0.069
7	0.050	0.052	0.054	0.056	0.058	0.060	0.062	0.064	0.066	0.069	0.071	0.071
8	0.050	0.052	0.054	0.056	0.058	0.061	0.063	0.065	0.068	0.070	0.073	0.073
9	0.050	0.052	0.054	0.057	0.059	0.061	0.064	0.066	0.069	0.072	0.075	0.075
10	0.050	0.052	0.055	0.057	0.060	0.062	0.065	0.068	0.070	0.073	0.076	0.076
11	0.050	0.052	0.055	0.057	0.060	0.063	0.066	0.069	0.072	0.075	0.078	0.078
12	0.050	0.052	0.055	0.058	0.061	0.064	0.066	0.070	0.073	0.076	0.080	0.080
13	0.050	0.053	0.055	0.058	0.061	0.064	0.067	0.071	0.074	0.078	0.081	0.081
14	0.050	0.053	0.056	0.058	0.062	0.065	0.068	0.072	0.075	0.079	0.083	0.083
15	0.050	0.053	0.056	0.059	0.062	0.065	0.069	0.072	0.076	0.080	0.084	0.084
16	0.050	0.053	0.056	0.059	0.062	0.066	0.070	0.073	0.077	0.081	0.086	0.086
17	0.050	0.053	0.056	0.060	0.063	0.066	0.070	0.074	0.078	0.083	0.087	0.087
18	0.050	0.053	0.056	0.060	0.063	0.067	0.071	0.075	0.079	0.084	0.088	0.088
19	0.050	0.053	0.057	0.060	0.064	0.068	0.072	0.076	0.080	0.085	0.090	0.090
20	0.050	0.053	0.057	0.060	0.064	0.068	0.072	0.077	0.081	0.086	0.091	0.091
21	0.050	0.053	0.057	0.061	0.065	0.069	0.073	0.078	0.082	0.087	0.092	0.092
22	0.050	0.054	0.057	0.061	0.065	0.069	0.074	0.078	0.083	0.088	0.094	0.094
23	0.050	0.054	0.057	0.061	0.065	0.070	0.074	0.079	0.084	0.089	0.095	0.095
24	0.050	0.054	0.058	0.062	0.066	0.070	0.075	0.080	0.085	0.090	0.096	0.096
25	0.050	0.054	0.058	0.062	0.066	0.071	0.076	0.081	0.086	0.092	0.098	0.097
50	0.050	0.055	0.061	0.067	0.074	0.081	0.089	0.097	0.106	0.116	0.125	0.125
100	0.050	0.058	0.066	0.076	0.086	0.098	0.111	0.125	0.140	0.156	0.174	0.174
200	0.050	0.061	0.074	0.089	0.106	0.126	0.148	0.172	0.199	0.228	0.259	0.259
400	0.050	0.066	0.087	0.111	0.140	0.174	0.213	0.256	0.304	0.355	0.408	0.408

TABLE 3 (*Continued*)

	Standardized effect size											
0.2	0.3	0.4	0.5	0.6	0.7	0.8	0.9	1.0	1.1	1.2	1.3	1.4
0.067	0.077	0.088	0.099	0.112	0.126	0.141	0.157	0.174	0.191	0.210	0.229	0.250
0.075	0.091	0.109	0.129	0.152	0.177	0.205	0.235	0.267	0.301	0.337	0.374	0.413
0.082	0.103	0.127	0.155	0.187	0.222	0.261	0.304	0.348	0.396	0.444	0.493	0.543
0.088	0.113	0.144	0.179	0.219	0.264	0.313	0.366	0.422	0.478	0.536	0.592	0.647
0.093	0.123	0.159	0.201	0.250	0.304	0.362	0.424	0.488	0.552	0.614	0.674	0.730
0.098	0.132	0.174	0.223	0.279	0.341	0.408	0.477	0.548	0.616	0.681	0.741	0.795
0.103	0.142	0.189	0.244	0.308	0.377	0.451	0.527	0.602	0.673	0.738	0.796	0.845
0.108	0.150	0.203	0.265	0.335	0.412	0.492	0.573	0.650	0.722	0.786	0.840	0.884
0.112	0.159	0.217	0.285	0.362	0.445	0.530	0.615	0.694	0.765	0.825	0.875	0.914
0.117	0.167	0.230	0.304	0.388	0.476	0.566	0.653	0.732	0.801	0.858	0.903	0.936
0.121	0.176	0.244	0.324	0.413	0.507	0.600	0.688	0.767	0.833	0.885	0.925	0.953
0.125	0.184	0.257	0.342	0.437	0.536	0.632	0.720	0.797	0.860	0.908	0.942	0.966
0.130	0.192	0.270	0.361	0.461	0.563	0.661	0.750	0.824	0.883	0.926	0.956	0.975
0.133	0.200	0.282	0.379	0.483	0.589	0.689	0.776	0.848	0.902	0.941	0.966	0.982
0.138	0.207	0.295	0.396	0.506	0.614	0.714	0.800	0.868	0.918	0.953	0.974	0.987
0.141	0.215	0.307	0.414	0.527	0.638	0.738	0.822	0.886	0.932	0.962	0.981	0.991
0.145	0.223	0.320	0.431	0.547	0.660	0.760	0.842	0.902	0.944	0.970	0.985	0.993
0.149	0.230	0.332	0.447	0.567	0.682	0.781	0.859	0.916	0.954	0.976	0.989	0.995
0.153	0.238	0.344	0.463	0.586	0.702	0.799	0.875	0.928	0.962	0.981	0.992	0.997
0.157	0.245	0.356	0.479	0.605	0.721	0.817	0.889	0.938	0.968	0.985	0.994	0.998
0.161	0.253	0.367	0.495	0.623	0.739	0.833	0.902	0.947	0.974	0.988	0.995	0.998
0.164	0.260	0.379	0.509	0.640	0.756	0.848	0.913	0.955	0.979	0.991	0.996	0.999
0.168	0.267	0.390	0.525	0.656	0.772	0.861	0.923	0.961	0.982	0.993	0.997	0.999
0.172	0.275	0.401	0.539	0.672	0.787	0.874	0.932	0.967	0.986	0.994	0.998	0.999
0.257	0.438	0.634	0.799	0.909	0.966	0.990	0.998	1.000	1.000	1.000	1.000	1.000
0.407	0.681	0.880	0.970	0.995	1.000	1.000	1.000	1.000	1.000	1.000	1.000	1.000
0.637	0.911	0.990	1.000	1.000	1.000	1.000	1.000	1.000	1.000	1.000	1.000	1.000
0.881	0.995	1.000	1.000	1.000	1.000	1.000	1.000	1.000	1.000	1.000	1.000	1.000

where $\bar{Y}_i^E$ and $\bar{Y}_i^C$ are the experimental and control sample means, and s_i is the pooled within-group sample standard deviation in the ith experiment. The (unobserved) test statistic t_i is the two-sample t-statistic given by

$$t_i = g_i\sqrt{n/2}. \tag{9}$$

To estimate an effect size we count the number of times that $t_i > C_\alpha$, i.e., that the t_i value is "significant" at level α. Thus the probability $p_C(\delta)$ of success is the probability that a noncentral t-variate with $2n - 2$ degrees of freedom and noncentrality parameter

$$\delta\sqrt{n/2}$$

exceeds C_α. When $\alpha = 0.50$ the critical value $C_{0.50} = 0.0$ and the vote-counting procedure consists in counting the number of times the mean difference, and consequently the g_i, are positive.

Table 2 gives values of $p_0(\delta)$ as a function of δ and the common sample size n in each group. Only values of $p_0(\delta)$ equal to or greater than 0.50 are tabulated because proportions less than 0.50 correspond to negative values of δ and these can be obtained by symmetry. For example, if $n = 20$ and $p_0(\delta) = 0.376$, the value of δ cannot be obtained directly, since $p_0(\delta)$ is less than 0.50. The identity

$$p_0(-\delta) = 1 - p_0(\delta)$$

yields values for $p_0(\delta)$ less than 0.50. Thus,

$$p_0(-\delta) = 1 - p_0(\delta) = 0.624,$$

which is tabulated. Entering Table 2 with $n = 20$, we find that $p_0(-\delta) = 0.624$ corresponds to $\delta = -0.20$. Table 3 gives $p_C(\delta)$ when $\alpha = 0.05$.

Example

Seven studies of the effects of open education on independence and self-reliance are described in Section C of Chapter 2, and the results (including effect size estimates) are given in Table 7 of Chapter 2. The common experimental and control group sample size is $n^E = n^C = 30$ in each study. In this case, the actual estimates of effect size from each study are available to us. Suppose, however, that these estimates were not available and we merely observed the number of studies in which the open education group mean exceeded the control group mean. Two of the seven studies show positive mean differences, so that $p_0(\delta) = \frac{2}{7} = 0.29$. Entering Table 2 and using linear interpolation, we see that $p_0(-\delta) = 1 - 0.29$ corresponds to the point estimate $\hat{\delta} = -0.15$. The average of the parametric point estimates for each study is $\bar{d} = -0.09$, so that the vote-count estimate is quite close to the fully parametric estimate.

Confidence intervals for δ can be obtained by either of the two methods presented in the previous section. Using the first method, we obtain from (3) a 90-percent confidence interval (p_L, p_U):

$$p_L = 0.29 - 1.645\sqrt{0.29(1.0 - 0.29)/7} = 0.01,$$
$$p_U = 0.29 + 1.645\sqrt{0.29(1.0 - 0.29)/7} = 0.57. \qquad (10)$$

From Table 2, with $n = 30$ and the values $p_L = 0.01$ and $p_U = 0.57$, we obtain a 90-percent confidence interval (δ_L, δ_U):

$$\delta_L = -0.69, \qquad \delta_U = 0.05$$

for δ.

Using the second method, we obtain a 90-percent confidence interval for $p_0(\delta)$ by finding $\tilde{p}_L$ and $\tilde{p}_U$ given by (6) with $\hat{p} = 0.29$ and $b = 2.706/7 = 0.387$:

$$\tilde{p}_L = \frac{2(0.29) + 0.387 - \sqrt{(0.387)^2 + 4(0.387)(0.29)(0.71)}}{2(1.387)} = 0.10,$$

$$\tilde{p}_U = \frac{2(0.29) + 0.387 + \sqrt{(0.387)^2 + 4(0.387)(0.29)(0.71)}}{2(1.387)} = 0.59.$$

Using linear interpolation in Table 2 with $n = 30$, $\tilde{p}_L = 0.10$ and $\tilde{p}_U = 0.59$, we obtain the 90-percent confidence interval $(\tilde{\delta}_L, \tilde{\delta}_U)$ for δ:

$$\tilde{\delta}_L = -0.37, \qquad \tilde{\delta}_U = 0.02.$$

In this example, we see that the second method produces a shorter confidence interval, and would be the preferable method.

Using the nomograph (Appendix E) with $k = 7$, the exact 90-percent confidence interval for p is (0.06, 0.70). This is to be compared with the two approximations (0.01, 0.57) and (0.10, 0.59). Note that the exact upper confidence limit is larger than either of the approximations.

F. ESTIMATING A CORRELATION

Consider a collection of k independent studies with common sample size n in which each study provides a correlation coefficient between variables having a bivariate normal distribution. Denote the sample and population correlations in the ith study by r_i and ρ_i, respectively, and suppose that all the studies share the same population correlation ρ:

$$\rho_1 = \rho_2 = \cdots = \rho_k = \rho.$$

If the sample correlations $r_1, \ldots, r_k$ are observed, then there are procedures for estimating ρ. The problem here is that we do not directly observe the r's, but only the vote count.

Obtaining vote-count estimates of a correlation ρ entails letting $T_i = r_i$ in the development of Section B. The probability $p_C(\rho)$ is then the probability that a sample correlation based on a sample of size n exceeds C when the population correlation is ρ. When the critical value C is zero, we count the number of times that r_i is positive.

Values of $p_0(\rho)$ were tabulated by Olkin (1971). Tables 4 and 5 give the value of $p_0(\rho)$ as a function of ρ and sample size n. Only values of $p_0(\rho)$

TABLE 4

Probability That a Correlation Coefficient from a Sample of Size n Will Be Positive. When the Population Correlation Coefficient Is ρ

	ρ									
n	0	0.1	0.2	0.3	0.4	0.5	0.6	0.7	0.8	0.9
3	0.500	0.550	0.600	0.650	0.700	0.762	0.811	0.858	0.906	0.953
4	0.500	0.564	0.626	0.688	0.748	0.805	0.858	0.906	0.948	0.981
5	0.500	0.575	0.648	0.718	0.784	0.844	0.896	0.939	0.972	0.993
6	0.500	0.584	0.666	0.743	0.813	0.873	0.923	0.960	0.985	0.997
7	0.500	0.593	0.683	0.765	0.837	0.896	0.942	0.973	0.991	0.999
8	0.500	0.601	0.697	0.784	0.857	0.915	0.956	0.982	0.995	1.000
9	0.500	0.608	0.710	0.800	0.874	0.929	0.967	0.988	0.997	
10	0.500	0.615	0.722	0.815	0.889	0.941	0.975	0.992	0.998	
11	0.500	0.621	0.733	0.828	0.901	0.951	0.980	0.994	0.999	
12	0.500	0.627	0.744	0.840	0.912	0.959	0.985	0.996	0.999	
13	0.500	0.633	0.754	0.851	0.922	0.966	0.988	0.997	1.000	
14	0.500	0.639	0.763	0.861	0.930	0.971	0.991	0.998		
15	0.500	0.644	0.771	0.871	0.938	0.976	0.993	0.999		
16	0.500	0.649	0.779	0.879	0.944	0.980	0.995	0.999		
17	0.500	0.654	0.787	0.887	0.950	0.983	0.996	0.999		
18	0.500	0.658	0.794	0.899	0.955	0.985	0.997	1.000		
19	0.500	0.663	0.801	0.901	0.960	0.988	0.997			
20	0.500	0.667	0.808	0.907	0.964	0.990	0.998			
21	0.500	0.671	0.814	0.913	0.967	0.991	0.998			
22	0.500	0.675	0.819	0.918	0.971	0.992	0.999			
23	0.500	0.679	0.826	0.923	0.974	0.994	0.999			
24	0.500	0.683	0.831	0.927	0.976	0.995	0.999			
25	0.500	0.687	0.836	0.932	0.979	0.995	0.999			
50	0.500	0.757	0.920	0.984	0.998	0.999	1.000			
100	0.500	0.840	0.978	0.999	1.000	1.000				
200	0.500	0.921	0.998	1.000						
400	0.500	0.977	1.000							

TABLE 5

Probability That a Correlation Coefficient from a Sample of Size n Will Be Positive When the Population Correlation Coefficient Is ρ

n	0.00	0.01	0.02	0.03	0.04	0.05	0.06	0.07	0.08	0.09	0.10
3	0.500	0.505	0.510	0.515	0.520	0.525	0.530	0.535	0.540	0.545	0.550
4	0.500	0.506	0.513	0.519	0.525	0.532	0.538	0.545	0.551	0.557	0.564
5	0.500	0.508	0.515	0.522	0.530	0.537	0.545	0.552	0.560	0.567	0.575
6	0.500	0.508	0.517	0.525	0.534	0.542	0.551	0.559	0.568	0.576	0.584
7	0.500	0.509	0.519	0.528	0.537	0.547	0.556	0.565	0.575	0.584	0.593
8	0.500	0.510	0.520	0.531	0.541	0.551	0.561	0.571	0.581	0.591	0.601
9	0.500	0.511	0.522	0.533	0.544	0.555	0.565	0.576	0.587	0.598	0.608
10	0.500	0.512	0.523	0.535	0.546	0.558	0.570	0.581	0.592	0.604	0.615
11	0.500	0.512	0.525	0.537	0.549	0.561	0.573	0.586	0.598	0.610	0.621
12	0.500	0.513	0.526	0.539	0.552	0.564	0.577	0.590	0.602	0.615	0.627
13	0.500	0.514	0.527	0.541	0.554	0.567	0.581	0.594	0.607	0.620	0.633
14	0.500	0.514	0.528	0.542	0.556	0.570	0.584	0.598	0.612	0.625	0.639
15	0.500	0.515	0.529	0.544	0.558	0.573	0.587	0.602	0.616	0.630	0.644
16	0.500	0.515	0.530	0.546	0.561	0.576	0.590	0.605	0.620	0.634	0.649
17	0.500	0.516	0.531	0.547	0.563	0.578	0.593	0.609	0.624	0.639	0.654
18	0.500	0.516	0.532	0.549	0.565	0.581	0.596	0.612	0.628	0.643	0.658
19	0.500	0.517	0.533	0.550	0.566	0.583	0.599	0.615	0.631	0.647	0.663
20	0.500	0.517	0.534	0.551	0.568	0.585	0.602	0.618	0.635	0.651	0.667
21	0.500	0.518	0.535	0.553	0.570	0.587	0.605	0.622	0.638	0.655	0.671
22	0.500	0.518	0.536	0.554	0.572	0.590	0.607	0.625	0.642	0.659	0.675
23	0.500	0.518	0.537	0.555	0.574	0.592	0.610	0.627	0.645	0.662	0.679
24	0.500	0.519	0.538	0.557	0.575	0.594	0.612	0.630	0.648	0.666	0.683
25	0.500	0.519	0.539	0.558	0.577	0.596	0.615	0.633	0.651	0.669	0.687
50	0.500	0.528	0.555	0.583	0.610	0.636	0.662	0.687	0.712	0.735	0.757
100	0.500	0.540	0.579	0.617	0.654	0.690	0.724	0.757	0.787	0.815	0.840
200	0.500	0.556	0.611	0.664	0.714	0.760	0.801	0.838	0.871	0.898	0.921
400	0.500	0.579	0.655	0.725	0.788	0.841	0.885	0.919	0.945	0.964	0.977

equal to or greater than 0.50 are tabulated because proportions less than 0.5 correspond to negative values of ρ and these can be obtained by symmetry. For example, if $n = 17$ and $p_0(\rho) = 0.050$, the value of ρ cannot be obtained directly, since $p_0(\rho)$ is less than 0.50. The identity

$$p_0(-\rho) = 1 - p_0(\rho)$$

yields values for $p_0(\rho)$ less than 0.50. Thus,

$$p_0(-\rho) = 1 - p_0(\rho) = 0.950,$$

which is tabulated. Entering Table 4 with $n = 17$, we find that $p_0(-\rho) = 0.950$ corresponds to $\rho = -0.4$.

Example

Seven studies of the relationship between teacher indirectness and achievement are described in Section D of Chapter 2, including the results (effect size estimates) of the studies. The sample sizes vary between 15 and 17, but we treat the studies as if each had a sample size of 16 for the purposes of our analysis. In this case, the actual values of the sample correlation are available to us, but we illustrate the vote-counting method by assuming that only the number U of studies yielding positive correlations is observed. Since the number of positive correlations is $U = 6$, we obtain $U/k = \frac{6}{7} = 0.86$. Entering Table 4 with $n = 16$, we obtain the solution $\hat{\rho}$ to the estimation equation $p_0(\hat{\rho}) = 0.86$ as $\hat{\rho} = 0.28$. The average of the seven sample correlations is $\bar{r} = 0.29$, which is quite close to the point estimate based on the vote count.

Confidence intervals for ρ can be obtained by either of the two methods presented in Section C. In both methods, we begin by obtaining a confidence interval for $p_0(\rho)$. Using the first method, we obtain a 90-percent confidence interval (p_L, p_U) for $p_0(\rho)$ using (3):

$$p_L = 0.86 - 1.645\sqrt{0.86(0.14)/7} = 0.64,$$

$$p_U = 0.86 + 1.645\sqrt{0.86(0.14)/7} = 1.07,$$

From Table 5 with $n = 16$, we obtain the value $\rho_L = 0.09$ corresponding to $p_L = 0.64$. Because $p_U > 1$ it provides no information about the upper confidence limit for ρ. Hence our 90-percent confidence interval using the first method is

$$0.09 \leq \rho \leq 1.$$

Using the second method, we obtain a 90-percent confidence interval for $p_0(\rho)$ by finding the confidence limits $\tilde{p}_L$ and $\tilde{p}_U$ for $p_0(\rho)$ using (6):

$$\tilde{p}_L = \frac{14.746 - \sqrt{7.322 + 9.122}}{19.412} = 0.55,$$

$$\tilde{p}_U = \frac{14.746 + \sqrt{7.322 + 9.122}}{19.412} = 0.97.$$

Using Tables 3 and 4 with $n = 16$, $\tilde{p}_L = 0.55$ and $\tilde{p}_U = 0.97$, so that we obtain $\tilde{\rho}_L = 0.03$ and $\tilde{\rho}_U = 0.48$. Hence the 90-percent confidence interval for ρ using the second method is

$$0.03 \leq \rho \leq 0.48.$$

Remark: The two methods yield 90-percent confidence intervals of (0.09, 1.00) and (0.03, 0.48). Since neither interval is totally included in the other, we cannot state that one interval is superior. However, when confidence intervals differ in length, one would normally prefer the shorter interval. Although in this example the second method provides a shorter confidence interval, this will not always happen.

G. LIMITATIONS OF VOTE-COUNTING ESTIMATORS

The vote-counting estimators described in this chapter are very simple and intuitively appealing. These estimators can sometimes provide a quick estimate of an effect size or a correlation based on minimal data from a series of studies. The estimators have several limitations that restrict their applicability in practice, however.

One limitation of counting estimators is that they require a relatively large number of studies (counts) to obtain an accurate estimate of $p_C(\theta)$. Moreover, the estimate $\hat{\theta}$ of θ obtained by these methods can be very sensitive to variations in U/k, especially when $p_C(\theta)$ is close to zero or unity. Thus a large sample of studies is even more important when $p_C(\theta)$ is close to zero or unity. Formally, the asymptotic theory that underlies vote-counting estimators holds as k gets large. Thus, vote-counting estimators (unlike other procedures described herein) depend on a reasonably large number of studies.

Another limitation stems from the logic of the derivation of the counting estimators. The development in this section depends on the fact that each study has an identical sample size n. Few, if any, collections of research studies consist entirely of studies with identical sample sizes. If the sample sizes are not very different, then a reasonable procedure is to treat the studies as if all had (the same) sample size equal to some "average value" (Hedges & Olkin, 1980).

The question is, which average value should be used? Using either the minimum or maximum sample size would be too conservative or too optimistic. The unweighted mean is subject to undue influence by extremes. Of course, the median sample size avoids this problem, but could be troublesome computationally. There is a general class of means

$$\bar{n}_m = \left(\frac{1}{k}\sum_{i=1}^{k} n_i^{1/m}\right)^m, \qquad m = 0, 1, 2, \ldots,$$

that is very useful. When $m = 0$ we obtain the maximum; when $m = 1$ we obtain the arithmetic mean; when $m = \infty$ we obtain the minimum. Indeed, $\bar{n}_m$ is monotone in m. On the basis of some computational evidence in another context, we recommend use of $m = 2$, which leads to the square mean root (SMR)

$$n_{\text{SMR}} = \left(\frac{\sqrt{n_1} + \cdots + \sqrt{n_k}}{k}\right)^2.$$

This "average" will not be as responsive to extremes as is the arithmetic mean and will be slightly more conservative. However, it is not as conservative as the minimum.

The methods given in this chapter can then be applied to obtain an estimate of θ and associated confidence intervals. In many cases, the sample sizes of the studies may differ substantially, and it may therefore be unreasonable to treat them as equal. Because unequal sample sizes are the rule in research synthesis rather than the exception, the counting estimators are likely to be most useful for providing quick approximate estimates rather than as the analytic tool for final analyses. In the next section we develop vote-counting methods that are designed to handle studies with unequal sample sizes. These estimators are considerably more complex than the vote-counting estimators previously discussed.

A different limitation of vote-counting estimators is that they cannot usually be used to estimate θ when $T_i > C$ for all or none of the studies, that is, when $U = 0$ or k. This is because the estimate of $p_C(\theta)$ is zero or unity. Note that the value of unity occurs in tables of $p_C(\theta)$, but $p_C(\theta) = 1$ does not define a unique θ in this case, since if $p_C(\theta_0) = 1$ for some θ_0 then $p_C(\theta) = 1$ for all $\theta \geq \theta_0$. In this special case we can use other estimation procedures to obtain an estimate of $p_C(\theta)$. It should be noted that the vote-counting estimator of θ discussed in Section A is consistent and asymptotically efficient. However, when we obtain other estimates of $p_C(\theta)$ and transform these estimates to obtain estimates of θ, these properties for the estimates of θ are no longer guaranteed.

The situation in which $U = 0$ or $U = k$ usually occurs when the true value of $p_C(\theta)$ or $1 - p_C(\theta)$ is large. This suggests that we may actually have prior knowledge of the magnitude of $p_C(\theta)$. One estimation procedure is to incorporate this prior knowledge into a Bayesian estimate of $p_C(\theta)$. For example, we can assume that $p_C(\theta)$ has a truncated uniform prior distribution, i.e., that there is a p_0 such that all values of $p_0(\theta)$ in the interval $(p_0, 1)$ are equally likely. The Bayes estimate $\hat{p}$ of $p_C(\theta)$ based on the truncated uniform prior distribution with $U = k$ is

$$\hat{p} = (k + 1)(1 - p_0^{k+2})/(k + 2)(1 - p_0^{k+1}).$$

The estimate of $p_C(\theta)$ when $U = 0$ is just $1 - \hat{p}$.

For large k, this estimate will be relatively insensitive to the choice of p_0, and if an a priori cutoff value can be determined, the Bayes estimator will give reasonable results. A complete discussion of Bayes estimation of a population proportion when the sample proportion is zero or unity is given in Chew (1971).

H. VOTE-COUNTING METHODS FOR UNEQUAL SAMPLE SIZES

The vote-counting methods described can be extended to handle unequal sample sizes explicitly, although much of the simplicity of the methods is lost. Extending the notation of Section B, let $T_1, \ldots, T_k$ be independent estimates of parameters $\theta_1, \ldots, \theta_k$ obtained from experiments with sample sizes $n_1, \ldots, n_k$. Suppose that we observe for each study whether the statistic T_i exceeds some critical value C_i. Note that the critical value C_i may differ from study to study. The probability that $T_i > C_i$ is a function of θ_i and n_i, so denote this function by

$$p(\theta_i, n_i) \equiv \text{Prob}\{T_i > C_i | \theta_i, n_i\}.$$

The fundamental idea is that for $\theta_1 = \cdots = \theta_k = \theta$ the probability that $T_i > C_i$ is a function of both θ and n_i. Because n_i is known, it should be possible to estimate θ from a knowledge of whether $T_i > C_i$ in each study. More specifically, we should be able to use the method of maximum likelihood to estimate θ from the data (whether $T_i > C_i$) in each of the k studies.

It is convenient to summarize the observed data via a dichotomous "scoring variable" X for each study, so that $X = 1$ if $T > C$ and $X = 0$ otherwise. That is, define

$$X_i = \begin{cases} 1 & \text{if} \quad T_i > C_i \\ 0 & \text{if} \quad T_i \leq C_i. \end{cases}$$

Then the observed data from the k studies are summarized by $X_1, \ldots, X_k$. Note that X_i is an integer, and that

$$\text{Prob}\{X_i = 1 | \theta, n_i\} = \text{Prob}\{T_i > C_i | \theta, n_i\} = p(\theta, n_i).$$

Consequently the log likelihood associated with the k studies is

$$\begin{aligned} L(\theta | X_1, \ldots, X_k) = {} & X_1 \log p(\theta, n_i) + (1 - X_1) \log[1 - p(\theta, n_i)] \\ & + \cdots + X_k \log p(\theta, n_k) + (1 - X_k) \log[1 - p(\theta, n_k)]. \end{aligned} \tag{11}$$

Because $n_1, \ldots, n_k$ are known and the data $X_1, \ldots, X_k$ are observed, the likelihood $L(\theta | X_1, \ldots, X_k)$ is a function of θ alone and this function can be maximized over θ to obtain the maximum likelihood estimator $\hat{\theta}$. There is generally no closed form expression for $\hat{\theta}$, so that $\hat{\theta}$ must be obtained numerically. One way to obtain $\hat{\theta}$ is to calculate the value of $L(\theta \mid X_1, \ldots, X_k)$ for a grid of θ values and select $\hat{\theta}$ as the value that gives the greatest value of the likelihood. If a good initial guess about the value of $\hat{\theta}$ is available, then the evaluation of a relatively small grid of trial values can give $\hat{\theta}$ to two decimal places or more. Of course more elaborate numerical methods such as the Newton–Raphson method can also be used to estimate $\hat{\theta}$, but these methods require a specialized computer program.

The vote-counting estimate for unequal sample sizes is formally equivalent to the item response models used in test theory (see Lord & Novick, 1968). In item response models a test item is scored dichotomously, and the probability of a "correct" response is a function (at least) of an "item" parameter that is different for each item and a "person" parameter that is the same across items for a given individual. In the present model studies are analogous to items. The event $T_i > C_i$ is analogous to a correct item response, the sample size n_i corresponds to the item parameter, and θ corresponds to the person parameter. Thus the process of estimating θ by this method is analogous in test theory to estimating person abilities when the item parameters are known. For example, when C_i is the $\alpha = 0.50$ significance level and T_i is the two-sample t-statistic, the vote-counting estimation procedure is equivalent to a so-called one-parameter normal ogive model in item response theory.

H.1 THE LARGE SAMPLE VARIANCE OF THE MAXIMUM LIKELIHOOD ESTIMATOR

When the number k of studies is large, $\hat{\theta}$ will be approximately normally distributed. The theory of maximum likelihood estimation can be used to obtain an expression for the large sample variance of $\hat{\theta}$:

$$\mathrm{Var}(\hat{\theta}) = \left\{ \sum_{i=1}^{k} \frac{[D_i^{(1)}]^2}{p_i(1 - p_i)} \right\}^{-1} \tag{12}$$

where

$$p_i = p(\theta, n_i), \qquad D_i^{(1)} = \partial p(\theta, n_i)/\partial \theta,$$

and the derivative is evaluated at $\theta = \hat{\theta}$. When k is large a $100(1 - \alpha)$-percent confidence interval $(\theta_{\mathrm{L}}, \theta_{\mathrm{U}})$ for θ is given by

$$\theta_{\mathrm{L}} = \hat{\theta} - C_{\alpha/2}\sqrt{\mathrm{Var}(\hat{\theta})}, \qquad \theta_{\mathrm{U}} = \hat{\theta} + C_{\alpha/2}\sqrt{\mathrm{Var}(\hat{\theta})}, \tag{13}$$

where $C_{\alpha/2}$ is the two-tailed critical value of the standard normal distribution.

H.2 ESTIMATING EFFECT SIZE

Consider a collection of k studies each of which compares the outcome of an experimental (E) group and a control (C) group. Use the same model and notation as are used in Section E, except that now the experimental and control group sample sizes of the ith study (n_i^E and n_i^C, respectively) need not be equal. We observe for each study whether the sample mean $\bar{Y}_i^E$ of the experimental group, exceeds the sample mean $\bar{Y}_i^C$ of the control group. That is, we observe whether $\bar{Y}_i^E - \bar{Y}_i^C > 0$ for each study. If the effect sizes are homogeneous, that is, if $\delta_1 = \cdots = \delta_k = \delta$, our model implies that

$$\bar{Y}_i^E - \bar{Y}_i^C \sim \mathcal{N}(\delta\sigma_i, \sigma_i^2/\tilde{n}_i),$$

where $\tilde{n}_i = n_i^E n_i^C/(n_i^E + n_i^C)$. Therefore the probability of a positive result is

$$p(\delta, \tilde{n}_i) = \text{Prob}\{\bar{Y}_i^E - \bar{Y}_i^C > 0\} = 1 - \Phi(-\sqrt{\tilde{n}_i}\delta).$$

Using the scoring variable X_i for each study, the likelihood (11) becomes

$$L(\delta \mid X_1, \ldots, X_k) = \sum_{i=1}^{k} \{X_i \log[1 - \Phi(-\sqrt{\tilde{n}_i}\delta)] + (1 - X_i) \log \Phi(-\sqrt{\tilde{n}_i}\delta)\}. \quad (14)$$

This likelihood is relatively easy to compute, but it must be maximized numerically to obtain the maximum likelihood estimate $\hat{\delta}$ of δ. Perhaps the simplest way to accomplish this is to compute the likelihood for a coarse grid of trial values of δ to determine the region where $\hat{\delta}$ lies, and then compute the likelihood for a finer grid of values to obtain a more precise estimate of $\hat{\delta}$.

Example

For the results of 18 studies of the effects of open versus traditional education on student self-concept given in Table 8 of Chapter 2, the sample sizes n^E, n^C and $\tilde{n}$, and the parametric effect size estimate d are given for each study in Table 6. Here we ignore the parametric effect size estimates and estimate the common effect size δ from only the *signs* of the estimates (which are the same as the signs of the mean differences). We first compute the likelihood (14) for δ between -0.50 and 0.50 in steps of 0.10. The values of the likelihood for this coarse grid of δ values are given on the left-hand side of Table 7. Because the likelihood is larger at $\delta = 0.00$ than at $\delta = \pm 0.10$, we see that the maximum occurs between -0.10 and $+0.10$. To refine our estimate of $\hat{\delta}$ we evaluate the likelihood between $\delta = \pm 0.10$ in steps of 0.01. The values of the likelihood for this finer grid of δ values are given on the right-hand side of Table 7. Because the likelihood is largest at $\delta = 0.03$, the maximum likelihood estimate of δ is $\hat{\delta} = 0.03$ (to two decimal places). It is interesting to note that the estimate derived by pooling the parametric

TABLE 6
Data from 18 Studies of the Effect of Open versus Traditional Education on Self-Concept

Study	n^E	n^C	$\tilde{n}$	X	Unbiased estimate of effect size, d
1	100	180	64.286	1	0.100
2	131	138	67.204	0	−0.162
3	40	40	20.000	0	−0.090
4	40	40	20.000	0	−0.049
5	97	47	31.660	0	−0.046
6	28	61	19.191	0	−0.010
7	60	55	28.696	0	−0.431
8	72	102	42.207	0	−0.261
9	87	45	29.659	1	0.134
10	80	49	30.388	1	0.019
11	79	55	32.425	1	0.175
12	70	109	29.262	1	0.056
13	36	93	25.954	1	0.045
14	9	18	6.000	1	0.103
15	14	16	7.467	1	0.121
16	21	22	10.744	0	−0.482
17	133	124	64.171	1	0.290
18	83	45	29.180	1	0.342

estimates of effect size from each study (see Section B.1 of Chapter 6) is $d_+ = 0.011$.

The large sample variance of $\hat{\delta}$ is evaluated using (12). Using the fact that

$$p(\delta, \tilde{n}_i) = 1 - \Phi(-\sqrt{\tilde{n}_i}\,\delta),$$

we obtain

$$D_i^{(1)} = \frac{\partial p_i(\hat{\delta}, \tilde{n}_i)}{\partial \delta} = \sqrt{\frac{\tilde{n}_i}{2\pi}} \exp(-\tfrac{1}{2}\,\tilde{n}_i \hat{\delta}^2).$$

Inserting the expressions for $p_i = p(\hat{\delta}, \tilde{n}_i)$ and $D_i^{(1)}$ into (12) yields an estimate of the large sample variance of $\hat{\delta}$.

TABLE 7

The Likelihood Function for δ Based on 18 Studies of the Effects of Open Education on Self-Concept

Coarse grid		Fine grid	
δ	$L(\delta \mid X_1, \ldots, X_k)$	δ	$L(\delta \mid X_1, \ldots, X_k)$
−0.50	−59.236	−0.10	−15.153
−0.40	−43.518	−0.09	−14.726
−0.30	−30.842	−0.08	−14.334
−0.20	−21.327	−0.07	−13.978
−0.10	−15.153	−0.06	−13.658
0.00	−12.477	−0.05	−13.372
0.10	−13.294	−0.04	−13.122
0.20	−17.367	−0.03	−12.908
0.30	−24.318	−0.02	−12.729
0.40	−33.813	−0.01	−12.585
0.50	−45.665	0.00	−12.477
		0.01	−12.403
		0.02	−12.365
		0.03	−12.361
		0.04	−12.393
		0.05	−12.458
		0.06	−12.558
		0.07	−12.692
		0.08	−12.859
		0.09	−13.060
		0.10	−13.294

Example

In our previous example of 18 studies of the effect of open education on self-concept, values of $p_i, D_i^{(1)}, D_i^{(2)}$, and $D_i^{(1)} + D_i^{(2)} + (D_i^{(1)})^2(1 - 2p_i)/p_i(1 - p_i)$ are given in Table 8. Using the values in Table 8, the estimated variance of $\hat{\delta}$ is

$$\text{Est Var}(\hat{\delta}) = 0.0122.$$

Consequently a 95-percent confidence interval for δ from (13) is given by $\delta_L = 0.03 - 1.96\sqrt{0.0122}$ and $\delta_U = 0.03 + 1.96\sqrt{0.0122}$ or

$$-0.19 \leq \delta \leq 0.25.$$

Note that a 95-percent confidence interval based on the parametric methods given in Chapter 6 is $-0.071 \leq \delta \leq 0.093$. Thus the confidence interval based on the parametric estimates is shorter than that based on the vote-counting estimator. The advantage of the vote-counting estimator is that it can be used when parametric estimates cannot be calculated.

TABLE 8

Computations for Obtaining the Large Sample Variance of the Estimator of Effect Size

Study	p_i	$D_i^{(1)}$	$[D_i^{(1)}]^2/[p_i(1-p_i)]$
1	0.5948	3.1009	9.9242
2	0.5970	3.1685	10.4847
3	0.5534	1.7681	2.8809
4	0.5533	1.7681	2.8809
5	0.5663	2.1904	4.4161
6	0.5520	1.7241	2.7493
7	0.5631	2.0846	3.9815
8	0.5771	2.5370	6.1093
9	0.5642	2.1205	4.1250
10	0.5653	2.1558	4.2699
11	0.5674	2.2245	4.5636
12	0.5642	2.1205	4.1250
13	0.5596	1.9724	3.5588
14	0.5293	0.9746	1.1263
15	0.5316	1.0522	1.2499
16	0.5378	1.2559	1.6163
17	0.5948	3.1009	9.9242
18	0.5642	2.1205	4.1250
			82.1085

CHAPTER 5

Estimation of a Single Effect Size: Parametric and Nonparametric Methods

This chapter is devoted to the study of parametric and nonparametric methods for estimating the effect size (standardized mean difference) from a single experiment. Imagine that a number of studies have been conducted each of which measures the effect of the same treatment on a psychological construct such as self-concept. The studies vary in that different psychological tests are used to measure self-concept. The reviewer will first want to estimate the size of that effect for each study. If there is reason to believe that the underlying (population) effect size is the same across the studies, then we estimate this common effect size using information from all the studies. It is important to recognize that estimating and interpreting a common effect size is based on the belief that the population effect size is actually the same across studies. Otherwise, estimating a "mean" effect may obscure important differences between the studies.

In Sections A and B we examine parametric point estimates and confidence intervals for effect size, and in Section C discuss robust and nonparametric estimators.

A. ESTIMATION OF EFFECT SIZE FROM A SINGLE EXPERIMENT

In this section we discuss several alternative point estimators of the effect size δ from a single two-group experiment. These estimators are based on the sample standardized mean difference, but differ by multiplicative constants that depend on the sample sizes involved. Although the estimates have identical large sample properties, they generally differ in terms of small sample properties.

The statistical properties of estimators of effect size depend on the model for the observations in the experiment. A convenient and often realistic model is to assume that the observations are independently normally distributed within groups of the experiment. That is, we assume that the experimental observations $Y_1^E, \ldots, Y_{n^E}^E$ are normally distributed with mean μ^E and common variance σ^2. Similarly, the control group observations $Y_1^C, \ldots, Y_{n^C}^C$ are normally distributed with mean μ^C and variance σ^2. More succinctly,

$$Y_j^E \sim \mathcal{N}(\mu^E, \sigma^2), \qquad j = 1, \ldots, n^E, \tag{1}$$

and

$$Y_j^C \sim \mathcal{N}(\mu^C, \sigma^2), \qquad j = 1, \ldots, n^C.$$

In this notation the *effect size* δ is defined as

$$\delta = (\mu^E - \mu^C)/\sigma. \tag{2}$$

A.1 INTERPRETING EFFECT SIZES

Effect sizes are natural parameters for use in the synthesis of experimental results. The most direct interpretation of the population effect size $\delta = (\mu^E - \mu^C)/\sigma$ is that δ is the mean difference that would obtain if the dependent variable were scaled to have unit variance within the groups of the experiment. Thus the effect size is just the mean difference reexpressed in units scaled so that $\sigma = 1$ to remove the dependence on the arbitrary scale factor σ.

When the observations in the experimental and control groups are normally distributed, δ can be used to quantify the degree of overlap between the distributions of observations in the experimental and control groups (Glass, 1976). Because $\delta = (\mu^E - \mu^C)/\sigma$ is the standardized score (z score) of the experimental group mean in the control group distribution, the quantity $\Phi(\delta)$ represents the proportion of *control group scores* that are less than the *average score in the experimental group*. Thus an effect size of $\delta = 0.5$ implies that the score of the average individual in the experimental group exceeds that of 69 percent of the individuals in the control group. Similarly an effect size of -0.5 implies that the score of the average individual in the experimental group exceeds that of only 31 percent of the individuals in the control group.

Another interpretation of effect size is obtained by converting δ to an estimate of a correlation coefficient. This, in effect, treats the experimental/control group distinction as a dichotomization of a continuous variable. In this situation

$$\rho^2 \cong \frac{\delta^2}{\delta^2 + (n^E + n^C - 2)/\tilde{n}},$$

where $\tilde{n} = n^E n^C/(n^E + n^C)$. If $n^E = n^C = n$ this formula reduces to

$$\rho^2 \cong \frac{\delta^2}{\delta^2 + 4(n-1)/n} \cong \frac{\delta^2}{\delta^2 + 4}.$$

Consequently, ρ^2 and δ^2 are related in a very simple manner.

The above expression as a squared correlation coefficient permits an interpretation of the effect size as a percentage of variance in the dependent variable that is accounted for by the effect of the experimental treatment.

The transformation of δ into a percentage of variance accounted for often leads investigators to conclude that what appears as a large effect size appears to be a trivial proportion of variance accounted for (see Rosenthal & Rubin, 1979a, 1982c). These authors suggest the use of a "binomial effect size display" as a simple way to illustrate the magnitude of effects expressed as correlation coefficients. They dichotomize the independent variable into experimental and control groups and the dependent variable into success and failure. Then they note that there is a simple relation between the correlation coefficient and the proportion of successes and failures in the experimental and control groups. Suppose that the overall proportion of successes is 0.50. Then the experimental success rate is $0.50 + \rho/2$ and the control group success rate is $0.50 - \rho/2$ as summarized in Table 1.

TABLE 1

The Relationship between Correlation Coefficients and the Proportion of Success in the Binomial Effect Size Display

	Outcome	
Group	Failure	Success
Control	$0.50 + \rho/2$	$0.50 - \rho/2$
Experimental	$0.50 - \rho/2$	$0.50 + \rho/2$

Note that the difference between the success rates in the experimental and control groups is always $\rho/2 + \rho/2 = \rho$. Thus even a rather small proportion of variance accounted for may correspond to a substantial change in the success rates between the experimental and control groups. For example, a value of $\rho^2 = 0.10$, which might seem rather small, corresponds (since $\rho = 0.32$) to a sizable increase in the success rate from 34 to 66 percent.

A.2 AN ESTIMATOR OF EFFECT SIZE BASED ON THE STANDARDIZED MEAN DIFFERENCE

Glass (1976) proposed an estimator of δ based on the sample value of the standardized mean difference. If $\bar{Y}^{E}$ and $\bar{Y}^{C}$ are the respective experimental and control group means, there are several ways to create a standardized mean difference:

$$(\bar{Y}^{E} - \bar{Y}^{C})/s^*,$$

where s^* is a standard deviation. Different choices of s^* yield different estimators. For example, we could define s^* as s^{C} (the standard deviation of the control group) or s^{E} (the standard deviation of the experimental group). Alternatively, we could use a pooled standard deviation that combines s^{E} and s^{C}.

Glass (1976) proposed an estimator that uses s^{C} to standardize the mean difference:

$$g' = (\bar{Y}^{E} - \bar{Y}^{C})/s^{C}.$$

His argument was that pooling pairs of variances could lead to different standardized values of identical mean differences within an experiment where several treatments were compared to a control. This argument depends on the fact that sample standard deviations of each treatment group will almost surely differ.

In many cases, however, the assumption of equal *population* variances is reasonable, which suggests that the most precise estimate of the population variance is obtained by pooling. Since we consider only two groups per experiment, with equal population variances, Glass's argument does not apply. Hence, a modified estimator is obtained by using a pooled estimate of the standard deviation.

One modification of g' is

$$g = (\bar{Y}^{E} - \bar{Y}^{C})/s, \tag{3}$$

where s is the pooled sample standard deviation,

$$s = \sqrt{\frac{(n^E - 1)(s^E)^2 + (n^C - 1)(s^C)^2}{n^E + n^C - 2}},$$

and n^E and n^C are the experimental and control group sample sizes, respectively.

Because g and g' are *sample statistics*, each has a sampling distribution. The sampling distributions of both statistics are closely related to the noncentral t-distribution. More specifically, if

$$\tilde{n} = \frac{n^E n^C}{n^E + n^C}, \tag{4}$$

then $\sqrt{\tilde{n}}g$ and $\sqrt{\tilde{n}}g'$ each have a noncentral t-distribution with noncentrality parameter $\sqrt{\tilde{n}}\delta$ and respective degrees of freedom $n^E + n^C - 2$ and $n^C - 1$.

As shown in the next sections, the bias and variance of g are both smaller than the bias and variance of the corresponding g'. Consequently, g is a (uniformly) better estimator than g' regardless of the value of δ, and we will not consider g' further. Other sampling properties of g can be obtained, and we list several of these. (For a more detailed discussion of these results, see Hedges, 1981.)

Fact: The mean (expected value) of g defined by (3) is approximately

$$E(g) \cong \delta + \frac{3\delta}{4N - 9}, \tag{5}$$

where $N = n^E + n^C$.

The exact mean is

$$E(g) = \delta / J(N - 2), \tag{6}$$

where $J(m)$ is a constant that is tabulated in Table 2 for values of m from 2 to 50. [The constant $J(m)$ appears as $c(m)$ in much of the published literature, e.g., Hedges, 1981. We changed the notation in order not to conflict with other notation used later.] The constant $J(m)$ is less than unity and approaches unity when m is large. It is closely approximated by

$$J(m) = 1 - \frac{3}{4m - 1}. \tag{7}$$

TABLE 2

Exact Values of the Bias Correction Factor $J(m)$

m	$J(m)$	m	$J(m)$	m	$J(m)$	m	$J(m)$
2	0.5642	15	0.9490	27	0.9719	39	0.9806
3	0.7236	16	0.9523	28	0.9729	40	0.9811
4	0.7979	17	0.9551	29	0.9739	41	0.9816
5	0.8408	18	0.9577	30	0.9748	42	0.9820
6	0.8686	19	0.9599	31	0.9756	43	0.9824
7	0.8882	20	0.9619	32	0.9764	44	0.9828
8	0.9027	21	0.9638	33	0.9771	45	0.9832
9	0.9139	22	0.9655	34	0.9778	46	0.9836
10	0.9228	23	0.9670	35	0.9784	47	0.9839
11	0.9300	24	0.9684	36	0.9790	48	0.9843
12	0.9359	25	0.9699	37	0.9796	49	0.9846
13	0.9410	26	0.9708	38	0.9801	50	0.9849
14	0.9453						

The variance of g is approximately

$$\mathrm{Var}(g) \cong \frac{1}{\tilde{n}} + \delta^2 \frac{1}{2(N - 3.94)}. \tag{8}$$

The expression for the exact variance is given in the Technical Commentary.

From (5) we see that the bias in estimating δ by g is approximately

$$\mathrm{Bias}(g) \cong 3\delta/(4N - 9),$$

so that for N as small as 12 the bias is only 0.08δ, which is small unless δ is quite large. A graphic representation of the bias of g as a function of the degrees of freedom $N - 2$ is given in Fig. 1.

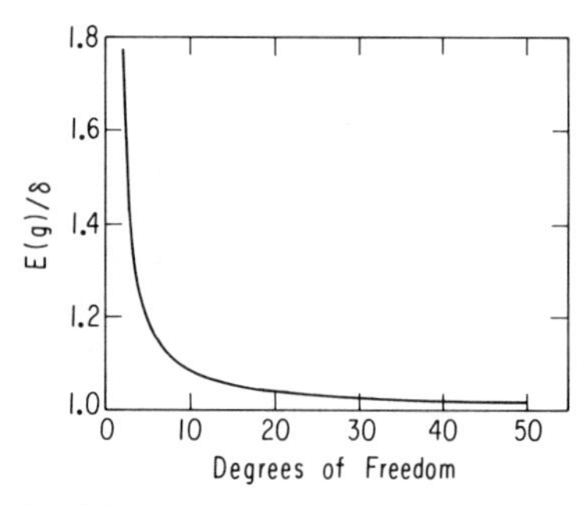

Fig. 1 The ratio, $E(g)/\delta$, of the expectation of the estimator $g = (\bar{Y}^E - \bar{Y}^C)/s$ to the true effect size δ as a function of the degrees of freedom, $N - 2$, of s^2 is used to estimate the common variance of the experimental and control groups.

A.3 AN UNBIASED ESTIMATOR OF EFFECT SIZE

Examination of Fig. 1 reveals that g has a small sample bias. It is easy to remove this bias by defining a new estimator

$$d = J(N-2) \quad g = J(N-2)\frac{\bar{Y}^{E} - \bar{Y}^{C}}{s} \tag{9}$$

or, approximately,

$$d \cong \left(1 - \frac{3}{4N-9}\right)g. \tag{10}$$

Because both the bias and the variance of d are smaller than the bias and variance of g, d has a smaller mean-squared error than g and dominates g as an estimator of δ.

When $n^{E} = n^{C}$, d is not only an unbiased estimator but also the unique minimum variance unbiased estimator of δ (see Hedges, 1981).

If we compare d and g, we find that

$$d/g = J(N-2),$$

which depends on the total sample size. As N gets large, $J(N-2)$ approaches unity, so that the distributions of both d and g tend to a normal distribution with identical means and variances in large samples. Since d tends to g in probability as N gets large, d and g are essentially the same estimator in large samples.

A.4 THE MAXIMUM LIKELIHOOD ESTIMATOR OF EFFECT SIZE

Maximum likelihood estimates have the advantage of being consistent and asymptotically efficient, both of which are large sample characteristic. However, maximum likelihood estimation is also widely used even when the requirement of having a large sample is not met. We now derive the maximum likelihood estimator of δ based on a single experiment.

In the case of a single experiment, the parameter δ is the ratio of the mean difference $\mu^{E} - \mu^{C}$ and the within-group standard deviation σ, and the maximum likelihood estimator of δ is the ratio of the maximum likelihood estimators of $\mu^{E} - \mu^{C}$ and σ. The maximum likelihood estimator of $\mu^{E} - \mu^{C}$ is $\bar{Y}^{E} - \bar{Y}^{C}$, but the maximum likelihood estimator of the standard deviation

σ is *not* s, the square root of the pooled within-group variance. In fact, the maximum likelihood estimator of σ is $\sqrt{N/(N-2)}s$, which corresponds to dividing the sum of squares about the mean by $n^E + n^C = N$ rather than by $N - 2$.

The maximum likelihood estimator $\hat{\delta}$ of δ is therefore given by

$$\hat{\delta} = \sqrt{\frac{N}{N-2}} \frac{\bar{Y}^E - \bar{Y}^C}{s} = \sqrt{\frac{N}{N-2}}\, g. \tag{11}$$

Note that $\hat{\delta}$ is biased in small samples and that $\hat{\delta}$ has larger absolute bias than Glass's estimator g'. Also, the maximum likelihood estimator $\hat{\delta}$ has larger variance than d. However, since $N/(N-2)$ is close to unity in large samples, all three estimators $\hat{\delta}$, g, and d of effect size from a single experiment have the same asymptotic distribution and are equivalent in large samples.

A.5 SHRUNKEN ESTIMATORS OF EFFECT SIZE

It is well known that the minimum variance unbiased estimator need not be the minimum mean-squared error estimator. For example, the mean-squared error of the maximum likelihood estimator of the variance of the normal distribution can be reduced by "shrinking" the estimates toward zero (Stein, 1964). The principle in these shrunken estimates is that the increase in the bias term of the mean-squared error is more than compensated for by the reduction of the variance term of the mean-squared error. Indeed, the minimum variance unbiased estimator of δ is dominated by (has uniformly larger mean-squared error than) a shrunken estimator that we denote by $\tilde{g}$.

Fact: Regardless of the value of the parameter δ,

$$\tilde{g} = \frac{N-4}{N-2} \frac{g}{J(N-2)} = \frac{N-4}{N-2} \frac{d}{[J(N-2)]^2} \tag{12}$$

has smaller mean-squared error than d.

A.6 COMPARING PARAMETRIC ESTIMATORS OF EFFECT SIZE

Four candidates have been proposed as estimators of effect size:

$$g = \frac{\bar{Y}^E - \bar{Y}^C}{s}, \quad d = J(N-2)\,g, \quad \hat{\delta} = \sqrt{\frac{N}{N-2}}\, g, \quad \tilde{g} = \frac{N-4}{N-2} \frac{g}{J(N-2)}. \tag{13}$$

Because the constants in the definitions of d, $\hat{\delta}$, and g approach unity as the sample sizes become large, all four estimators have the same large sample distribution, and therefore the same large sample properties.

Example

Suppose that a study with $n^E = n^C = 10$ yields a sample effect size value of $g = 0.60$. Using (13) and Table 2, the unbiased estimator d for this study is

$$d = J(18)\,0.60 = (0.9577)0.60 = 0.57.$$

The maximum likelihood estimator of the effect size for this study is

$$\hat{\delta} = \sqrt{20/18}\;0.60 = 0.63.$$

The shrunken estimator $\tilde{\delta}$ of the effect size for this study is

$$\tilde{\delta} = [16/(18)(0.9577)]0.60 = 0.51.$$

Note that even in this example, where $n^E = n^C = 10$, the values 0.60, 0.57, 0.63, and 0.51 of the four estimators are not very different.

The four estimators of δ are ordered, except for sign, so that

$$\hat{\delta}^2 \geq g^2 \geq d^2 \geq \tilde{g}^2.$$

From the definition (13) it follows that the variances are ordered in the same way:

$$\mathrm{Var}(\hat{\delta}) \geq \mathrm{Var}(g) \geq \mathrm{Var}(d) \geq \mathrm{Var}(\tilde{g}).$$

However, the mean-squared errors may not be ordered in the same way.

Table 3 is a tabulation of the variance, bias, and mean-squared error of the four estimators for selected sample sizes and effect sizes δ. Bias($\hat{\delta}$) and bias(g) are positive when δ is positive, and negative when δ is negative; bias($\tilde{g}$) is negative when δ is positive, and positive when δ is negative. The best estimator by the criterion of mean-squared error is $\tilde{g}$, and the differences among the estimators are largest when the sample size is small.

Figure 2 presents a graphic illustration of the mean-squared error of each of the four estimators as a function of the degrees of freedom. The figure illustrates that although the estimators differ in finite samples, the differences in performance of the estimators are small for 16 or more degrees of freedom. Indeed, the differences are appreciable only for very small degrees of freedom, which is unrealistic in most applications. Many applications involve sample sizes of at least 10 subjects per group, and in these cases the differences among the four estimators are likely to be negligible. Because d has good properties for small sample sizes, we use it as the basic estimator of effect size from a single study.

TABLE 3

Variance, Bias, and Mean-Squared Error of Four Estimators of Effect Size for Effect Sizes $\delta = 0.0$, 0.2, 0.5, and 1.0 and Various Degrees of Freedom

Degrees of freedom	Variance				Bias				Mean-squared error			
	$\hat{\delta}$	g	d	$\tilde{g}$	$\hat{\delta}$	g	d	$\tilde{g}$	$\hat{\delta}$	g	d	$\tilde{g}$
						$\delta = 0.0$						
4	2.000	1.333	0.849	0.524	0.000	0.000	0.000	0.000	2.000	1.333	0.849	0.524
6	1.000	0.750	0.566	0.442	0.000	0.000	0.000	0.000	1.000	0.750	0.566	0.442
8	0.667	0.533	0.435	0.368	0.000	0.000	0.000	0.000	0.667	0.533	0.435	0.368
10	0.500	0.417	0.355	0.313	0.000	0.000	0.000	0.000	0.500	0.417	0.355	0.313
15	0.308	0.271	0.244	0.226	0.000	0.000	0.000	0.000	0.308	0.271	0.244	0.226
20	0.222	0.202	0.187	0.177	0.000	0.000	0.000	0.000	0.222	0.202	0.187	0.177
25	0.174	0.161	0.151	0.145	0.000	0.000	0.000	0.000	0.174	0.161	0.151	0.145
50	0.083	0.080	0.078	0.076	0.000	0.000	0.000	0.000	0.083	0.080	0.078	0.076
100	0.041	0.040	0.039	0.039	0.000	0.000	0.000	0.000	0.041	0.040	0.039	0.039
						$\delta = 0.2$						
4	2.026	1.351	0.860	0.530	0.107	0.051	0.000	−0.043	2.037	1.353	0.860	0.532
6	1.009	0.757	0.571	0.446	0.066	0.030	0.000	−0.023	1.014	0.758	0.571	0.446
8	0.672	0.538	0.438	0.371	0.048	0.022	0.000	−0.016	0.674	0.538	0.438	0.371
10	0.504	0.420	0.357	0.315	0.037	0.017	0.000	−0.012	0.505	0.420	0.357	0.316
15	0.310	0.273	0.246	0.228	0.024	0.011	0.000	−0.008	0.310	0.273	0.246	0.228
20	0.224	0.203	0.188	0.178	0.018	0.008	0.000	−0.005	0.224	0.203	0.188	0.178
25	0.175	0.162	0.152	0.146	0.014	0.006	0.000	−0.004	0.175	0.162	0.152	0.146
50	0.084	0.081	0.078	0.076	0.007	0.003	0.000	−0.002	0.084	0.081	0.078	0.076
100	0.041	0.040	0.039	0.039	0.004	0.002	0.000	−0.001	0.041	0.040	0.039	0.039
						$\delta = 0.5$						
4	2.161	1.441	0.917	0.566	0.267	0.127	0.000	−0.107	2.233	1.457	0.917	0.577
6	1.058	0.794	0.599	0.468	0.165	0.076	0.000	−0.058	1.085	0.799	0.599	0.471
8	0.670	0.560	0.456	0.386	0.119	0.054	0.000	−0.040	0.714	0.563	0.456	0.388
10	0.523	0.436	0.371	0.327	0.094	0.042	0.000	−0.030	0.531	0.437	0.371	0.328
15	0.320	0.282	0.254	0.235	0.061	0.027	0.000	−0.019	0.324	0.283	0.254	0.236
20	0.231	0.210	0.194	0.184	0.045	0.020	0.000	−0.014	0.233	0.210	0.194	0.184
25	0.180	0.167	0.157	0.150	0.036	0.016	0.000	−0.011	0.182	0.167	0.157	0.150
50	0.087	0.083	0.080	0.079	0.018	0.008	0.000	−0.005	0.086	0.083	0.083	0.079
100	0.042	0.041	0.041	0.040	0.009	0.004	0.000	−0.002	0.042	0.041	0.041	0.040
						$\delta = 1.0$						
4	2.644	1.763	1.122	0.692	0.535	0.253	0.000	−0.215	2.930	1.827	0.122	0.73[illegible]
6	1.233	0.925	0.698	0.545	0.329	0.151	0.000	−0.116	1.341	1.240	0.698	0.55[illegible]
8	0.799	0.639	0.521	0.441	0.239	0.108	0.000	−0.080	0.856	0.770	0.521	0.44[illegible]
10	0.591	0.492	0.419	0.370	0.187	0.084	0.000	−0.060	0.626	0.565	0.419	0.37[illegible]
15	0.357	0.315	0.284	0.263	0.122	0.054	0.000	−0.038	0.372	0.448	0.284	0.26[illegible]
20	0.256	0.232	0.215	0.203	0.090	0.040	0.000	−0.027	0.264	0.318	0.215	0.20[illegible]
25	0.199	0.184	0.173	0.166	0.072	0.031	0.000	−0.022	0.204	0.185	0.173	0.16[illegible]
50	0.094	0.090	0.088	0.086	0.054	0.015	0.000	−0.010	0.096	0.091	0.088	0.08[illegible]
100	0.046	0.045	0.044	0.044	0.018	0.008	0.000	−0.005	0.046	0.045	0.044	0.04[illegible]

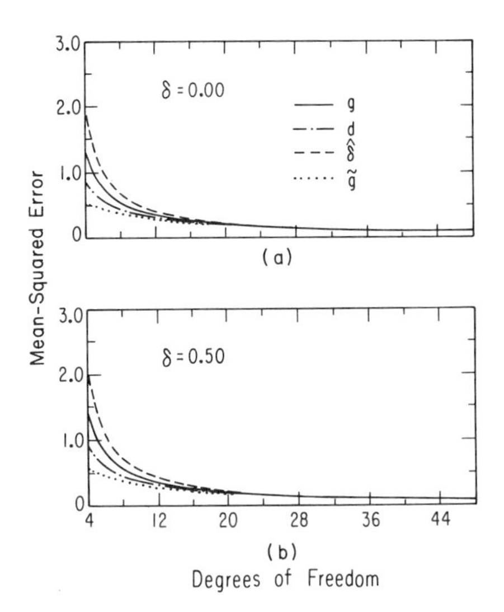

Fig. 2 Exact mean squared error of four estimators of effect size as a function of degrees of freedom for effect sizes (a) $\delta = 0.00$ and (b) $\delta = 0.50$.

B. DISTRIBUTION THEORY AND CONFIDENCE INTERVALS FOR EFFECT SIZES

Whenever we estimate a parameter we need to know the distribution of the estimator. In addition to providing a point estimate of the parameter, we wish to obtain a confidence interval. Frequently the exact distribution of parametric estimators of effect size is of a rather complicated analytic form. Fortunately several excellent approximations to the distribution of effect size estimates can be used to compute confidence intervals when the combined sample size $n^E + n^C$ is moderate to large. These are given in Sections B.1 and B.2. In Section B.3 we supply exact results for the case where $n^E + n^C \leq 20$, the situation in which approximations may not be very accurate.

B.1 THE ASYMPTOTIC DISTRIBUTION OF ESTIMATORS OF EFFECT SIZE

Recall from the last section that the four parametric estimators of effect size based on a single study have the same large sample distribution. This distribution and the large sample approximation based on the distribution are given below.

Fact: The large sample distribution of $d = J(N-2)(\bar{Y}^{\mathrm{E}} - \bar{Y}^{\mathrm{C}})/s$ tends to normality. Specifically, if n^{E} and n^{C} increase at the same rate, that is, if n^{E}/N and n^{C}/N remain fixed, then the asymptotic distribution of d is normal with mean δ and asymptotic variance

$$\sigma_{\infty}^{2}(d) = \frac{n^{\mathrm{E}} + n^{\mathrm{C}}}{n^{\mathrm{E}} n^{\mathrm{C}}} + \frac{\delta^{2}}{2(n^{\mathrm{E}} + n^{\mathrm{C}})}. \tag{14}$$

This formal asymptotic distribution can be used to obtain an excellent large sample approximation to the distribution of d. In practice, this approximation is used by substituting d for δ in the expression (14) for the variance. The estimated variance $\hat{\sigma}^{2}(d)$ is given by

$$\hat{\sigma}^{2}(d) = \frac{n^{\mathrm{E}} + n^{\mathrm{C}}}{n^{\mathrm{E}} n^{\mathrm{C}}} + \frac{d^{2}}{2(n^{\mathrm{E}} + n^{\mathrm{C}})}. \tag{15}$$

A $100(1-\alpha)$-percent confidence interval $(\delta_{\mathrm{L}}, \delta_{\mathrm{U}})$ for δ is given by

$$\delta_{\mathrm{L}} = d - C_{\alpha/2}\hat{\sigma}(d), \qquad \delta_{\mathrm{U}} = d + C_{\alpha/2}\hat{\sigma}(d), \tag{16}$$

where $C_{\alpha/2}$ is the two-tailed critical value of the standard normal distribution.

Example

Suppose that we observe an effect size of $g = 0.60$ in a study with $n^{\mathrm{E}} = n^{\mathrm{C}} = 10$, and that we wish to obtain a 95-percent confidence interval for δ. Recall that

$$d = J(18)\, g = 0.57.$$

Using (15) to calculate the variance $\hat{\sigma}^{2}(d)$ of d, we obtain

$$\hat{\sigma}^{2}(d) = \frac{10 + 10}{(10)(10)} + \frac{(0.57)^{2}}{2(10 + 10)} = 0.208,$$

and using (16) a 95-percent confidence interval for δ is $0.57 \pm 1.96\sqrt{0.208}$ or

$$-0.32 \leq \delta \leq 1.46.$$

This approximation has been studied in detail by Johnson and Welch (1939). They examined the behavior of the first three moments of both the exact and the approximate distribution and concluded that the approximation is quite good when δ is small and N is moderately large.

Another way to study the accuracy of the large sample approximation to the distribution of d is by simulation (see Hedges, 1982a). The distribution of the estimator d was simulated for $n^{\mathrm{E}} = n^{\mathrm{C}} = 10, 20, 30, 40, 50, 100$ and $\delta = 0.25, 0.50, 1.00, 1.50$. For each value of $n^{\mathrm{E}} = n^{\mathrm{C}}$ and δ, either 4000 or 10,000 d's were generated, and a confidence interval using (16) was constructed for each estimate. The empirical proportions of confidence intervals containing δ for various significance levels were tabulated. A reproduction of those results is given in Table 4.

The expected proportions of the confidence intervals that contain δ are very close to the observed probability content of the confidence intervals. The large sample approximation to the distribution of the effect size estimator seems reasonably accurate for sample sizes that exceed 10 per group, when the effect sizes are between 0.25 and 1.50 in absolute magnitude. (The approximation might be good outside of these ranges for sample size and effect size, but we have no empirical evidence concerning this.) Values from quantitative research syntheses usually fall within the ranges examined. The large sample approximation is likely to be less accurate for studies with small sample sizes

TABLE 4

The Small Sample Accuracy of Confidence Intervals for δ Based on the Normal Approximation to the Distribution of d

Sample size $n^E = n^C$	δ	Mean of d	Variance of d	Proportion of confidence intervals containing δ with nominal significance level					
				0.60	0.70	0.80	0.90	0.95	0.99
10	0.25	0.252	0.206	0.621	0.714	0.813	0.910	0.955	0.991
20	0.25	0.255	0.103	0.604	0.704	0.806	0.903	0.951	0.990
30	0.25	0.246	0.067	0.608	0.708	0.809	0.908	0.954	0.989
40	0.25	0.248	0.051	0.601	0.703	0.800	0.900	0.950	0.991
50	0.25	0.250	0.039	0.612	0.709	0.809	0.904	0.952	0.991
100	0.25	0.251	0.020	0.600	0.705	0.808	0.903	0.949	0.990
10	0.50	0.504	0.214	0.620	0.714	0.807	0.906	0.954	0.990
20	0.50	0.497	0.105	0.609	0.709	0.807	0.904	0.955	0.990
30	0.50	0.500	0.070	0.603	0.699	0.800	0.903	0.952	0.990
40	0.50	0.499	0.052	0.599	0.700	0.803	0.903	0.952	0.991
50	0.50	0.496	0.042	0.597	0.696	0.799	0.904	0.953	0.990
100	0.50	0.498	0.021	0.606	0.698	0.799	0.906	0.954	0.989
10	1.00	0.993	0.241	0.602	0.703	0.801	0.901	0.952	0.992
20	1.00	0.993	0.112	0.609	0.707	0.808	0.907	0.955	0.992
30	1.00	0.993	0.078	0.594	0.697	0.799	0.897	0.952	0.991
40	1.00	0.995	0.057	0.607	0.709	0.805	0.901	0.953	0.992
50	1.00	0.996	0.046	0.603	0.706	0.806	0.900	0.951	0.989
100	1.00	1.000	0.023	0.607	0.702	0.805	0.905	0.953	0.990
10	1.50	1.498	0.282	0.599	0.696	0.797	0.901	0.953	0.990
20	1.50	1.501	0.137	0.601	0.697	0.797	0.898	0.950	0.989
30	1.50	1.505	0.091	0.600	0.699	0.797	0.899	0.950	0.991
40	1.50	1.502	0.069	0.594	0.693	0.796	0.899	0.949	0.990
50	1.50	1.500	0.056	0.593	0.691	0.794	0.897	0.949	0.991
100	1.50	1.500	0.026	0.603	0.705	0.810	0.906	0.955	0.990

and very large effect sizes, but little is known about the exact decrease in accuracy under these conditions.

B.2 CONFIDENCE INTERVALS FOR EFFECT SIZES BASED ON TRANSFORMATIONS

In some instances the estimator d can be transformed to a new variate that has a simpler distribution. After obtaining a confidence interval for the transformed parameter we can then transform back to obtain a confidence interval for δ.

In the present case, because the variance of d depends on the unknown parameter δ, we use the variance-stabilizing transformation

$$h(d) = \sqrt{2}\,\sinh^{-1}\frac{d}{a} = \sqrt{2}\,\log\left(\frac{d}{a} + \sqrt{\frac{d^2}{a^2} + 1}\right), \tag{17}$$

where $\sinh^{-1}$ is the inverse hyperbolic sine function, and

$$a = \sqrt{4 + 2(n^E/n^C) + 2(n^C/n^E)}.$$

Note that the exact form of the transformation depends on the balance n^E/n^C in the experiment. For simplicity of notation, denote the transformed value of the estimate by h and the transformed value of the parameter by η:

$$h = h(d) \qquad \text{and} \qquad \eta = h(\delta).$$

Then the transformed estimate h, when appropriately normalized, has an approximate standard normal distribution. More explicitly,

$$\sqrt{N}(h - \eta) \sim N(0, 1).$$

Therefore a $100(1 - \alpha)$-percent confidence interval (η_L, η_U) for η is given by

$$\eta_L = h - C_{\alpha/2}/\sqrt{N}, \qquad \eta_U = h + C_{\alpha/2}/\sqrt{N}, \tag{18}$$

where $C_{\alpha/2}$ is the two-tailed critical value of the standard normal distribution. The confidence limits η_L and η_U are inverted to produce a confidence interval (δ_L, δ_U) for δ by finding the values δ_L and δ_U that correspond to η_L and η_U:

$$\delta_L = h^{-1}(\eta_L) \qquad \text{and} \qquad \delta_U = h^{-1}(\eta_U), \tag{19}$$

where $h^{-1}(x) = a\,\sinh(x/\sqrt{2})$.

Example

Return to our example of an experiment with $n^E = n^C = 10$ yielding a sample effect size of $g = 0.60$ and an unbiased estimate of effect size of

$d = 0.57$. We calculate a 95-percent confidence interval for δ using (17) with $a = \sqrt{4 + 2(10/10) + 2(10/10)} = \sqrt{8}$,

$$h = h(d) = \sqrt{2}\,\sinh^{-1}(0.57/\sqrt{8}) = 0.283.$$

Expression (18) gives the confidence limits η_L and η_U for η as

$$\eta_L = 0.283 - 1.96\sqrt{1/(10 + 10)} = -0.155,$$

$$\eta_U = 0.283 + 1.96\sqrt{1/(10 + 10)} = 0.721.$$

Using $h^{-1}(x) = \sqrt{8}\,\sinh(x/\sqrt{2})$, we obtain the confidence limits

$$\delta_L = \sqrt{8}\,\sinh(-0.155/\sqrt{2}) = -0.31,$$
$$\delta_U = \sqrt{8}\,\sinh(0.721/\sqrt{2}) = 1.50,$$

and the confidence interval $-0.31 \le \delta \le 1.50$.

This variance-stabilizing transformation was originally studied by Hedges and Olkin (1983a,b) in the balanced case $n^E/n^C = 1$, and is closely related to a variance-stabilizing transformation for the noncentral t-distribution that was studied by Laubscher (1960). Available evidence in the form of simulations indicates that, in most circumstances, the confidence intervals produced by the variance-stabilizing transformation are as accurate as those based on the large sample approximation given in Section B.

A different transformation was suggested by Kraemer (1983). This transformation exploits the fact that the distribution of the effect size estimate (the noncentral t-distribution) and the distribution of the product–moment correlation coefficient are related (see Kraemer & Paik, 1979). Kraemer (1983) shows that if

$$r = d/(d^2 + v)^{1/2} \qquad \text{and} \qquad \rho = \delta/(\delta^2 + v)^{1/2},$$

where $v = N(N - 2)/n^E n^C$, then the transformed variate

$$u = u(r, \rho) = (\rho - r)/(1 - r\rho) \tag{20}$$

has approximately the null distribution of a product–moment correlation. Confidence limits for u can be computed from a table of percentage points of the null distribution of the correlation coefficient distribution (e.g., Fisher & Yates, 1957). Kraemer and Paik (1979) show that the variate

$$\sqrt{N - 2}\,u/(1 - u^2)^{1/2} \tag{21}$$

has an approximate Student's t-distribution with $N - 2$ degrees of freedom, which can be used to obtain confidence limits for u from percentage points of the (central) t-distribution.

After obtaining (by either method) confidence limits u_L and u_U, we obtain confidence limits ρ_L and ρ_U by finding the values of ρ that correspond to u_L and u_U:

$$\rho_L = (u_L - r)/(u_L r - 1), \qquad \rho_U = (u_U - r)/(u_U r - 1). \tag{22}$$

Finally, we obtain the confidence interval for δ by transforming ρ_L and ρ_U to δ_L and δ_U:

$$\delta_L = \rho_L\sqrt{v}/(1 - \rho_L^2)^{1/2}, \qquad \delta_U = \rho_U\sqrt{v}/(1 - \rho_U^2)^{1/2}. \tag{23}$$

Example

Return to our example of a study with $n^E = n^C = 10$, an observed effect size of $g = 0.60$, and an unbiased estimate of effect size of $d = 0.57$. To calculate a 95-percent confidence interval for δ, we first calculate

$$v = 18/20(\tfrac{1}{2})(\tfrac{1}{2}) = 3.6,$$

then calculate

$$r = 0.57/(0.57^2 + 3.6)^{1/2} = 0.288.$$

Using the fact that $u = (r - \rho)/(1 - r\rho)$ has approximately the null distribution of a correlation coefficient from $n = 20$ observations, we can obtain the 2.5 and 97.5 percentile points from a table of the critical values of the correlation. These critical values are $U_L = -0.444$ and $U_U = 0.444$. Inserting these values into (22), we obtain

$$\rho_L = (0.444 - 0.288)/[(0.444)(0.288) - 1] = -0.179,$$

$$\rho_U = (-0.444 - 0.288)/[(-0.444)(0.288) - 1] = 0.649.$$

Inserting the values of ρ_L and ρ_U into (23) yields

$$\delta_L = -0.179\sqrt{3.6}/[1 - (0.179)^2]^{1/2} = -0.34,$$

$$\delta_U = 0.649\sqrt{3.6}/[1 - (0.649)^2]^{1/2} = 1.62.$$

Thus the 95-percent confidence interval for δ using the present method is

$$-0.34 \leq \delta \leq 1.62.$$

Relatively little is known about the comparative accuracy of these transformation methods. For small to moderate effect sizes, namely, $|\delta| \leq 1.5$, the two transformations seem to produce very similar results (Kraemer, 1983). There may be more substantial differences for larger effect sizes or for very small sample sizes. Neither method for generating confidence intervals based on transformations seems to be superior to the method based on the asymptotic distribution of d given in Section B.1.

B.3 EXACT CONFIDENCE INTERVALS FOR EFFECT SIZES

Although some methods yield exact confidence intervals for δ, we discuss approximations because the exact methods are rather complicated. Exact results involve either extensive calculations or the use of special tables, whereas the approximations are relatively simple. We recommend the use of the approximate confidence interval given in (16) when the lesser of n^E and n^C is moderate to large (greater than 10).

For smaller sample sizes, these asymptotic confidence intervals may not be sufficiently accurate. Exact confidence intervals for δ are obtained by using the exact distribution of the effect size estimator g. The statistic $g\sqrt{n^E n^C/N}$ has a noncentral t-distribution with $N - 2$ degrees of freedom and noncentrality parameter $\delta\sqrt{n^E n^C/N}$. Denote the cumulative distribution function of the estimator g by $F(g; N - 2, \delta)$. Then the confidence limits δ_L and δ_U are defined as the solutions of the equations

$$F(g; N - 2, \delta_L) = \alpha/2, \qquad F(g; N - 2, \delta_U) = 1 - \alpha/2. \tag{24}$$

Unfortunately, the cumulative distribution function of g and that of the noncentral t-distribution have rather complicated analytic forms. There are various integral and series representations of the distribution functions, but they are not easily used for hand calculation (see, e.g., Johnson & Kotz, 1970, Chapter 31). Therefore we provide nomographs (Appendix F) that give 90-, 95-, and 99-percent confidence intervals for δ when $0 \le g \le 1.5$ and $n^E = n^C = 2$ to 10. These nomographs are used by locating the value of g on the abscissa and moving vertically until the curves corresponding to the appropriate sample size are intersected. The ordinate values at these intersections are the confidence limits.

Example

Suppose that we observe an effect size of $g = 0.60$ in a study where $n^E = n^C = 10$ and wish to obtain a 95-percent confidence interval for δ. Entering the appropriate nomograph at the abscissa value of $g = 0.60$, we move upward until we intersect the lower confidence band for $n = 10$. Reading the ordinate value -0.33 at this point of intersection, we conclude that $\delta_L = -0.33$. Moving upward along the abscissa value of $g = 0.60$ until we intersect the upper confidence band for $n = 10$, we read the ordinate value of $\delta_U = 1.50$. Hence an exact 95-percent confidence interval for δ is $-0.33 \le \delta \le 1.50$.

C. ROBUST AND NONPARAMETRIC ESTIMATION OF EFFECT SIZE

In Section A we used the parametric model (1) to derive estimators of effect size. The usual model (that the assumptions of the t-test are met in each study) is often convenient, but not always realistic. Several examples of real data that do not meet the assumptions of the usual model are cited by Kraemer and Andrews (1982). They suggest that in many cases data are skewed or otherwise require a transformation to achieve even approximate normality.

When the assumption of normality is not met, effect sizes calculated from the data before and after transformation may differ substantially. One approach to estimating δ is to use the effect size calculated on the transformed data, which, presumably, are reasonably normal. An alternative is to develop an estimator of the effect size that is unaffected by monotonic transformations of the observations. This was the motivation for the nonparametric estimator of effect size proposed by Kraemer and Andrews (1982) and for extensions of this estimator proposed by Hedges and Olkin (1984).

The standardized mean difference provides a guide to the overlap between the distributions of scores in the experimental and control groups. But when the data are not normally distributed the interpretation of effect size in terms of the percentiles of the standard normal distribution is faulty. If, however, the data can be transformed monotonically in such a way that they become more normally distributed, then it is important to have an estimator that remains unchanged in the transforming process.

The usual model for the observations in each study (assumptions of the t-test) may also be invalid if some observations are outliers. The presence of outliers distorts the values of test statistics and estimates of effect size. We may eliminate outliers before any analysis takes place, and calculate the effect size estimate on the outlier-free data. Alternatively, we can develop methods for estimating effect size that are insensitive to the presence of outliers.

Below we examine some alternatives to parametric estimation of effect size. Although the estimators are nonparametric in the sense that they do not explicitly depend on the parametric structural model for the observations, they may not be distribution-free (see Krauth, 1983). We begin by discussing estimators of effect size derived by using robust estimators of μ^E, μ^C, and σ, and then examine a nonparametric estimator of effect size proposed by Kraemer and Andrews (1982). Their method is suggestive of a number of alternative estimators, and these are illustrated on the data provided by Kraemer and Andrews.

C.1 ESTIMATES OF EFFECT SIZE THAT ARE ROBUST AGAINST OUTLIERS

The effect size δ is the ratio of a mean difference to a standard deviation. To obtain an estimator of effect size that is robust against outliers, we estimate both the mean difference and the standard deviation by robust estimators. Many such estimators are possible. For example, the population mean of each group can be estimated by the median, and the standard deviation of each group can be estimated by the range or other linear combinations of order statistics. We can let

$$\tilde{\delta} = [\text{Med}(Y^E) - \text{Med}(Y^C)]/\tilde{\sigma}, \tag{25}$$

where $\text{Med}(Y^E)$ and $\text{Med}(Y^C)$ are the medians of the observations in the experimental and control groups, respectively, and

$$\tilde{\sigma} = a_2 y_{(2)}^C + \cdots + a_{n-1} y_{(n-1)}^C, \tag{26}$$

where $y_{(1)}^C \leq y_{(2)}^C \leq \cdots \leq y_{(n-1)}^C \leq y_{(n)}^C$ are the ordered values in the control group. The optimal coefficients $a_2, \ldots, a_{n-1}$ are chosen to minimize the variance of $\tilde{\sigma}$. They are not easily described but are tabulated (see, e.g., Sarhan & Greenberg, 1962, p. 218–251). In order to diminish sensitivity to extreme values, the smallest and largest values have been omitted in the estimator of σ.

Since the estimate $\tilde{\delta}$ of δ given in (25) does not involve the two extreme observations, it is robust against shifts of the most extreme observations.

C.2 NONPARAMETRIC ESTIMATORS OF EFFECT SIZE

A nonparametric approach to estimating effect size was suggested by Kraemer and Andrews (KA). Their technique requires that both the pretest (x) and posttest (y) scores be available for each individual in the experimental and control groups of an experiment (Table 5).

We first review the rationale for the KA estimator; this serves to clarify how additional estimators can be constructed.

For each group determine the sample proportion $\hat{p}$ of x scores that lie below the median y score. This proportion corresponds to a standard normal deviate $\hat{\delta}$:

$$\Phi(\hat{\delta}) = \hat{p} \qquad \text{or} \qquad \hat{\delta} = \Phi^{-1}(\hat{p}), \tag{27}$$

where $\Phi(x)$ is the standard normal cumulative distribution function. An application of this procedure to each of the experimental and control groups

TABLE 5
Systolic Blood Pressure Data on Pre and Post Experimental and Control Data from 40 Hypertensives

	Experimental group		Control group	
	Pre	Post	Pre	Post
	134	130	139	130
	135	131	140	131
	135	135	141	144
	136	136	143	146
	145	136	151	128
	147	138	152	156
	148	124	152	161
	150	126	153	162
	151	104	153	160
	153	142	154	131
	153	114	154	158
	155	166	159	166
	156	153	160	150
	158	169	160	186
	162	127	162	188
	165	130	163	153
	167	120	165	144
	168	121	169	147
	179	149	175	169
	180	150	176	170
Median	153	133	154	154.5

yields posttest versus pretest effect size estimates $\hat{\delta}^E$ and $\hat{\delta}^C$ for the experimental and control groups, respectively. From these an overall effect size $\hat{\delta}$ is obtained as

$$\hat{\delta} = \hat{\delta}^E - \hat{\delta}^C.$$

The key idea in the above development is to estimate effect size by using the proportion of x scores that lie below the "middle" of the y score distribution. However, different definitions of the middle of the y score distribution can be used, and these would result in slightly different estimators. For simplicity, we describe the procedure using the median as the definition of the center of the posttest score distribution.

Remark: Kraemer and Andrews do not use the median as the center, because they are especially interested in the pre- to posttest change of individuals near the center of the pretest score distribution. They define a "critical prerange" as the x median plus the nearest two x scores on either side of the median and then define the center of y scores as the median of the five posttest scores of individuals with x scores in the critical prerange. Once the middle of the y score distribution is defined, their procedure proceeds as described in Section C.2.

Example

Data on systolic blood pressure given in Kraemer and Andrews (1982) are reproduced in Table 5. The pretreatment (x) and posttreatment (y) scores of experimental and control group subjects are arranged in ascending order of pretest scores. The median of the y scores in the control group is 154.5. The number of x scores less than 154.5 is 11, so that

$$\hat{\delta}^{C} = \Phi^{-1}(\tfrac{11}{20}) = 0.126.$$

In a similar manner we find that $\hat{\delta}^{E} = \Phi^{-1}(0)$. In such a case, Kraemer and Andrews recommend using $p = 1/(n + 1) = \frac{1}{21} = 0.048$ instead of zero, so that $\hat{\delta}^{E} = \Phi^{-1}(0.048) = -1.665$. Combining $\hat{\delta}^{E}$ and $\hat{\delta}^{C}$, we obtain an estimate of the effect size:

$$\hat{\delta} = \hat{\delta}^{E} - \hat{\delta}^{C} = -1.665 - 0.126 = -1.79.$$

Remark: The result obtained by Kraemer and Andrews using slightly different definitions of $\hat{\delta}^{E}$ and $\hat{\delta}^{C}$ is -1.85.

C.3 ESTIMATORS BASED ON DIFFERENCES OF CONTROL VERSUS TREATMENT PROPORTIONS

The logic of nonparametric estimators of effect size is best illustrated by examining bivariate plots of the pretest (x) scores versus posttest (y) scores (see Fig. 3). The KA procedure consists in looking at the proportion p of points (x, y) such that x is less than the median of y, i.e., calculating the proportion of points to the left of line L_1 for the control group. We then use the inverse normal cumulative distribution function Φ^{-1} to obtain an estimate of pre–post effect size from p. The KA estimator is

$$\hat{\delta}_1 = \Phi^{-1}(\hat{p}_1^{E}) - \Phi^{-1}(\hat{p}_1^{C}). \tag{28}$$

When the observations are normally distributed, $\hat{\delta}_1$ is an estimate of

$$\frac{\mu_y^{E} - \mu_x^{E}}{\sigma_x^{E}} - \frac{\mu_y^{C} - \mu_x^{C}}{\sigma_x^{C}},$$

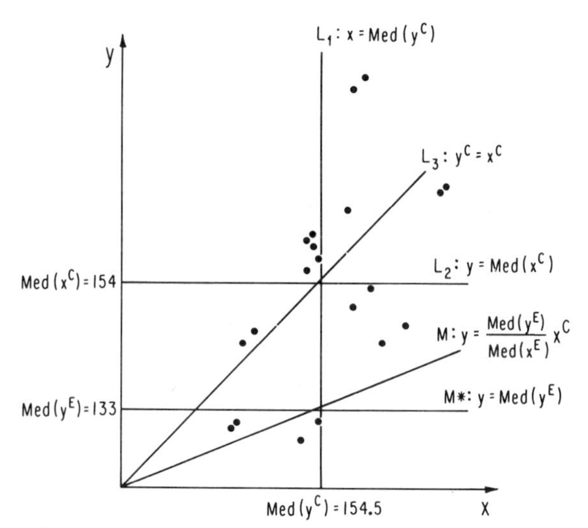

Fig. 3 Plot of control group Pretest and Posttest scores to exhibit computation of nonparametric estimators.

where μ_y^E, μ_x^E, and σ_x^E are the experimental group posttest mean, pretest mean, and pretest standard deviation, respectively, and μ_y^C, μ_x^C, and σ_x^C are the analogous parameters for the control group.

Other regions of the bivariate plot can be used to derive estimates of effect size. Indeed, the KA procedure is only one of several natural estimators of effect size obtained by using proportions of points falling into different regions of the bivariate plot. The rationale for the KA procedure is that the larger the treatment effect, the smaller the proportion of treatment group pretest scores that exceed the posttest median.

A related estimator $\hat{\delta}_2$ is based on the proportion $\hat{p}_2^C$ of the posttest scores in the control group that exceed the pretest median in the control group, and the proportion $\hat{p}_2^E$ of the posttest scores in the experimental group that exceed the pretest median in the experimental group. For the control group the relevant proportion corresponds to the proportion of observations above the line L_2. (The relevant proportion for the experimental group is calculated in a similar fashion.) The above description is translated to the formula for $\hat{\delta}_2$:

$$\hat{\delta}_2 = \Phi^{-1}(\hat{p}_2^E) - \Phi^{-1}(\hat{p}_2^C). \tag{29}$$

For the Kraemer–Andrews data $\hat{p}_2^E = \frac{2}{20} = 0.10$ and $p_2^C = \frac{10}{20} = 0.50$. Therefore $\hat{\delta}_2^E = \Phi^{-1}(0.10) = -1.282$, $\hat{\delta}_2^C = \Phi^{-1}(0.50) = 0.0$, and $\hat{\delta}_2 = \hat{\delta}_2^E - \hat{\delta}_2^C = -1.282$.

When the observations are normally distributed, $\hat{\delta}_2$ is an estimate of

$$\frac{\mu_y^E - \mu_x^E}{\sigma_y^E} - \frac{\mu_y^C - \mu_x^C}{\sigma_y^C},$$

where σ_y^{E} and σ_y^{C} are experimental and control group posttest standard deviations.

Still another alternative can be considered. Let $\hat{p}_3^{\mathrm{C}}$ denote the proportion of individuals in the control group whose scores increase from pretest to posttest. This corresponds to the proportion of observations above the line L_3. If $\hat{p}_3^{\mathrm{E}}$ is the proportion of individuals whose scores increase from pretest to posttest in the experimental group, the estimator $\hat{\delta}_3$ is given by

$$\hat{\delta}_3 = \Phi^{-1}(\hat{p}_3^{\mathrm{E}}) - \Phi^{-1}(\hat{p}_3^{\mathrm{C}}). \tag{30}$$

In the experimental group in Table 5 there are two ties in the pretest scores and two ties in the posttest scores. We recommend counting each tie as half an observation in each direction, that is, one-half an observation for which $y > x$ and one-half an observation for which $y < x$.

Using this correction for ties, $\hat{p}_3^{\mathrm{E}} = \frac{3}{20} = 0.15$ and $\hat{p}_3^{\mathrm{C}} = \frac{10}{20} = 0.50$, so that $\hat{\delta}_3^{\mathrm{E}} = \Phi^{-1}(0.15) = -1.038$, $\hat{\delta}_3^{\mathrm{C}} = \Phi^{-1}(0.50) = 0.00$, and

$$\hat{\delta} = \hat{\delta}_3^{\mathrm{E}} - \hat{\delta}_3^{\mathrm{C}} = -1.038.$$

When the scores are normally distributed, $\hat{\delta}$ is an estimate of

$$\frac{\mu_y^{\mathrm{E}} - \mu_x^{\mathrm{E}}}{\sigma_*^{\mathrm{E}}} - \frac{\mu_y^{\mathrm{C}} - \mu_x^{\mathrm{C}}}{\sigma_*^{\mathrm{C}}},$$

where μ_y^{E}, μ_y^{C}, μ_x^{E}, and μ_x^{C} are experimental and control group population means on the posttest and pretest, respectively, and σ_*^{E} and σ_*^{C} are the gain score standard deviations. Note that $\hat{\delta}_3$ estimates the difference between standardized mean gains in the experimental and control groups.

The main distinction between the estimators $\hat{\delta}_1$, $\hat{\delta}_2$, $\hat{\delta}_3$ lies in the standardizations. Each estimator estimates a parameter of the form

$$\frac{\mu_y^{\mathrm{E}} - \mu_x^{\mathrm{E}}}{\sigma^{\mathrm{E}}} - \frac{\mu_y^{\mathrm{C}} - \mu_x^{\mathrm{C}}}{\sigma^{\mathrm{C}}},$$

where the standard deviations σ are $\sigma_x, \sigma_y, \sigma_*$ for $\hat{\delta}_1, \delta_2, \delta_3$, respectively. Thus, the particular $\hat{\delta}$ that should be used depends on the standardization desired.

C.4 ESTIMATORS BASED ON GAIN SCORES IN THE EXPERIMENTAL GROUP RELATIVE TO THE CONTROL GROUP

A different attack makes use of the proportion of scores in the control group that gain less than would be expected in the experimental group. This idea is illustrated graphically in Fig. 3 as follows.

Determine the medians of the pretest and posttest scores in the experimental group, and on the control group data draw a line M from the origin (0, 0) through the point [Med(x^E), Med(y^E)]. The equation of the line M is

$$y^C = \frac{\text{Med}(y^E)}{\text{Med}(x^E)} x^C. \tag{31}$$

The proportion $\hat{q}_1$ of points above line M represents the proportion of individuals in the control group whose scores increased relatively less than those of their experimental group counterparts. Then the estimator $\hat{\gamma}_1$ of effect size is given by

$$\hat{\gamma}_1 = \Phi^{-1}(\hat{q}_1). \tag{32}$$

For the data in Table 5, $\hat{q}_1 = \frac{2}{20} = 0.10$ and $\hat{\gamma}_1 = \Phi^{-1}(0.10) = -1.282$. If the scores are normally distributed, $\hat{\gamma}_1$ estimates $(\mu_y^E - \mu_y^C)/\sigma_y^C$.

A symmetric version is obtained by drawing on the plot of experimental data a line between the origin (0, 0) and the point [Med(x^C), Med(y^C)] representing the medians of the posttest scores in the control group. Define $\hat{q}_2$ as the proportion of scores in the experimental group that lie above the line, that is, for which

$$y^E > \frac{\text{Med}(y^C)}{\text{Med}(x^C)} x^E. \tag{33}$$

The estimator $\hat{\gamma}_2$ of effect size is

$$\hat{\gamma}_2 = \Phi^{-1}(\hat{q}_2). \tag{34}$$

For the data in Table 5, $\hat{q}_2 = \frac{2}{20} = 0.10$ and $\hat{\gamma}_2 = \Phi^{-1}(0.10) = -1.282$. If the scores are normally distributed then $\hat{\gamma}_2$ estimates $(\mu_y^E - \mu_y^E)/\sigma_y^E$.

Remark: The estimators based on relative gain are *not* invariant under all monotonic transformations of the x and y scores. They are included primarily to illustrate a class of robust estimators that can be constructed using the principle of relative gain in experimental and control groups.

C.5 NONPARAMETRIC ESTIMATORS INVOLVING ONLY POSTTEST SCORES

If random assignment is made to the control and treatment groups, then one expects the distribution of pretest scores to be similar in both groups. This suggests using only the posttest scores in the experimental and control groups. (This procedure does have the advantage that it can be used when pretest scores are unavailable.) Calculate the proportion $\hat{q}_1^*$ of control group scores less than the experimental group median (below line M^*) and transform the proportion to an estimate $\hat{\gamma}_1^*$ of effect size,

$$\hat{\gamma}_1^* = \Phi^{-1}(\hat{q}_1^*). \tag{35}$$

Examining Table 5, we see that $\hat{q}_1^* = \frac{4}{20} = 0.20$ for the Kraemer–Andrews data, which corresponds to $\hat{\gamma}_1^* = \Phi^{-1}(0.20) = -1.038$. When the observations are independently normally distributed, $\hat{\gamma}$ is an estimator of $(\mu_y^{\mathrm{E}} - \mu_y^{\mathrm{C}})/\sigma_y^{\mathrm{C}}$, the mean difference standardized by the standard deviation σ_y^{C} of the control group scores.

A symmetric estimator is obtained by using the proportion $\hat{q}_2^*$ of posttest scores in the experimental group that exceed the median score in the control group. The proportion $\hat{q}_2^*$ is transformed into an estimator of effect size via

$$\hat{\gamma}_2^* = \Phi^{-1}(\hat{q}_2^*).$$

The Kraemer–Andrews data in Table 5 have $\hat{q}_2^* = \frac{2}{20} = 0.10$, which corresponds to $\hat{\gamma}_2^* = \Phi^{-1}(0.10) = -1.282$.

When the observations are independently normally distributed, $\hat{\gamma}_2^*$ is an estimator of $(\mu_y^{\mathrm{E}} - \mu_y^{\mathrm{C}})/\sigma_y^{\mathrm{E}}$, the experimental–control group mean difference standardized by the standard deviation σ_y^{E} of the scores in the experimental group. Therefore if the scores are normally distributed and the experimental and control group standard deviations are equal, both $\hat{\gamma}_1^*$ and $\hat{\gamma}_2^*$ estimate the same quantity. We would use $\hat{\gamma}_1$ rather than $\hat{\gamma}_1^*$ when both pre- and posttest data are available, and use $\hat{\gamma}_1^*$ when only posttest data are available.

C.6 RELATIONSHIPS BETWEEN ESTIMATORS

The seven nonparametric estimators of effect size are different in the sense that they use the data differently and estimate slightly different parameters when the data are normally distributed. The proportions used to calculate each of the estimators are given in Table 6 along with the quantity estimated under the assumption that the observations are normally distributed. Note that $\hat{\gamma}_1$ and $\hat{\gamma}_1^*$ estimate the same quantity, and that $\hat{\gamma}_2$ and $\hat{\gamma}_2^*$ estimate the same quantity. When $\sigma_y^{\mathrm{E}} = \sigma_y^{\mathrm{C}}$, the four estimators $\hat{\gamma}_1, \hat{\gamma}_1^*, \hat{\gamma}_2$, and $\hat{\gamma}_2^*$ estimate identical quantities. Similarly, $\hat{\delta}_1$ and $\hat{\delta}_2$ estimate the same parameter if $\sigma_x^{\mathrm{E}} = \sigma_y^{\mathrm{C}}$ and $\sigma_y^{\mathrm{E}} = \sigma_y^{\mathrm{C}}$.

Little is known about the sampling properties of nonparametric estimators of effect size. However, they probably are less efficient than their parametric counterparts when the assumptions of the parametric procedures are satisfied. Thus the nonparametric estimators should be used only when there is reason to believe that parametric assumptions are seriously violated. Nonparametric estimators of effect size can be computed only when all of

TABLE 6
Summary of Nonparametric Estimators

Estimator	Proportion of points to be calculated	Quantity estimated under normality assumption
$\hat{\delta}_1^C$	$x^C < \text{Med}(y^C)$	$(\mu_y^C - \mu_x^C)/\sigma_x^C$
$\hat{\delta}_1^E$	$x^E < \text{Med}(y^E)$	$(\mu_y^E - \mu_x^E)/\sigma_x^E$
$\hat{\delta}_1 = \hat{\delta}_1^E - \hat{\delta}_1^C$		$\delta_1^E - \delta_1^C$
$\hat{\delta}_2^C$	$y^C > \text{Med}(x^C)$	$(\mu_y^C - \mu_x^C)/\sigma_y^C$
$\hat{\delta}_2^E$	$y^E > \text{Med}(x^E)$	$(\mu_y^E - \mu_x^E)/\sigma_y^E$
$\hat{\delta}_2 = \hat{\delta}_2^E - \hat{\delta}_2^C$		$\delta_2^E - \delta_2^C$
$\hat{\delta}_3^C$	$y^C > x^C$	$(\mu_y^C - \mu_x^C)/\sigma_*^C$
$\hat{\delta}_3^E$	$y^E > x^E$	$(\mu_y^E - \mu_x^E)/\sigma_*^E$
$\hat{\delta}_3 = \hat{\delta}_3^E - \hat{\delta}_3^C$		$\delta_3^E - \delta_3^C$
$\hat{\gamma}_1$	$y^C < \dfrac{\text{Med}(y^E)}{\text{Med}(x^E)} x^C$	$(\mu_y^E - \mu_y^C)/\sigma_y^C$
$\hat{\gamma}_2$	$y^E > \dfrac{\text{Med}(y^C)}{\text{Med}(x^C)} x^E$	$(\mu_y^E - \mu_y^C)/\sigma_y^E$
$\hat{\gamma}_1^*$	$y^C < \text{Med}(y^E)$	$(\mu_y^E - \mu_y^C)/\sigma_y^C$
$\hat{\gamma}_2^*$	$y^E > \text{Med}(y^C)$	$(\mu_y^E - \mu_y^C)/\sigma_y^E$

the original observations (the raw data as opposed to means and standard deviations) are available. This seriously limits the applicability of these estimators in practice, since the raw data are rarely available in meta-analysis.

D. OTHER MEASURES OF EFFECT MAGNITUDE

An alternative approach to the estimation of a scale-free effect magnitude is based on the idea of variance accounted for due to the introduction of an explanatory variable. Measures of effect magnitude in this category are all constructed via the relationship

$$\text{effect magnitude} = (\text{explained variance})/(\text{total variance}),$$

where the total variance is the variance of all observations and the explained variance is the reduction in variance due to the introduction of an explanatory variable. The explained "variance" is often not formally a variance at all but the difference between the total variance and a conditional variance.

Measures of effect magnitude based on variance accounted for have some serious deficiencies (discussed in Section D.4). One reason is that such measures are not directional and are therefore ill-suited for use in combining the results of different studies. However, because these measures do appear in the literature, for the sake of completeness we provide a brief overview of the basic principles and two examples. The first is the squared correlation coefficient and its generalization, the correlation ratio. The second is a measure of effect magnitude introduced by Hays (1963) that is quite similar to the correlation ratio.

D.1 THE CORRELATION COEFFICIENT AND CORRELATION RATIO

The best-known index based on variance accounted for is the squared correlation coefficient

$$\rho^2 = \frac{\sigma_{\mathrm{T}}^2 - \sigma_{\mathrm{RL}}^2}{\sigma_{\mathrm{T}}^2} = \frac{\text{variance accounted for}}{\text{total variance}},$$

where σ_{T}^2 is the total variance of the dependent variable and σ_{RL}^2 is the variance about the regression line.

The squared correlation coefficient can sometimes be used as an index of effect magnitude even when the independent variable is not continuous. In multigroup experiments where the levels of the independent variable can be assigned meaningful numerical values, the squared correlation can be calculated, and reflects only the proportion of variance accounted for by the *linear* relationship between the independent and dependent variables. If the number of discrete values of the independent variable exceeds two, then nonlinear components of the relationship are possible, which are not necessarily reflected in the size of the squared correlation.

For such cases a generalization of the squared correlation, the correlation ratio η^2, was introduced by Pearson (1905). It is defined as

$$\eta^2 = \frac{\sigma_Y^2 - \sigma_{Y|X}^2}{\sigma_Y^2},$$

where σ_Y^2 is the total variance of the dependent variable Y and $\sigma_{Y|X}^2$ is the conditional variance of Y given the value of the discrete independent variable X. The correlation ratio generalizes the squared correlation by incorporating the reduction in the variance of Y due to both the linear and the nonlinear components of the relationship between Y and X. Note that if the number of discrete values of X is two, the correlation ratio is identical to the squared correlation coefficient. If X takes on more than two values, η^2 will be at least as large as ρ^2.

The usual estimator of η^2 in a multigroup experiment is

$$\hat{\eta}^2 = \frac{SS_B}{SS_B + SS_W},$$

where SS_B is the usual between-group sum of squares and SS_W is the usual within-group sum of squares. An alternative estimator E^2 of η^2 was proposed by Kelley (1935):

$$E^2 = 1 - \frac{N-1}{N-k}(1 - \hat{\eta}^2) = \frac{MS_T - MS_W}{MS_T},$$

where N is the total sample size, k is the number of groups in the experiment (discrete values of X), MS_W and MS_T are the usual mean square within groups and mean square total in the analysis of variance. The rationale for E^2 is that MS_W and MS_T provide unbiased estimators of $\sigma^2_{Y|X}$ and σ^2_Y, respectively, so that the ratio of unbiased estimates might be expected to be a "good" estimator of η^2. Although the estimator E^2 is sometimes referred to as unbiased, this is untrue in the technical sense. However, there is some evidence that E^2 is less biased than the usual estimator $\hat{\eta}^2$ of η^2 (Winer, 1971). Note that the exact sampling properties of $\hat{\eta}^2$ have been studied extensively. Both Fisher (1922) and Hotelling (1925) independently obtained the sampling distribution of $\hat{\eta}^2$ when the null hypothesis $\hat{\eta}^2 = 0$ holds. Fisher (1928) and Wishart (1932) obtained the general (noncentral) distribution of the squared multiple correlation coefficient and correlation ratio, respectively.

D.2 THE INTRACLASS CORRELATION COEFFICIENT

One variant of the correlation ratio is the analogue to η^2 for the random effects model. The intraclass correlation coefficient ρ_I is defined as

$$\rho_I = \frac{\sigma^2_R}{\sigma^2_R + \sigma^2_E},$$

where σ^2_R is the variance component of the random group effects and σ^2_E is the error variance component in the one-way random effects model analysis of variance. Note that the intraclass correlation is defined *only* in the random effects model. There are many common estimators of ρ_I (see Shrout & Fleiss, 1979), but one of the simplest estimators is

$$\hat{\rho}_I = \frac{MS_B - MS_W}{MS_B + (N-1)MS_W},$$

where MS_B and MS_W are the usual mean square between groups and mean square within groups, respectively, for the random effects analysis of variance, and N is the total sample size. The distribution of $\hat{\rho}_I$ was first obtained by Fisher (1921), and the estimator has been used extensively since then, particularly as an index of reliability in test theory.

D.3 THE OMEGA-SQUARED INDEX

Hays (1981, p. 349) introduced an index of effect magnitude (in the fixed effects analysis of variance) which he called ω^2 to distinguish it from other similar measures such as η^2, ρ_I^2, and ρ^2. The parameter ω^2 is motivated by the idea that ω^2 is the relative reduction in the variance of Y due to X:

$$\omega^2 = \lambda^2/(N + \lambda^2),$$

where N is the total sample size and λ^2 is the noncentrality parameter of the F-statistic for testing the difference among group means.

The usual estimator of ω^2 in the one-way analysis of variance is

$$\hat{\omega}^2 = \frac{(k-1)(MS_B - MS_W)}{(k-1)MS_B + (N-k+1)MS_W},$$

where MS_B and MS_W are the usual mean squares between and within groups, and k is the number of groups in the experiment. The distribution of $\hat{\omega}^2$ was derived by Abu Libdeh (1984), who also provided a study of the bias of $\hat{\omega}^2$, and nomographs for interval estimation of ω^2 from $\hat{\omega}^2$.

D.4 PROBLEMS WITH VARIANCE-ACCOUNTED-FOR MEASURES

Indices of effect magnitude based on the proportion of variance accounted for are intuitively appealing but are not well suited for combination across studies. These indices are inherently nondirectional, and can have the same value even though the research studies exhibit substantively different results. They can even have the same value for conflicting patterns of results. For example, consider two studies, one in which the treatment group mean exceeds the control group mean by one standard deviation and one in which the reverse holds; namely, the control group mean exceeds the treatment group mean by one standard deviation. Both studies would yield identical values of $\hat{\rho}^2$, $\hat{\eta}^2$, or $\hat{\omega}^2$, even though the results of the two studies are exactly opposite. An examination of only the $\hat{\omega}^2$ values would lead to the conclusion that the two experiments obtained the same results—a very misleading conclusion.

A more subtle difficulty with indices of effect magnitude based on variance accounted for is that these indices can be sensitive to the definition of groups or patterns of X values (Glass & Hakstian, 1969). When the treatment groups of an experiment are formed by choosing arbitrarily from a range of (theoretically continuous) X values, the number and range of X values used in an experiment can differ. The value of $\hat{\rho}^2$, $\hat{\eta}^2$, or $\hat{\omega}^2$ will, in general, depend on the particular levels chosen in a study as well as on the relationship between X and Y. Consequently indices of variance accounted for depend on functions of arbitrary design decisions as well as the underlying relationship between theoretical constructs.

Because of these inherent difficulties we do not discuss indices of effect magnitude derived from the percentage of variance accounted for in subsequent chapters.

E. TECHNICAL COMMENTARY

Section A.2: The exact mean and variance of g are

$$E(g) = \delta / J(n^*),$$

$$\operatorname{Var}(g) = \frac{n^*}{(n^* - 2)\tilde{n}} + \delta^2\left(\frac{n^*}{n^* - 2} - \frac{1}{[J(n^*)]^2}\right), \tag{36}$$

where $n^* = n^E + n^C - 2$, $\tilde{n} = n^E n^C/(n^E + n^C)$,

$$J(a) = \frac{\Gamma(a/2)}{\sqrt{a/2}\,\Gamma((a-1)/2)}, \tag{37}$$

and $\Gamma(x)$ is the gamma function.

Section A.2: We now prove that

$$J(m) = \frac{\Gamma(m/2)}{\sqrt{m/2}\,\Gamma((m-1)/2)} < 1$$

Let $x = m/2$, so that we need to prove that

$$\Gamma^2(x) < x\Gamma^2(x - \tfrac{1}{2}).$$

Since $\Gamma(x) = (x - 1)\Gamma(x - 1)$,

we need to show that

$$(x - 1)\Gamma(x - 1)\Gamma(x) < x\Gamma^2(x - \tfrac{1}{2}).$$

Since $0 < x - 1 < x$, it suffices to show that

$$\Gamma(x-1)\Gamma(x) < \Gamma^2(x-\tfrac{1}{2})$$

or, by taking logarithms,

$$\frac{\log \Gamma(x-1) + \log \Gamma(x)}{2} < \log \Gamma(x-\tfrac{1}{2}). \tag{38}$$

But the function $\log \Gamma(x)$ is convex for x positive. Its derivative

$$\psi(x) = d \log \Gamma(x)/dx$$

is called the digamma function and is monotone increasing for x positive. Since $x - \frac{1}{2} = \frac{1}{2}(x-1) + \frac{1}{2}(x)$, the convexity of $\log \Gamma(x)$ proves (38), which completes the proof.

Section A.3: If $d = J(n^*)\,\delta$, then $E(d) = \delta$ and

$$\text{Var}(d) = [J(n^*)]^2 \text{ Var}(g) < \text{Var}(g).$$

The inequality holds because $J(n^*) < 1$. Consequently, the mean-squared errors of g and d are

$$\text{MSE}(g) = \text{Var}(g) + [\text{Bias}(g)]^2 > \text{Var}(d) + [\text{Bias}(g)]^2 = \text{MSE}(d) > \text{Var}(d).$$

An explicit expression for the variance of d is

$$\text{Var}(d) = c_1 + c_2\delta^2,$$

where

$$c_1 = n^*[J(n^*)]^2/(n^*-2)\tilde{n}, \qquad c_2 = c_1\tilde{n} - 1,$$

and $n^* = n^{\text{E}} + n^{\text{C}} - 2$.

Section A.4: The mean-squared error of g^* is

$$\text{MSE}(g^*) = \text{Var}(g^*) + [\text{Bias}(g^*)]^2.$$

If we confine ourselves to linear combinations of d:

$$g^* = ad + b,$$

then

$$\text{Var}(g^*) = a^2 \text{ Var}(d) = a^2(c_1 + c_2\delta^2), \qquad \text{Bias}(g^*) = (a-1)\delta + b,$$

and

$$\text{MSE}(g^*) = a^2(c_1 + c_2\delta^2) + [(a-1)\delta + b]^2,$$

which is to be minimized with respect to a and b. This implies that $b = 0$, for otherwise the choice of $\delta = 0$ would increase the MSE(g^*). Then, with $b = 0$,

$$\text{MSE}(g^*) = a^2 c_1 + \delta^2[a^2 c_2 + (a - 1)^2].$$

Since we want to minimize this for all δ, it implies that we need to

$$\operatorname*{Min}_a \{a^2 c_2 + (a - 1)^2\},$$

which yields the minimizer

$$a_0 = 1/(c_2 + 1).$$

CHAPTER 6

Parametric Estimation of Effect Size from a Series of Experiments

If a series of k studies can reasonably be expected to share a common effect size δ, then we should pool the information from all the studies to obtain a combined estimate of δ. However, there are alternative ways to pool information. We now present several methods for obtaining pooled estimates of δ from a series of studies with moderate to large sample sizes.

One simple pooled estimate is the average of the estimates obtained from each study, and because of its simplicity, this method is used frequently. As might be expected, if the studies do not have a common sample size, there should be some weighting. This becomes clearer when we note that the variance of the estimator depends on the sample size, so that estimates from studies with larger sample sizes are more precise than those from studies with smaller sample sizes.

Two methods are presented for obtaining "optimal" combinations of estimates of effect size from a series of studies: (i) a direct weighted linear combination of estimators from different studies and (ii) a maximum likelihood estimator. Both estimators are shown to have the same asymptotic distribution and hence are asymptotically equivalent. Two other methods are suggested that involve a transformation of the effect size estimates. As we shall show, the method based on linear combinations of estimators is simpler, perhaps more intuitively appealing, and involves less computation.

A. MODEL AND NOTATION

Statistical properties of procedures for combining results from a series of experiments depend on the structural model for the results of the experiments. Here, the structural model requires that each experiment be based on a response scale from a collection of congeneric measures; that is, each response scale is a linear transformation of a response scale with unit variance within groups. This assumption is satisfied if the tests are linearly equatable.

The population standardized mean difference was proposed by Cohen (1969) as an index of effect magnitude in connection with an assessment of the power of the two-sample t-test. We denote this population parameter by δ and its unbiased estimate by d. In one of the early quantitative syntheses of studies involving experimental and control groups, Glass (1976) estimated δ for each study and then combined the estimates across studies. The statistical analyses in such studies typically involve the use of t- or F-tests to test for differences between the group means. If the assumptions for the validity of the t-test are met, it is possible to derive the properties of estimators of δ exactly. We start by stating these assumptions explicitly.

Suppose that the data arise from a series of k independent studies, in which each study is a comparison of an experimental group (E) with a control group (C) as indicated below.

Data base:

	Observations	
Study	Experimental	Control
1	$Y_{11}^E, \ldots, Y_{1n_1^E}^E$	$Y_{11}^C, \ldots, Y_{1n_1^C}^C$
$\vdots$	$\vdots$	$\vdots$
k	$Y_{k1}^E, \ldots, Y_{kn_k^E}^E$	$Y_{k1}^C, \ldots, Y_{kn_k^C}^C$

Parameters:

	Experimental		Control		
Study	Mean	Variance	Mean	Variance	Effect size
1	μ_1^E	σ_1^2	μ_1^C	σ_1^2	$\delta_1 = (\mu_1^E - \mu_1^C)/\sigma_1$
$\vdots$	$\vdots$	$\vdots$	$\vdots$	$\vdots$	$\vdots$
k	μ_k^E	σ_k^2	μ_k^C	σ_k^2	$\delta_k = (\mu_k^E - \mu_k^C)/\sigma_k$

Assume that for the ith study the experimental observations $Y_{i1}^{\mathrm{E}}, \ldots, Y_{in_i^{\mathrm{E}}}^{\mathrm{E}}$ are normally distributed with a common mean μ_i^{E} and a common variance σ_i^2. Similarly, the control observations $Y_{i1}^{\mathrm{C}}, \ldots, Y_{in_i^{\mathrm{C}}}^{\mathrm{C}}$ are normally distributed with a common mean μ_i^{C} and a common variance σ_i^2. By omitting a superscript E or C on the variance, we have tacitly assumed that the variances of the experimental and control groups, for the ith experiment, are identical. More succinctly,

$$Y_{ij}^{\mathrm{E}} \sim \mathcal{N}(\mu_i^{\mathrm{E}}, \sigma_i^2), \qquad j = 1, \ldots, n_i^{\mathrm{E}}, \quad i = 1, \ldots, k,$$

and (1)

$$Y_{ij}^{\mathrm{C}} \sim \mathcal{N}(\mu_i^{\mathrm{C}}, \sigma_i^2), \qquad j = 1, \ldots, n_i^{\mathrm{C}}, \quad i = 1, \ldots, k.$$

In this notation, the *effect size* δ_i for the ith study is defined as

$$\delta_i = (\mu_i^{\mathrm{E}} - \mu_i^{\mathrm{C}})/\sigma_i. \tag{2}$$

Although the values of the population means and the standard deviations may change under a linear transformation of the observations, the effect size δ remains unaltered; that is, δ is invariant under location and scale changes of the outcome variable. This implies that if the same population of test scores is represented on two tests that are linearly equatable, the effect sizes will be identical. The virtue of effect sizes is that they are comparable even though they may be derived from different but linearly equatable measures.

Conversely, if two measures are not linearly equatable, then the same population of test scores would, in general, yield different effect sizes when represented on the two measures. In particular, if two tests do not measure the same construct, there is little reason to believe that effect sizes based on those two tests would be the same. The implication is that even if a treatment produces a uniform effect size on measures of one construct (such as mathematics achievement), there is little reason to expect that effect size to be the same as the effect size for studies that measure the influence of the same treatment on another construct (such as attitude).

We now assume that each experiment in the series is a replication of the others, i.e., that all experiments measure the same outcome construct and differ only in response scale and sample sizes. That is, the population value of the treatment effect would be identical if all of the studies used the same response scale for the dependent variable. This assumption means that

$$\delta_1 = \delta_2 = \cdots = \delta_k = \delta^*.$$

B. WEIGHTED LINEAR COMBINATIONS OF ESTIMATES

When a series of k independent studies share a common effect size δ, it is natural to estimate δ by pooling estimates from each of the studies. If the sample sizes of the studies differ, then the estimates from the larger studies will be more precise than the estimates from the smaller studies. In this case it is reasonable to give more weight to the more precise estimates when pooling. This leads to weighted estimators of the form

$$d_w \equiv w_1 d_1 + \cdots + w_k d_k, \tag{3}$$

where $w_1, \ldots, w_k$ are nonnegative weights that sum to unity.

B.1 ESTIMATING WEIGHTS

The weights that minimize the variance of d_w give weight inversely proportional to the variance in each study. This is intuitively clear in that smaller variance, i.e., more precision, should lead to a larger weight. Consequently,

$$w_i = \frac{1}{\sigma^2(d_i)} \bigg/ \sum_{j=1}^{k} \frac{1}{\sigma^2(d_j)},$$

where $\sigma^2(d_i)$ is the variance of d_i. Because we use large sample theory, the weights are

$$w_i = \frac{1}{\sigma_\infty^2(d_i)} \bigg/ \sum_{j=1}^{k} \frac{1}{\sigma_\infty^2(d_j)}, \tag{4}$$

where $\sigma_\infty^2(d_i)$ is the large sample variance given in expression (14) of Chapter 5. The practical problem in calculating the most precise weighted estimate is that the ith weight depends on the variance of d_i, which in turn depends on the unknown parameter δ, and which must be estimated.

To resolve this difficulty weights that are based on an approximation of the variances that do not depend on δ can be used. This procedure results in a pooled estimator that is unbiased, and usually less precise than if optimal weights are used. For example, approximate weights are given by

$$w_i \cong \frac{\tilde{n}_i}{\sum_{j=1}^{k} \tilde{n}_j}, \tag{5}$$

where $\tilde{n}_i = n_i^E n_i^C / (n_i^E + n_i^C)$. The weights thus derived are close to optimal when δ is near zero and the $\tilde{n}_i$ are large.

Example

The results of 18 studies of the effects of open versus traditional education on student self-concept were described in Section C of Chapter 2. The sample sizes n_i^E and n_i^C, $\tilde{n}_i$, $1/\tilde{n}_i$, the weights w_i given in (5), and the effect size estimate

d_i for each study are listed in Table 1. Using the weights (5) from Table 1, the weighted average effect size from (3) is

$$d_w = -0.014.$$

If a nonzero a priori estimate of δ is available, then weights could be estimated by inserting that value of δ in expression (14) of Chapter 5 for the variance of d_i and using the formula (4) for w_i. In general, the result will be an unbiased pooled estimator of δ that is slightly less precise than the most precise weighted estimator.

The weighted estimator of δ based on using the sample estimate of δ to calculate the weights for each study is given by

$$d_+ = \sum_{i=1}^{k} \frac{d_i}{\hat{\sigma}^2(d_i)} \Big/ \sum_{i=1}^{k} \frac{1}{\hat{\sigma}^2(d_i)}, \tag{6}$$

where $\hat{\sigma}^2(d)$ is defined in expression (15) of Chapter 5. The estimator d_+ is therefore obtained by calculating the weights using d_i to estimate δ_i. Although the d_i are unbiased, d_+ is not. The bias of d_+ is small in large samples and tends to zero as the sample sizes become large.

TABLE 1

Effect Size Estimates from 18 Studies of the Effects of Open versus Traditional Education on Student Self-Concept

Study	n^E	n^C	$\tilde{n}$	w	d
1	100	180	64.286	0.115	0.100
2	131	138	67.205	0.120	−0.162
3	40	40	20.000	0.036	−0.090
4	40	40	20.000	0.036	−0.049
5	97	47	31.660	0.057	−0.046
6	28	61	19.191	0.034	−0.010
7	60	55	28.696	0.051	−0.431
8	72	102	42.207	0.076	−0.261
9	87	45	29.659	0.053	0.134
10	80	49	30.388	0.054	0.019
11	79	55	32.425	0.058	0.175
12	40	109	29.262	0.053	0.056
13	36	93	25.953	0.047	0.045
14	9	18	6.000	0.011	0.103
15	14	16	7.467	0.013	0.121
16	21	22	10.744	0.019	−0.482
17	133	124	64.171	0.115	0.290
18	83	45	29.180	0.052	0.342

This estimator d_+ could be modified by replacing d_i by d_+ in the expression for $\hat{\sigma}^2(d_i)$ and iterating. Define $d_+^{(0)}$ to be d_+ to obtain the first iterate:

$$d_+^{(1)} = \sum_{i=1}^{k} \frac{d_i}{\hat{\sigma}^2(d_i | d_+^{(0)})} \Bigg/ \sum_{i=1}^{k} \frac{1}{\hat{\sigma}^2(d_i | d_+^{(0)})},$$

and now define $d_+^{(2)}$ by

$$d_+^{(2)} = \sum_{i=1}^{k} \frac{d_i}{\hat{\sigma}^2(d_i | d_+^{(1)})} \Bigg/ \sum_{i=1}^{k} \frac{1}{\hat{\sigma}^2(d_i | d_+^{(1)})},$$

where $\hat{\sigma}^2(d_i | X)$ is an estimate of $\sigma_\infty^2(d_i)$ obtained by replacing d by X in expression (15) of Chapter 5 for the variance of d. For any stage l, the estimate at the next stage is obtained from

$$d_+^{(l+1)} = \sum_{i=1}^{k} \frac{d_i}{\hat{\sigma}^2(d_i | d_+^{(l)})} \Bigg/ \sum_{i=1}^{k} \frac{1}{\hat{\sigma}^2(d_i | d_+^{(l)})}, \qquad l = 0, 1, 2, \ldots. \tag{7}$$

The iterated estimator $d_+^{(l)}$ will tend to be less biased than $d_+^{(l-1)}$. If the effect sizes are homogeneous across experiments, the iteration process usually will not change the estimate very much (see Hedges 1983a).

The asymptotic distribution of d_+ can be used to obtain large sample confidence intervals for δ based on d_+. The definition of large sample in this case means that the individual sample sizes n_i^E and n_i^C, $i = 1, \ldots, k$, grow at the same rate and that $N = \sum (n_i^E + n_i^C)$ is large. We give the asymptotic distribution in the following.

Fact: If the sample sizes of the experimental and control groups in each of the k studies, $n_1^E, \ldots, n_k^E, n_1^C, \ldots, n_k^C$, become large at the same rates in the sense that n_i^E/N, n_i^C/N remain fixed, then the distribution of d_+ tends to normality with a mean

$$\delta_+ = \sum_{i=1}^{k} \frac{\delta_i}{\sigma_\infty^2(d_i)} \Bigg/ \sum_{i=1}^{k} \frac{1}{\sigma_\infty^2(d_i)} \tag{8}$$

and a variance

$$\sigma_\infty^2(d_+) = \left(\sum_{i=1}^{k} \frac{1}{\sigma_\infty^2(d_i)} \right)^{-1}, \tag{9}$$

where $\sigma_\infty^2(d_i)$ is given by expression (14) of Chapter 5.

When all the parameters $\delta_1, \ldots, \delta_k$ are equal to δ^*, say, as is the case in this chapter, then $\delta_+ = \delta^*$.

The large sample approximation derived from the asymptotic distribution above is that d_+ is normally distributed with a mean δ^* and an asymptotic variance (9). Consequently an approximate $100(1-\alpha)$-percent confidence interval (δ_L, δ_U) for δ^* is given by

$$\delta_L = d_+ - C_{\alpha/2}\hat{\sigma}(d_+), \qquad \delta_U = d_+ + C_{\alpha/2}\hat{\sigma}(d_+), \tag{10}$$

where $C_{\alpha/2}$ is the two-tailed critical value of the standard normal distribution and $\hat{\sigma}^2(d_+)$ is the sample estimate of the variance of d_+ given by

$$\hat{\sigma}^2(d_+) = \left(\sum_{i=1}^{k} \frac{1}{\hat{\sigma}^2(d_i)}\right)^{-1}. \tag{11}$$

Of course, if the interval (δ_L, δ_U) does not include zero, then we reject the hypothesis that $\delta^* = 0$ at the 100α-percent significance level.

The formal asymptotic distribution of the iterated estimator $d_+^{(l)}$ is the same as that of d_+. Therefore analogous procedures using $d_+^{(l)}$ exist for establishing confidence intervals for δ^* and a test of the hypothesis that $\delta^* = 0$. These procedures simply involve using $\hat{\sigma}^2(d_i | d_+^{(l)})$ in place of each $\hat{\sigma}^2(d_i)$ in (11) to compute the variance of $d_+^{(l)}$, and then using $d_+^{(l)}$ and the resulting variance of $d_+^{(l)}$ in (10).

Example

For the 18 studies of effects of open versus traditional education on student self-concept described in Section C of Chapter 2, the sample sizes and effect sizes are reported in Table 2. In order to find d_+ we have included the computations of $\hat{\sigma}^2(d_i)$, $1/\hat{\sigma}^2(d_i)$, and $d_i/\hat{\sigma}^2(d_i)$. The weighted mean using (4) is

$$d_+ = 5.881/555.600 = 0.011,$$

with estimated variance

$$\hat{\sigma}^2(d_+) = 1/555.600 = 0.00180.$$

If there is a common effect size δ^* for all 18 studies, then a 95-percent confidence interval (δ_L, δ_U) for δ^* is

$$\delta_L = 0.011 - 1.96\sqrt{0.00180} = -0.072,$$
$$\delta_U = 0.011 + 1.96\sqrt{0.00180} = 0.094.$$

Since this confidence interval contains zero, we do not reject the hypothesis that $\delta^* = 0$ at the $\alpha = 0.05$ level of significance.

B.2 EFFICIENCY OF WEIGHTED ESTIMATORS

The estimators discussed in this section are derived by finding the weights that minimize the variance of the resulting linearly weighted estimator.

TABLE 2

Data from 18 Studies of the Effects of Open versus Traditional Education on Student Self-Concept

Study	n^E	n^C	d	$\hat{\sigma}^2(d)$	$1/\hat{\sigma}^2(d)$	$d/\hat{\sigma}^2(d)$	$d^2/\hat{\sigma}^2(d)$
1	100	180	0.100	0.0156	64.212	6.421	0.642
2	131	138	−0.162	0.0149	66.985	−10.852	1.758
3	40	40	−0.090	0.0501	19.980	−1.798	0.162
4	40	40	−0.049	0.0500	19.994	−0.980	0.048
5	97	47	−0.046	0.0316	31.652	−1.456	0.067
6	28	61	−0.010	0.0521	19.191	−0.192	0.002
7	60	55	−0.431	0.0357	28.046	−12.088	5.210
8	72	102	−0.261	0.0239	41.861	−10.926	2.852
9	87	45	0.134	0.0338	29.599	3.966	0.531
10	80	49	0.019	0.0329	30.386	0.577	0.011
11	79	55	0.175	0.0309	32.306	5.653	0.989
12	40	109	0.056	0.0342	29.253	1.638	0.092
13	36	93	0.045	0.0385	25.948	1.168	0.053
14	9	18	0.103	0.1669	5.993	0.617	0.064
15	14	16	0.121	0.1342	7.453	0.902	0.109
16	21	22	−0.482	0.0958	10.441	−5.033	2.426
17	133	124	0.290	0.0157	63.504	18.416	5.341
18	83	45	0.342	0.0347	28.796	9.848	7.368
					555.600	5.881	23.725

However, there could be a better estimator that is not a linear combination of $d_1, \ldots, d_k$. This suggests that we investigate whether the best linearly weighted estimator is also the most precise in a larger class of estimators of effect size that includes those that are *not* weighted linear combinations of the d_i. Fortunately, it turns out that d_+ is actually best even in a larger class of estimators. It is the "best" in the sense that for large samples the variance of d_+ is as small as possible. Thus no other estimator has an asymptotic variance smaller than that of d_+. This result also implies that d_+ has the same asymptotic distribution as the maximum likelihood estimator of δ based on k experiments.

B.3 THE ACCURACY OF THE LARGE SAMPLE APPROXIMATION TO THE DISTRIBUTION OF WEIGHTED ESTIMATORS OF EFFECT SIZE

The accuracy of the confidence intervals and the test of the hypothesis that $\delta^* = 0$ depends on the accuracy of the large sample normal approximation to the distribution of d_+. The distribution of the estimator d_+, which is a

TABLE 3

The Small Sample Accuracy of Confidence Intervals for δ Based on the Normal Approximation to the Distribution of d_+ with 2000 Replications

Sample sizes	δ	Mean of d_+	Variance of d_+	Proportion of confidence intervals containing δ with nominal significance level					
				0.60	0.70	0.80	0.90	0.95	0.99
$k = 2$									
(10, 10)	0.25	0.243	0.096	0.620	0.721	0.821	0.920	0.964	0.994
(10, 20)	0.25	0.252	0.064	0.628	0.731	0.824	0.915	0.959	0.992
(10, 50)	0.25	0.245	0.033	0.616	0.719	0.819	0.909	0.949	0.991
(50, 50)	0.25	0.250	0.020	0.596	0.705	0.802	0.910	0.957	0.993
(10, 10)	0.50	0.485	0.097	0.620	0.731	0.826	0.912	0.959	0.992
(10, 20)	0.50	0.484	0.067	0.611	0.712	0.816	0.903	0.960	0.992
(10, 50)	0.50	0.487	0.035	0.589	0.684	0.791	0.896	0.948	0.990
(50, 50)	0.50	0.490	0.021	0.601	0.710	0.811	0.901	0.945	0.990
(10, 10)	1.00	0.959	0.111	0.613	0.709	0.808	0.908	0.946	0.992
(10, 20)	1.00	0.976	0.072	0.625	0.719	0.808	0.905	0.951	0.990
(10, 50)	1.00	0.985	0.038	0.598	0.693	0.790	0.904	0.951	0.991
(50, 50)	1.00	0.991	0.022	0.610	0.712	0.810	0.903	0.954	0.993
(10, 10)	1.50	1.458	0.122	0.591	0.706	0.814	0.917	0.959	0.992
(10, 20)	1.50	1.474	0.087	0.595	0.690	0.803	0.909	0.951	0.990
(10, 50)	1.50	1.487	0.042	0.595	0.694	0.804	0.902	0.950	0.992
(50, 50)	1.50	1.493	0.025	0.587	0.694	0.791	0.904	0.953	0.992
$k = 5$									
(10, 10, 10, 10, 10)	0.25	0.242	0.037	0.625	0.718	0.821	0.922	0.964	0.993
(10, 10, 10, 50, 50)	0.25	0.247	0.015	0.621	0.715	0.818	0.904	0.954	0.991
(20, 20, 20, 20, 20)	0.25	0.250	0.021	0.601	0.702	0.798	0.902	0.946	0.987
(50, 50, 50, 50, 50)	0.25	0.249	0.008	0.607	0.704	0.807	0.908	0.959	0.990
(10, 10, 10, 10, 10)	0.50	0.484	0.039	0.625	0.718	0.821	0.914	0.955	0.991
(10, 10, 10, 50, 50)	0.50	0.487	0.015	0.597	0.701	0.809	0.905	0.951	0.990
(20, 20, 20, 20, 20)	0.50	0.487	0.020	0.606	0.698	0.800	0.905	0.960	0.991
(50, 50, 50, 50, 50)	0.50	0.493	0.008	0.613	0.721	0.819	0.904	0.945	0.988
(10, 10, 10, 10, 10)	1.00	0.953	0.042	0.605	0.714	0.813	0.905	0.951	0.989
(10, 10, 10, 50, 50)	1.00	0.980	0.017	0.612	0.708	0.798	0.897	0.947	0.991
(20, 20, 20, 20, 20)	1.00	0.975	0.021	0.611	0.711	0.807	0.913	0.956	0.989
(50, 50, 50, 50, 50)	1.00	0.990	0.009	0.598	0.700	0.792	0.901	0.952	0.989
(10, 10, 10, 10, 10)	1.50	1.438	0.050	0.585	0.692	0.792	0.895	0.941	0.984
(10, 10, 10, 50, 50)	1.50	1.476	0.019	0.586	0.685	0.787	0.896	0.953	0.992
(20, 20, 20, 20, 20)	1.50	1.472	0.025	0.605	0.695	0.796	0.898	0.946	0.991
(50, 50, 50, 50, 50)	1.50	1.489	0.010	0.602	0.706	0.803	0.905	0.953	0.990

TABLE 4

The Small Sample Accuracy of Confidence Intervals for δ Based on the Normal Approximation to the Distribution of $d_{+}^{(1)}$

				Proportion of confidence intervals containing δ with nominal significance level					
Sample sizes	δ	Mean of $d_{+}^{(1)}$	Variance of $d_{+}^{(1)}$	0.60	0.70	0.80	0.90	0.95	0.99
			$k = 2$						
(10, 10)	0.25	0.250	0.102	0.605	0.705	0.803	0.912	0.957	0.992
(10, 20)	0.25	0.256	0.066	0.618	0.716	0.815	0.910	0.952	0.992
(10, 50)	0.25	0.247	0.033	0.610	0.714	0.814	0.906	0.947	0.988
(50, 50)	0.25	0.251	0.020	0.595	0.701	0.798	0.909	0.956	0.992
(10, 10)	0.50	0.498	0.103	0.613	0.716	0.811	0.902	0.954	0.989
(10, 20)	0.50	0.491	0.069	0.604	0.704	0.808	0.897	0.956	0.992
(10, 50)	0.50	0.490	0.036	0.586	0.678	0.783	0.897	0.947	0.988
(50, 50)	0.50	0.493	0.021	0.601	0.708	0.806	0.902	0.944	0.988
(10, 10)	1.00	0.986	0.119	0.604	0.698	0.791	0.898	0.939	0.989
(10, 20)	1.00	0.992	0.075	0.620	0.713	0.803	0.894	0.948	0.990
(10, 50)	1.00	0.994	0.039	0.593	0.685	0.788	0.903	0.949	0.989
(50, 50)	1.00	0.996	0.022	0.604	0.708	0.811	0.903	0.954	0.992
(10, 10)	1.50	1.499	0.133	0.590	0.688	0.795	0.904	0.952	0.990
(10, 20)	1.50	1.499	0.091	0.588	0.688	0.794	0.901	0.943	0.988
(10, 50)	1.50	1.499	0.044	0.590	0.691	0.796	0.898	0.947	0.989
(50, 50)	1.50	1.501	0.026	0.591	0.690	0.791	0.900	0.954	0.991
			$k = 5$						
(10, 10, 10, 10, 10)	0.25	0.252	0.041	0.600	0.689	0.799	0.909	0.957	0.990
(10, 10, 10, 50, 50)	0.25	0.251	0.015	0.609	0.710	0.812	0.900	0.947	0.989
(20, 20, 20, 20, 20)	0.25	0.255	0.022	0.595	0.694	0.785	0.894	0.940	0.984
(50, 50, 50, 50, 50)	0.25	0.250	0.008	0.606	0.697	0.802	0.903	0.956	0.989
(10, 10, 10, 10, 10)	0.50	0.504	0.043	0.605	0.697	0.794	0.896	0.942	0.986
(10, 10, 10, 50, 50)	0.50	0.494	0.016	0.591	0.693	0.805	0.903	0.947	0.987
(20, 20, 20, 20, 20)	0.50	0.497	0.021	0.591	0.686	0.793	0.897	0.951	0.989
(50, 50, 50, 50, 50)	0.50	0.497	0.008	0.603	0.710	0.813	0.903	0.946	0.988
(10, 10, 10, 10, 10)	1.00	0.994	0.047	0.596	0.696	0.794	0.898	0.944	0.987
(10, 10, 10, 50, 50)	1.00	0.995	0.018	0.611	0.707	0.791	0.893	0.947	0.993
(20, 20, 20, 20, 20)	1.00	0.994	0.022	0.614	0.709	0.806	0.906	0.954	0.989
(50, 50, 50, 50, 50)	1.00	0.997	0.009	0.595	0.693	0.790	0.899	0.954	0.988
(10, 10, 10, 10, 10)	1.50	1.499	0.058	0.582	0.669	0.773	0.878	0.938	0.982
(10, 10, 10, 50, 50)	1.50	1.500	0.020	0.579	0.681	0.793	0.897	0.947	0.991
(20, 20, 20, 20, 20)	1.50	1.503	0.027	0.585	0.697	0.797	0.894	0.948	0.987
(50, 50, 50, 50, 50)	1.50	1.501	0.010	0.610	0.710	0.800	0.901	0.952	0.992

combination of $d_1, \ldots, d_k$, is derived from the distributions of the individual terms $d_1, \ldots, d_k$. The accuracy of the large sample normal approximations to the distribution of the d_i suggests that the approximation to the distribution of d_+ will be reasonably accurate in many situations. We should expect d_+ to have a slight negative bias for $\delta > 0$, since fluctuations in the estimators are negatively correlated with fluctuations in the estimated weights. This is because the weights used to calculate $d_+ = \sum w_i d_i$ are of the approximate form $w_i = c_1/(c_2 + d_i^2)$. Larger values of d_i lead to smaller values of w_i, and smaller values of d_i lead to larger values of w_i. Consequently when $\delta > 0$ large values of d_i are "underweighted" and small values of d_i are "overweighted" in the calculation of d_+. We also expect that the iterated estimators $d_+^{(l)}$ will be less biased than d_+, although the very slight dependence of $\hat{\sigma}^2(d_i)$ on d_i suggests that iteration will not usually change the value of the estimate by very much.

As is the case for a single effect size, the accuracy of the large sample approximation to the distribution of d_+ and $d_+^{(1)}$ is studied by Monte Carlo (simulation) methods for two and five independent studies (Hedges, 1982a). Results for $k = 2$ and 5, some combinations of the sample sizes $n^E = n^C = 10$, 20, 30, 40, and 50, and the effect sizes $\delta = 0.25$, 0.50, 1.00, and 1.50 are given in Tables 3 and 4. These sample sizes and effect sizes were chosen because they are typical of values found in most meta-analyses. Two thousand replications of each sample size configuration were generated for each population effect size. The large sample approximation was used to calculate confidence intervals for δ. Means and variances of the estimates as well as empirical proportions of confidence intervals that contained δ were calculated for each sample size configuration for each effect size.

These results suggest that the large sample approximation to the distribution of the estimators is reasonably accurate for the range of δ examined, even when all the studies have a sample size of 10 per group. The accuracy of the approximation tends to improve as the sample sizes increase. As expected, the estimator d_+ has a slight negative bias, tending to underestimate δ. The simulation (Table 3) verifies that the iterated estimator $d_+^{(1)}$ is less biased than d_+, although it is not markedly superior.

C. OTHER METHODS OF ESTIMATION OF EFFECT SIZE FROM A SERIES OF EXPERIMENTS

Most statistical procedures for estimating an effect size from a series of experiments are derived by linearly combining estimates from individual experiments. However, other methods exist for combining estimates of effect sizes. In this section some of these alternative methods are reviewed. In particular, maximum likelihood estimation and methods based on transformations of the effect sizes are discussed.

C.1 THE MAXIMUM LIKELIHOOD ESTIMATOR OF EFFECT SIZE FROM A SERIES OF EXPERIMENTS

The method of maximum likelihood is a general method for obtaining estimates of parameters given a specific statistical model. It has the theoretical advantage of yielding estimates that become closer to the true value of the parameter as the sample size increases (consistency). It is also asymptotically efficient in the sense that no other estimator is more precise in large samples than is the maximum likelihood estimator.

If the k experiments share a common effect size δ, that is,

$$\delta_1 = \delta_2 = \cdots = \delta_k = \delta^*,$$

then the maximum likelihood estimator of δ^* can be obtained. Unfortunately it is expressible only as the solution of an equation and not explictly in a formula. The maximum likelihood estimator $\hat{\delta}$ based on observed effect sizes $g_1, \ldots, g_k$ is the solution of the equation

$$A\hat{\delta} + B_1\sqrt{\hat{\delta}^2 + c_1} + \cdots + B_k\sqrt{\hat{\delta}^2 + c_k} = 0, \tag{12}$$

where

$$A = \tilde{n}_1(2 - L_1) + \cdots + \tilde{n}_k(2 - L_k), \qquad B_i = (\text{sign } g_i)\tilde{n}_i L_i,$$
$$\tilde{n}_i = n_i^E n_i^C / N_i, \qquad N_i = n_i^E + n_i^C,$$
$$L_i = \tilde{n}_i g_i^2 / (\tilde{n}_i g_i^2 + N_i - 2),$$
$$c_i = 4N_i / \tilde{n}_i L_i, \qquad i = 1, \ldots, k.$$

Because the terms $A, B_1, \ldots, B_k, c_1, \ldots, c_k$ are determined by $g_1, \ldots, g_k$, Eq. (12) can (in principle) be solved for $\hat{\delta}$, the maximum likelihood estimator of δ^*.

In general, when $k > 2$, explicit formulas for $\hat{\delta}$ cannot be obtained. However, since the equation is well behaved, algorithms for the numerical solution of (12) can be implemented on a computer without great difficulty. The large sample properties of $\hat{\delta}$ and d_+ are the same (Hedges, 1982a), so that $\hat{\delta}$ has approximately a normal distribution in large samples with a mean of δ and a variance of

$$\hat{\sigma}^2(\hat{\delta}) = \left(\sum_{i=1}^{k} \frac{1}{\hat{\sigma}^2(d_i)}\right)^{-1},$$

where $\hat{\sigma}^2(d_i)$ is given by formula (15) of Chapter 5. The small sample properties of $\hat{\delta}$ have not been investigated in general. When $k = 1$ they are not the same as those of d_+. Given the poor small sample performance of $\hat{\delta}$ when $k = 1$ (see Section A.4 of Chapter 5), it is unlikely that $\hat{\delta}$ will perform better than d_+ when $k \geq 2$. In addition, $\hat{\delta}$ is more difficult to compute than d_+, so that there is little reason to recommend the use of the maximum likelihood estimator.

C.2 ESTIMATORS OF EFFECT SIZE BASED ON TRANSFORMED ESTIMATES

One of the problems in combining estimates of effect size from k studies is that the optimal linear combination of estimates depends on the variances of $d_1, \ldots, d_k$, which in turn depend on the unknown δ. In Section B we dealt with this difficulty by using estimates of δ to estimate the optimal weights for combining $d_1, \ldots, d_k$. Another alternative is to *transform* the estimates so that the resulting transformed variates have variances that are independent of δ. The transformed estimates can then be combined in a simple way to produce a pooled estimate in the transformed metric. By inverting the transformation the pooled estimate then yields an estimate of δ^* (see Section B.2 of Chapter 5). Note that the general transformation given in Section B.2 of Chapter 5 depends explicitly on the balance n^E/n^C. In order to use the same transformation for each study, the ratio n_i^E/n_i^C must be identical for each study. We discuss two alternative transformations and note that at this time little is known about comparative properties.

If each experiment is balanced in the sense that experimental and control sample sizes within studies are equal, $n_j^E = n_j^C = n_j$, $j = 1, \ldots, k$, then a variance-stabilizing transformation for d is given by

$$h(d) = \sqrt{2}\,\sinh^{-1}(d/2\sqrt{2}). \tag{13}$$

Let $h_1 = h(d_1), \ldots, h_k = h(d_k)$ be the transformed estimates and $\eta = h(\delta)$ be the transformed effect size parameter assumed to be the same for all studies. Then each of the transformed estimates h_i has an approximate normal distribution with a mean of η and a variance of $1/(2n_i)$. Consequently the linearly weighted estimate of η with the smallest variance is

$$h_+ = 2\sum_{i=1}^{k} \frac{n_i h_i}{N}, \tag{14}$$

where $N = 2\sum n_i$ is the total sample size. This estimate of η is transformed into an estimate of δ^* by the inverse transformation $\hat{\delta}^* = 2\sqrt{2}\,\sinh(h_+/\sqrt{2})$.

Moreover, h_+ will itself be normally distributed with a variance of approximately

$$\sigma_\infty^2(h_+) = 1/N.$$

Thus a 100(1 − α)-percent confidence interval (η_L, η_U) for η is

$$\eta_L = h_+ - C_{\alpha/2}/\sqrt{N}, \qquad \eta_U = h_+ + C_{\alpha/2}/\sqrt{N}, \tag{15}$$

where $C_{\alpha/2}$ is the two-tailed critical value of the standard normal distribution. The confidence interval is inverted to obtain a confidence interval (δ_L, δ_U) for δ from

$$\delta_L = 2\sqrt{2}\,\sinh(\eta_L/\sqrt{2}), \qquad \delta_U = 2\sqrt{2}\,\sinh(\eta_U/\sqrt{2}). \tag{16}$$

Example

The method described above was used to combine estimates of effect size from six studies of the effects of open versus traditional education on student creativity (see Section C of Chapter 2). The sample sizes $n_i^E = n_i^C = n_i$, effect size estimates d_i, and transformed estimates $h_i = \sqrt{2}\,\sinh^{-1}(d_i/2\sqrt{2})$ are given in Table 5. The weighted average (14) of the h_i is

$$h_+ = 3.502/2(141) = 0.012,$$

which corresponds to $\hat{\delta} = 2\sqrt{2}\,\sinh(h_+/\sqrt{2}) = 0.024$. A 95-percent confidence interval for η given by (15) is $0.012 \pm 1.96/\sqrt{282}$, so that $\eta_L = -0.105$ and $\eta_U = 0.129$. Hence a 95-percent confidence interval for δ from (16) is

$$-0.210 \le \delta \le 0.258.$$

An alternative procedure is to use the relationship between the distribution of the effect size and that of the product–moment correlation discussed in Section B.2 of Chapter 5. If the k experiments are balanced (i.e., $n_1^E = n_1^C, \ldots, n_k^E = n_k^C$) then we can transform the effect size estimates $d_1, \ldots, d_k$ to the corresponding variates $r_1, \ldots, r_k$ from the relation

$$r_i \cong d_i/(d_i^2 + 4)^{1/2}. \tag{17}$$

The corresponding transformation of the population effect size is

$$\rho \cong \delta/(\delta^2 + 4)^{1/2}. \tag{18}$$

Next we transform the variates $r_1, \ldots r_k$ to $z_1, \ldots, z_k$ by Fisher's z-transformation

$$z_i = \tfrac{1}{2}\log\frac{1 + r_i}{1 - r_i}. \tag{19}$$

TABLE 5

Effect Size Estimates from Six Studies of the Effects of Open versus Traditional Education on Student Creativity

Study	$n_i^E = n_i^C = n_i$	d_i	$h_i = \sqrt{2}\sinh^{-1}(d_i/2\sqrt{2})$	$2n_i h_i$	$2n_i(h_i - h_+)^2$
1	22	0.563	0.280	12.306	3.160
2	10	0.308	0.154	3.074	0.403
3	10	0.081	0.040	0.810	0.016
4	10	0.598	0.297	5.936	1.624
5	39	−0.178	−0.089	−6.937	0.796
6	50	−0.234	−0.117	−11.687	1.664
	141			3.502	7.663

The corresponding transformation of the parameter is

$$\zeta = \tfrac{1}{2}\log\frac{1+\rho}{1-\rho}. \tag{20}$$

Since each z_i is approximately normally distributed with a mean of ζ and a variance of $1/(2n_i - 3)$, a pooled estimate of ζ is obtained:

$$z_+ = \sum_{i=1}^{k} \frac{(2n_i - 3)z_i}{N - 3k}. \tag{21}$$

Moreover, z_+ is approximately normally distributed with a mean of ζ and an asymptotic variance $\sigma_\infty^2(z_+)$ given by

$$\sigma_\infty^2(z_+) = 1/(N - 3k).$$

We are now in a position to obtain a 100(1 − α)-percent confidence interval (ζ_L, ζ_U) for ζ:

$$\zeta_L = z_+ - C_{\alpha/2}/\sqrt{N - 3k}, \qquad \zeta_U = z_+ + C_{\alpha/2}/\sqrt{N - 3k}, \tag{22}$$

where $C_{\alpha/2}$ is the two-tailed critical value of the standard normal distribution. Transform ζ_L and ζ_U to ρ_L and ρ_U using the inverse transformation of (20):

$$\begin{aligned} \rho_L &= [\exp(2\zeta_L) - 1]/[\exp(2\zeta_L) + 1], \\ \rho_U &= [\exp(2\zeta_U) - 1]/[\exp(2\zeta_U) + 1]. \end{aligned} \tag{23}$$

Finally, a confidence interval (δ_L, δ_U) for δ^* is obtained from the inverse of (18):

$$\delta_L = 2\rho_L/(1 - \rho_L^2)^{1/2}, \qquad \delta_U = 2\rho_U/(1 - \rho_U^2)^{1/2}. \tag{24}$$

This procedure was suggested by Kraemer (1983).

TABLE 6

Effect Size Estimates from Six Studies of the Effects of Open versus Traditional Education on Student Creativity

Study	n	d	r	z	$(2n-3)z$
1	22	0.563	0.271	0.278	11.394
2	10	0.308	0.152	0.153	2.608
3	10	0.081	0.041	0.041	0.688
4	10	0.598	0.287	0.295	5.010
5	39	−0.178	−0.089	−0.089	−6.666
6	50	−0.234	−0.116	−0.117	−11.323
	141				1.711

Example

Return to the six studies of the effects of open versus traditional education on creativity examined in the last example. The effect size estimates d_i are transformed to r_i by (17), which in turn are transformed to z_i by (19). These values are given in Table 6. The weighted average (21) of the z-transformed correlations is

$$z_+ = 1.711/(242 - 18) = 0.006,$$

which corresponds to a point estimate of $\hat{\rho} = 0.006$ and $\hat{\delta} = 0.012$. A 95-percent confidence interval for ζ is given by (22) as $0.006 \pm 1.96/\sqrt{264}$, so that $\zeta_L = -0.115$ and $\zeta_U = 0.127$ and the corresponding confidence interval for ρ given by (23) is $-0.115 \leq \rho \leq 0.127$. Finally, from (24) the confidence interval for δ is

$$\delta_L = 2(-0.115)/\sqrt{1 - 0.013} = -0.232$$

and

$$\delta_U = 2(0.127)/\sqrt{1 - 0.016} = 0.256.$$

D. TESTING FOR HOMOGENEITY OF EFFECT SIZES

Before pooling the estimates of effect size from a series of k studies, it is important to determine whether the studies can reasonably be described as sharing a common effect size. A statistical test for the homogeneity of effect sizes is formally a test of the hypothesis

$$H_0: \quad \delta_1 = \delta_2 = \cdots = \delta_k$$

versus the alternative that at least one of the effect sizes δ_i differs from the remainder.

An exact small sample test of H_0 is not known, but large sample tests can be devised. One such large sample test is based on the statistic

$$Q = \sum_{i=1}^{k} \frac{(d_i - d_+)^2}{\hat{\sigma}^2(d_i)}, \tag{25}$$

where d_+ is the weighted estimator of effect size given in (6), and $\hat{\sigma}^2(d_i)$ is given by formula (15) of Chapter 5. The test statistic Q is the sum of squares of the d_i about the weighted mean d_+, where the ith square is weighted by the reciprocal of the estimated variance of d_i.

When each study has a large sample size, the asymptotic distribution of Q can be used as the basis for an approximate test of the homogeneity of the δ_i. If all k studies have the same population effect size (i.e., if H_0 is true) then the test statistic Q has an asymptotic chi-square distribution with $k - 1$ degrees of freedom (Hedges, 1982a). Therefore, if the obtained value of Q exceeds the $100(1 - \alpha)$-percent critical value of the chi-square distribution with $k - 1$ degrees of freedom, we reject the hypothesis that the δ_i are equal. If we reject this null hypothesis, we may decide not to pool the estimates of δ, since they do not estimate the same parameter. When the sample sizes are *very* large, however, it is worth studying the variation in the values of d_i, since rather small differences may lead to large values of the test statistic. If the d_i values do not differ much in an absolute sense, the investigator may elect to pool the estimates even though there is reason to believe that the underlying parameters are not identical.

Example

For the 18 studies of the effects of open versus traditional education on student self-concept discussed in Section B.1, the sample sizes, the effect size estimates, and the variances of the effect size estimates are given in Table 2. Using the computational formula (29) and the values of $1/\hat{\sigma}^2(d)$, $d/\hat{\sigma}^2(d)$, and $d^2/\hat{\sigma}^2(d)$ given in Table 2, we compute Q as

$$Q = 23.725 - (5.881)^2/555.600 = 23.663.$$

Comparing the value 23.663 of Q with the percentage points of the chi-square distribution with 17 degrees of freedom, we see that a value of Q as large as that obtained would occur between 10 and 25 percent of the time if $\delta_1 = \cdots = \delta_{18}$. Hence we do not reject the hypothesis of homogeneity of effect size for the present data, and we consider pooling the data to obtain an estimate of the common effect size in the 18 studies.

D.1 SMALL SAMPLE SIGNIFICANCE LEVELS FOR THE HOMOGENEITY TEST STATISTIC

The small sample behavior of the test for homogeneity of effect sizes depends on the accuracy of the asymptotic distribution of the statistic in finite samples. The results of simulation studies (Hedges, 1982a) show that the large sample distribution of the test statistic Q is reasonably accurate in moderate-sized samples (10 or larger per group). In particular, results for $k = 2$ and 5, and for combinations of the sample sizes $n^E = n^C = 10, 20, 30, 40$, and 50, are provided in Table 7. Two thousand replications were generated for each sample size configuration and for each effect size $\delta = 0.25, 0.50, 1.00$, and 1.50. The homogeneity statistic Q was computed for each replication, and the proportion of test statistics exceeding various critical values was tabulated.

The large sample approximation to the distribution was fairly accurate even for $n^E = n^C = 10$. The actual significance values of Q tend to be lower than the nominal significance levels. Thus the test for homogeneity may be slightly conservative. The large sample approximation to the distribution of Q also seems to be more accurate as δ and the sample sizes increase.

D.2 OTHER PROCEDURES FOR TESTING HOMOGENEITY OF EFFECT SIZES

An alternative to the homogeneity test statistic Q is the likelihood ratio test. However, the likelihood ratio test involves a rather difficult computation of the maximum likelihood estimate of δ from k experiments. Because of this difficulty, we recommend the use of the Q-statistic, which is easier to compute and more intuitively appealing than the analogous likelihood ratio test.

However, there are other alternatives. If the k experiments are balanced, that is, if $n_1^E = n_1^C, \ldots, n_k^E = n_k^C$, then the transformation methods discussed in Section B.2 of Chapter 5 can be used to obtain tests of homogeneity of effect size. Using (17) of Chapter 5 with $a = 1$, we transform $d_1, \ldots, d_k$ to $h_1, \ldots, h_k$ and $\delta_1, \ldots, \delta_k$ to $\eta_1, \ldots, \eta_k$ via

$$h_i = \sqrt{2} \sinh^{-1}(d_i/2\sqrt{2}), \qquad \eta_i = \sqrt{2} \sinh^{-1}(\delta_i/2\sqrt{2}). \tag{26}$$

The equality of $\delta_1, \ldots, \delta_k$ is equivalent to the equality of $\eta_1, \ldots, \eta_k$, so that a test of homogeneity of effect size is the same as a test of homogeneity of the η's. We reject the hypothesis

$$H_0: \qquad \delta_1 = \cdots = \delta_k$$

TABLE 7

The Small Sample Behavior of the Homogeneity Test Statistic Q Based on a Simulation with 2000 Replications

Sample sizes	δ	Mean of Q	Variance of Q	Proportion of test statistics exceeding the nominal significance value					
				0.40	0.30	0.20	0.10	0.05	0.01
			$k = 2$						
(10, 10)	0.25	0.988	2.241	0.380	0.286	0.195	0.100	0.052	0.011
(10, 20)	0.25	0.992	2.154	0.402	0.290	0.194	0.090	0.044	0.011
(10, 50)	0.25	0.980	2.086	0.389	0.284	0.190	0.102	0.051	0.009
(50, 50)	0.25	0.951	1.937	0.383	0.274	0.184	0.089	0.048	0.012
(10, 10)	0.50	0.954	1.825	0.386	0.289	0.186	0.092	0.045	0.008
(10, 20)	0.50	0.907	1.756	0.366	0.262	0.178	0.083	0.041	0.007
(10, 50)	0.50	0.915	1.742	0.369	0.271	0.179	0.094	0.045	0.008
(50, 50)	0.50	1.043	2.130	0.412	0.305	0.199	0.114	0.058	0.012
(10, 10)	1.00	1.006	2.205	0.393	0.302	0.199	0.098	0.049	0.013
(10, 20)	1.00	0.953	1.730	0.389	0.291	0.192	0.090	0.042	0.008
(10, 50)	1.00	1.015	1.956	0.418	0.313	0.201	0.093	0.051	0.011
(50, 50)	1.00	0.990	2.056	0.390	0.288	0.194	0.099	0.055	0.007
(10, 10)	1.50	1.034	2.107	0.403	0.311	0.208	0.104	0.056	0.010
(10, 20)	1.50	0.992	1.944	0.394	0.304	0.202	0.095	0.046	0.010
(10, 50)	1.50	0.992	1.841	0.400	0.301	0.206	0.098	0.043	0.006
(50, 50)	1.50	1.059	2.383	0.402	0.295	0.211	0.109	0.058	0.011
			$k = 5$						
(10, 10, 10, 10, 10)	0.25	3.828	7.327	0.378	0.282	0.184	0.085	0.040	0.009
(10, 10, 10, 50, 50)	0.25	3.788	7.602	0.377	0.274	0.179	0.079	0.042	0.010
(20, 20, 20, 20, 20)	0.25	3.896	8.199	0.392	0.285	0.184	0.091	0.048	0.011
(50, 50, 50, 50, 50)	0.25	3.858	7.297	0.397	0.298	0.191	0.092	0.035	0.005
(10, 10, 10, 10, 10)	0.50	3.846	7.490	0.377	0.287	0.177	0.090	0.048	0.008
(10, 10, 10, 50, 50)	0.50	3.900	7.883	0.389	0.290	0.189	0.095	0.045	0.009
(20, 20, 20, 20, 20)	0.50	3.900	7.320	0.395	0.288	0.190	0.093	0.048	0.007
(50, 50, 50, 50, 50)	0.50	3.992	8.078	0.394	0.299	0.198	0.097	0.045	0.012
(10, 10, 10, 10, 10)	1.00	3.933	8.035	0.397	0.293	0.192	0.087	0.044	0.013
(10, 10, 10, 50, 50)	1.00	3.950	7.807	0.394	0.295	0.187	0.092	0.047	0.011
(20, 20, 20, 20, 20)	1.00	3.821	7.489	0.361	0.268	0.183	0.091	0.050	0.007
(50, 50, 50, 50, 50)	1.00	3.926	7.729	0.395	0.289	0.195	0.086	0.043	0.010
(10, 10, 10, 10, 10)	1.50	3.932	7.292	0.396	0.300	0.205	0.093	0.041	0.009
(10, 10, 10, 50, 50)	1.50	4.068	8.362	0.412	0.308	0.211	0.106	0.057	0.010
(20, 20, 20, 20, 20)	1.50	4.019	8.024	0.400	0.304	0.195	0.099	0.054	0.009
(50, 50, 50, 50, 50)	1.50	4.121	8.784	0.412	0.316	0.222	0.112	0.054	0.013

if the statistic

$$Q_1 = 2 \sum_{i=1}^{k} n_i(h_i - h_+)^2 \tag{27}$$

is too large, where h_+ is the weighted mean of $h_1, \ldots, h_k$ given in (14). When H_0 is true the statistic Q_1 has a chi-square distribution with $k - 1$ degrees of freedom.

Example

In our example of six studies of the effects of open education on creativity the values of h_i are given in Table 5. The value of the homogeneity statistic (27) is

$$Q_1 = 7.663.$$

Comparing 7.663 with the percentage points of the chi-square distribution with five degrees of freedom, we see that a value this large or larger will occur between 10 and 25 percent of the time if the null hypothesis is true. Consequently we do not reject the hypothesis that $\delta_1 = \cdots = \delta_6$.

Although the computations for Q_1 were obtained directly from (27), we show in Section E that a simpler computational formula is

$$Q_1 = 2 \sum n_i h_i^2 - N h_+^2.$$

The relationship between the distribution of the effect size estimate and the product-moment correlation can also be exploited to yield a test for homogeneity of effect size in k balanced experiments. Using (17), transform $d_1, \ldots, d_k$ to $r_1, \ldots, r_k$ and $\delta_1, \ldots, \delta_k$ to $\rho_1, \ldots, \rho_k$; then, using (19) and (20), transform $r_1, \ldots, r_k$ to $z_1, \ldots, z_k$ and $\rho_1, \ldots, \rho_k$ to $\zeta_1, \ldots, \zeta_k$. Because $\delta_1 = \cdots = \delta_k$ is equivalent to $\zeta_1 = \cdots = \zeta_k$, a test for homogeneity of the ζ's is equivalent to a test of homogeneity of the δ's. Hence a test of

$$H_0: \quad \delta_1 = \cdots = \delta_k$$

can be based on the statistic

$$Q_2 = \sum_{i=1}^{k} (2n_i - 3)(z_i - z_+)^2, \tag{28}$$

where z_+ is the weighted mean of $z_1, \ldots, z_k$ given by (21). When H_0 is true the statistic Q_2 has a chi-square distribution with $k - 1$ degrees of freedom, and large values of Q_2 lead to rejection of H_0.

Another test for the homogeneity of effect sizes uses the fact that a one-tailed p-value (see Section A of Chapter 3) is a function of the sample size and sample effect size. If p_i is the one-tailed p-value associated with the t-test for the difference between means of the ith experiment, and if the sample sizes n_i^E and n_i^C are relatively large, then

$$z_i = \Phi^{-1}(p_i)$$

has approximately a unit normal distribution with mean

$$\lambda_i = \delta_i \sqrt{n_i^E n_i^C / (n_i^E + n_i^C)}$$

(Rosenthal and Rubin, 1979).

If the experimental and control group sample sizes in k studies are equal, that is, if $n_i^E = \cdots = n_k^E$ and $n_i^C = \cdots = n_k^C$, the condition $\lambda_1 = \cdots = \lambda_k$ is equivalent to $\delta_1 = \cdots = \delta_k$, so that we can test the homogeneity of the λ_i's instead of the δ_i's. To test the hypothesis $\lambda_1 = \cdots = \lambda_k$, compute the statistic

$$Q_3 = \sum_{i=1}^{k} (z_i - \bar{z})^2,$$

where $\bar{z}$ is the unweighted mean of $z_1, \ldots, z_k$. When the hypothesis of homogeneity is true, Q_3 has a chi-square distribution with $k - 1$ degrees of freedom, and we reject the hypothesis for large values of Q_3.

When the sample sizes are unequal this method still provides a test of homogeneity of the λ_i, but now the equality of the λ's does not imply equality of the δ's. Therefore the test based on Q_3 should be used only if there is a common experimental group sample size and a common control group sample size. In most research syntheses, studies generally do not have common sample sizes, so that the test described in this section may have limited applicability.

E. COMPUTATION OF HOMOGENEITY TEST STATISTICS

Computational formulas facilitate the calculation of the homogeneity statistics given in Section D. These formulas rely on the identity

$$\sum a_i(x_i - x_+)^2 = \sum a_i x_i^2 - \left(\sum a_i x_i\right)^2 / \sum a_i,$$

where $x_+ = \sum a_i x_i / \sum a_j$. Applying this identity with $a_i = 1/\hat{\sigma}^2(d_i)$, $x_i = d_i$, and $x_+ = d_+$, we obtain a computational formula for the homogeneity statistic Q,

$$Q = \sum_{i=1}^{k} \frac{d_i^2}{\hat{\sigma}^2(d_i)} - \left(\sum_{i=1}^{k} \frac{d_i}{\hat{\sigma}^2(d_i)} \right)^2 \Bigg/ \sum_{i=1}^{k} \frac{1}{\hat{\sigma}^2(d_i)}. \tag{29}$$

This formula has the advantage that Q can be calculated from the sums of three variables: $1/\hat{\sigma}^2(d_i)$, $d_i/\hat{\sigma}^2(d_i)$, and $d_i^2/\hat{\sigma}^2(d_i)$. It is important to note, however, that it is not always numerically advantageous to use the computational formula (29). (For a discussion of the numerical issues, see Chan, Golub, & LeVeque 1983.)

The weighted mean d_+ and its variance $\hat{\sigma}^2(d_+)$ are also calculated from the sums of two of these variables: $1/\hat{\sigma}^2(d)$ and $d/\hat{\sigma}^2(d)$. Consequently, a direct calculation of these statistics is obtained from the values of the three variables for each study and their totals. Any program such as SAS Proc Means that computes means or sums can be used for these calculations.

The computations in the second example given in Section B can be obtained using the sums given in Table 2 and the computational formula, so that

$$Q = 23.725 - (5.581)^2/555.600.$$

F. ESTIMATION OF EFFECT SIZE FOR SMALL SAMPLE SIZES

Large sample statistical theory for estimating effect size from a series of studies was presented in Section B. The results of simulation studies show that this large sample theory is reasonably accurate when effect sizes are less than 1.5 in absolute magnitude and the sample sizes in each group are at least 10. Although the accuracy of the large sample theory has not been studied for sample sizes less than 10, we suspect that it may not be very accurate.

One situation that occurs in practice is that the data available to the research synthesizer consist of a *large number* of studies each with a relatively small sample size. Thus we have another notion of largeness, namely, many studies with small sample sizes in contrast to few studies with large sample sizes. This new notion of largeness requires a different analysis.

In the case of a large number of small sized studies, it is profitable to consider a different version of large sample theory—one in which the number k of studies becomes large. The large sample normal approximation in this theory is applicable whenever the number k of studies is large. Note that the two different kinds of asymptotic theory do not give the same results. The results are very close, however, when both the number of studies and the sample sizes are large.

When the sample sizes n_i^E and n_i^C of the ith study are small, the bias in the estimators $\hat{\delta}_i$ and g_i is not negligible. For example, if $n_i^E = n_i^C = 3$, the bias of g_i is over 25 percent. This bias does not necessarily disappear when a large number of estimators are averaged, because the biases can all be in the same direction. Thus if $n_i^E = n_i^C = 3$ for a series of k studies, the mean $\bar{g}$ of the k estimators will have a bias of 25 percent regardless of the size of k. Although the variance of $\bar{g}$ tends to zero as k becomes large, the mean $\bar{g}$ tends to the wrong value, $1.25\delta^*$, instead of δ^* as $k \to \infty$. This emphasizes the need to use the unbiased estimator d_i when studies have very small sample sizes.

There are several alternative methods for estimating the effect size from a large series of studies, each of which has a small sample size. One method is based on weighted linear combinations of estimators d_i. The second method is analogous to maximum likelihood and is based on a suggestion of Neyman and Scott (1948). Both methods have the advantage that the estimates approach δ^* when the number of studies increases, in contrast to the maximum likelihood estimate, which does not.

F.1 ESTIMATION OF EFFECT SIZE FROM A LINEAR COMBINATION OF ESTIMATES

One of the simplest methods for estimating effect size from a series of studies with small sample sizes is to use a weighted mean. The weighted mean with the smallest variance is given by

$$\bar{d}_w = w_1 d_1 + \cdots + w_k d_k, \tag{30}$$

where the optimal weights w_i are estimated by

$$w_i = \frac{1}{v_i(\bar{d})} \bigg/ \sum_{j=1}^{k} \frac{1}{v_j(\bar{d})}, \tag{31}$$

$\bar{d}$ is the unweighted mean of $d_1, \ldots, d_k$, and

$$v_i(\bar{d}) = a_i + b_i \bar{d}^2, \tag{32}$$

where

$$\begin{aligned} a_i &= (N_i - 2)[J(N_i - 2)]^2/[\tilde{n}_i(N_i - 4)] \\ b_i &= \{(N_i - 2)[J(N_i - 2)]^2 - (N_i - 4)\}/(N_i - 4), \end{aligned} \tag{33}$$

and $J(m)$ is given in Table 2 of Chapter 5.

Note that $v_i(\delta)$ is the exact variance of d_i and that the weights $w_1, \ldots, w_k$ derived from the $v_i(\bar{d})$ are calculated by using $\bar{d}$ to estimate δ.

An analysis of the distribution of $\bar{d}$ shows that it approaches normality as the number of studies becomes large, with a mean of δ^* and an estimated variance of

$$v = \left(\sum_{i=1}^{k} \frac{1}{v_i(\bar{d})} \right)^{-1}. \tag{34}$$

Consequently, we can obtain a $100(1 - \alpha)$-percent confidence interval (δ_L, δ_U) for δ:

$$\delta_L = \bar{d}_w - C_{\alpha/2}\sqrt{v}, \qquad \delta_U = \bar{d}_w + C_{\alpha/2}\sqrt{v}, \tag{35}$$

where $C_{\alpha/2}$ is the two-tailed critical value of the standard normal distribution.

Example

The sample size $n^E = n^C = n$ and sample standardized mean differences for 20 studies with $n_i \leq 10$ are given in Table 8. The exact value $J(2n - 2)$ of the correlation factor is obtained from Table 2 of Chapter 5, and the unbiased estimator d of effect size is computed from Eq. (9) of Chapter 5. To obtain the weighted mean (30), the constants a and b are given in (33), from which the variances $v(\bar{d})$ and weights can be determined from (32) and (31). The weighted average effect size is given by (30);

$$\bar{d}_w = w_1 d_1 + \cdots + w_{20} d_{20} = 0.320,$$

and the variance of $\bar{d}_w$ from (34) is approximately

$$v = 1/63.909 = 0.0156.$$

TABLE 8
Effect Size Estimates from 20 Studies with $n^E = n^C = n \leq 10$

Study	n	g	$J(2n-2)$	d	a	b	v	w
1	5	0.461	0.903	0.416	0.435	0.087	0.442	0.035
2	5	0.782	0.903	0.706	0.435	0.087	0.442	0.035
3	7	0.513	0.936	0.480	0.300	0.051	0.305	0.051
4	10	0.612	0.958	0.586	0.207	0.033	0.209	0.075
5	8	−0.131	0.945	−0.124	0.261	0.042	0.264	0.059
6	10	−0.018	0.958	−0.017	0.207	0.033	0.209	0.075
7	6	0.774	0.923	0.714	0.355	0.065	0.360	0.043
8	6	0.138	0.923	0.127	0.355	0.065	0.360	0.043
9	4	−0.482	0.869	−0.419	0.566	0.133	0.577	0.027
10	9	0.333	0.952	0.317	0.230	0.036	0.233	0.067
11	9	0.701	0.952	0.667	0.230	0.036	0.233	0.067
12	7	−0.222	0.936	−0.208	0.300	0.051	0.305	0.051
13	7	0.399	0.936	0.373	0.300	0.051	0.305	0.051
14	5	0.538	0.903	0.486	0.435	0.087	0.442	0.035
15	5	−0.190	0.903	−0.172	0.435	0.087	0.442	0.035
16	4	0.833	0.869	0.724	0.566	0.133	0.577	0.027
17	10	0.512	0.958	0.491	0.207	0.033	0.209	0.075
18	8	0.601	0.945	0.568	0.261	0.042	0.264	0.059
19	4	−0.366	0.869	−0.318	0.566	0.133	0.577	0.027
20	8	0.510	0.945	0.482	0.261	0.042	0.264	0.059

A 95-percent confidence interval for δ^* is obtained from (35) as

$$0.320 \pm 1.96\sqrt{0.0156},$$

so that

$$0.075 \leq \delta^* \leq 0.565.$$

F.2 MODIFIED MAXIMUM LIKELIHOOD ESTIMATION OF EFFECT SIZE*

The method of maximum likelihood produces consistent estimators of the effect size δ^* if the number k of experiments is fixed and the sample sizes $n_1^E, n_1^C, \ldots, n_k^E, n_k^C$ become large. However, the maximum likelihood estimator need not be consistent if the sample sizes remain small even though $k \to \infty$ (Neyman & Scott, 1948). A modified form of the maximum likelihood estimator that is consistent can be obtained (Hedges, 1980) as the solution of an estimation equation of the form

$$\sum_{i=1}^{k} A(\delta; n_i^E, n_i^C) = B(\delta; n_1^E, n_1^C, \ldots, n_k^E, n_k^C),$$

where A is a function of δ and the sample sizes n^E and n^C, and B is a function of δ and all of the sample sizes. Unfortunately both A and B have complicated analytic forms, so that this estimator is rather difficult to compute (see Hedges, 1980).

The distribution of the estimator has been obtained under the condition that k is large, and examination of that distribution suggests that the modified maximum likelihood estimator is usually less efficient than the weighted estimator given in Section F.1 in the sense that the weighted estimator of δ^* usually has smaller variance than that of the modified maximum likelihood estimator. Because the weighted estimator is so much easier to compute, there is little to recommend the modified maximum likelihood estimator, and we will not pursue it further.

G. THE EFFECTS OF MEASUREMENT ERROR AND INVALIDITY

Previously the assumption that $\delta_1, \ldots, \delta_k$ are identical is derived from the notion that the treatment produces the same effect on the underlying construct in each study. If the measuring devices (tests) used in the studies are identical, then raw data from the studies can be pooled directly and analyzed using standard methods. Few, if any, collections of studies use precisely the same measuring device in each of the studies. A more realistic assumption is that studies use different but linearly equatable tests as measures of the underlying construct. If different studies produce the same effect on the construct, then the effect size will be the same when a population of scores is represented on different but linearly equatable tests. Thus if different studies produce the same effect on the construct, and use different (but linearly equatable) tests, the population effect sizes will be the same.

The assumption of perfect linear equatability between tests that measure the same characteristic may not always be tenable. Two factors that could lead to violations of this assumption are (i) measurement error and (ii) the presence of unique factors (invalidity). In this section we study the effects of these departures from linear equatability on population effect sizes and estimators of effect sizes, and propose corrections for the effects of measurement error and invalidity.

G.1 THE EFFECTS OF MEASUREMENT ERROR

The standardized mean difference δ defined as in Section A of Chapter 5 is a measure of the magnitude of the treatment effect compared to the variability within the two groups of an experiment. An implicit assumption is that the variability within the experimental and control groups arises from stable differences between subjects (or more generally between experimental units). If the response measure is not perfectly reliable, that is, if errors of measurement are present, then measurement error also contributes to within-group variability. Measurement error therefore alters the population value of the standardized mean difference. If the object is to estimate the value δ of the standardized mean difference when no errors of measurement are present, some procedure to correct for measurement error is necessary. In Section G.1.a a structural model including measurement error is presented. In Section G.1.b the measurement error problem is formulated in terms of classical test theory. The reliability of the response measure is used to obtain an estimator of δ that is not biased by measurement error in Section G.1.c. Testing the homogeneity of disattenuated effect sizes is discussed in Section G.1.d.

G.1.a A Model for Errors of Measurement

The structural model (1) for combining experiments implies a single residual term for each subject. This residual term includes all sources of deviation from a perfect fit to the rest of the structural model. Two conceptually distinct sources of this residual term come to mind. One such source might be called *subject-treatment interaction*, and accounts for individual differences in the responses of subjects within groups. Subject-treatment interaction arises because subjects (or experimental units) are not identical, and hence the response to treatment will not be identical even if the treatment is the same for each of the subjects. Subject-treatment interaction can also be characterized as the disturbing effect of all unmeasured causal variables, that is, all causal variables except treatment.

Another contribution to the residual term arises whenever the measurements of the dependent variable are fallible. If the dependent variable is measured with error, then the difference between the "true" value and the observed value of the variable represents another contribution to the residual term. Errors of measurement are conceptually distinct from subject-treatment interactions. That is, errors of measurement, unlike subject-treatment interactions, are not stable individual differences, but are random fluctuations of the scores of individuals with a constant "true score."

Each of the observations in the ith study can be represented in a model containing a measurement error and a subject-treatment interaction for each individual as in Table 9.

TABLE 9
Models for Experimental and Control Groups

(a) Experimental group of the ith study

Observation		True score: Control group mean		True score: Treatment effect		True score: Subject-treatment interaction		Measurement error
Y_{i1}^{E}	$=$	μ_i^{C}	$+$	$\delta\sigma_i$	$+$	ξ_{i1}^{E}	$+$	ε_{i1}^{E}
	$\vdots$				$\vdots$			
$Y_{in_i^E}^{E}$	$=$	μ_i^{C}	$+$	$\delta\sigma_i$	$+$	$\xi_{in_i^E}^{E}$	$+$	$\varepsilon_{in_i^E}^{E}$

(b) Control group of the ith study

Observation		True score: Group mean		True score: Subject-treatment interaction		Measurement error
Y_{i1}^{C}	$=$	μ_i^{C}	$+$	ξ_{i1}^{C}	$+$	ε_{i1}^{C}
	$\vdots$				$\vdots$	
$Y_{in_i^C}^{C}$	$=$	μ_i^{C}	$+$	$\xi_{in_i^C}^{C}$	$+$	$\varepsilon_{in_i^C}^{C}$

The structural model incorporating both measurement error and subject-treatment interaction can be represented more succinctly as

$$\begin{aligned} Y_{ij}^{E} &= \mu_i^{C} + \delta\sigma_i + (\xi_{ij}^{E} + \varepsilon_{ij}^{E}), \quad && j = 1, \ldots, n_i^{E}, \quad i = 1, \ldots, k, \\ Y_{ij}^{C} &= \mu_i^{C} + (\xi_{ij}^{C} + \varepsilon_{ij}^{C}), && j = 1, \ldots, n_i^{C}, \quad i = 1, \ldots, k, \end{aligned} \tag{36}$$

where ξ_{ij} denotes the interaction of the jth subject with the treatment in the ith experiment, and ε_{ij} is the error of measurement. If $\varepsilon_{ij} = 0$ for each i and j, model (36) is equivalent to (1). Note that the two terms ξ and ε always occur together as $(\xi_{ij} + \varepsilon_{ij})$ and are therefore indistinguishable without further information. When information on the reliability of the response measure is available it is then possible to determine the variances of ξ and ε separately.

G.1.b Classical Test Theory for Errors of Measurement

The assumptions of classical test theory for model (36) are that each of ξ_{ij} and ε_{ij}, $j = 1, \ldots, n_i$, $i = 1, \ldots, k$, is normally distributed, that ξ_{ij} and ε_{ij} are independent of ξ_{st} (and ε_{st}) whenever $i \neq s$ and $j \neq t$, and that both ξ and ε have zero means within groups. (See, e.g., Lord and Novick 1968.) We further assume that the variances of ξ_{ij} and ε_{ij} (denoted by $\sigma_{\xi_i}^2$ and $\sigma_{\varepsilon_i}^2$, respectively) are the same for the experimental and control groups of the same experiment. These assumptions allow us to rewrite the within-group variance σ_i^2 of the ith experiment as $\sigma_i^2 = \sigma_{\xi_i}^2 + \sigma_{\varepsilon_i}^2$.

In classical test theory, the structural model is usually simplified into an equation with two terms, the true score and the error. The structural equation would therefore be written as

$$\begin{aligned} Y_{ij}^{E} &= \tau_{ij}^{E} + \varepsilon_{ij}^{E}, \quad && j = 1, \ldots, n_i^{E}, \quad i = 1, \ldots, k, \\ Y_{ij}^{C} &= \tau_{ij}^{C} + \varepsilon_{ij}^{C}, && j = 1, \ldots, n_i^{C}, \quad i = 1, \ldots, k, \end{aligned}$$

where

$$\tau_{ij}^{E} = \mu_i^{E} + \xi_{ij}^{E}, \qquad \tau_{ij}^{C} = \mu_i^{C} + \xi_{ij}^{C}.$$

The reliability $\rho(Y_i, Y_i')$ of the ith response measure is defined as the pooled within-group ratio of the true score variance $\sigma_{\tau_i}^2$ to total variance $\sigma_{\tau_i}^2 + \sigma_{\varepsilon_i}^2$.

$$\rho(Y_i, Y_i') = \sigma_{\tau_i}^2/(\sigma_{\tau_i}^2 + \sigma_{\varepsilon_i}^2) = \sigma_{\xi_i}^2/(\sigma_{\xi_i}^2 + \sigma_{\varepsilon_i}^2).$$

Consider the population value of the standardized mean difference in two cases, one in which the measurements are error free (that is, $\sigma_\varepsilon^2 = 0$) and one in which errors of measurement are present (that is $\sigma_\varepsilon^2 \neq 0$). For simplicity of notation, the subscript i denoting the particular experiment is omitted in the exposition that follows, but the results apply to each experiment when properly indexed.

If there are no errors of measurement, then the within cell standard deviation is simply σ_ξ^2. Let δ denote the population value of the standardized mean difference when there are no errors of measurement. Then

$$\delta = (\mu^{\mathrm{E}} - \mu^{\mathrm{C}})/\sigma_\xi, \tag{37}$$

where μ^{E} and μ^{C} are the population means of the experimental and control groups, respectively. Note that the use of the symbol δ in (37) is consistent with the definition of δ as used in expression (2) of Chapter 5.

In the second case, when errors of measurement are present, the population means μ^{E} and μ^{C} are unchanged, but the within-group variance is larger. If δ' denotes the value of the standardized mean difference when errors of measurement are present, then

$$\delta' = (\mu^{\mathrm{E}} - \mu^{\mathrm{C}})/\sqrt{\sigma_\xi^2 + \sigma_\varepsilon^2}.$$

The relationship between δ' and δ can be expressed as

$$\delta' = \delta\sqrt{\sigma_\xi^2/(\sigma_\xi^2 + \sigma_\varepsilon^2)} = \delta\sqrt{\rho(Y, Y')}, \tag{38}$$

where $\rho(Y, Y')$ is the reliability of the response measure. Note that in this context $\rho(Y, Y')$ is always nonnegative.

G.1.c Correcting Estimates for Measurement Error

We see from (38) that the population value of the standardized mean difference depends explicitly on the reliability of the response measure. If the object is to estimate the standardized mean difference when no errors of measurement are present, then estimating δ' instead of δ can result in a bias. Since reliabilities cannot exceed unity, the effect of measurement error is to *reduce* the magnitude of the parameter δ' as compared with δ. In particular, errors of measurement cause the estimator d to estimate δ' instead of δ, so that $E(d) = \delta' = \delta\sqrt{\rho(Y, Y')}$. Hence, errors of measurement result in *underestimates* of the parameter δ.

If the reliability $\rho(Y, Y')$ is known, the bias can be removed by dividing d by $\sqrt{\rho(Y, Y')}$. When we combine several estimates that use reponse scales with different reliabilities, each estimate can be corrected for measurement error separately. This leads to the estimators

$$d_1/\sqrt{\rho(Y_1, Y_1')}, d_2/\sqrt{\rho(Y_2, Y_2')}, \ldots, d_k/\sqrt{\rho(Y_k, Y_k')}, \tag{39}$$

which are unbiased estimators of the disattenuated effect size δ. Since each estimator in (39) is a constant times d_i, the variance of $d_i/\sqrt{\rho(Y_i, Y_i')}$ is

$1/\rho(Y_i, Y'_i)$ times the variance of d_i. Therefore the weighted estimate of δ corrected for reliability in response measures denoted by d^R is

$$d^R = \sum_{i=1}^{k} \frac{d_i\sqrt{\rho(Y_i, Y'_i)}}{\hat{\sigma}^2(d_i)} \Big/ \sum_{i=1}^{k} \frac{\rho(Y_i, Y'_i)}{\hat{\sigma}^2(d_i)}, \tag{40}$$

where $\hat{\sigma}^2(d_i)$ is given in (15) of Chapter 5. Note that d^R is obtained by replacing d_i by $d_i/\sqrt{\rho(Y_i, Y'_i)}$ and $\hat{\sigma}^2(d_i)$ by $\hat{\sigma}^2(d_i)/\rho(Y_i, Y'_i)$ in the expression (6) for d_+. The large sample distribution of d^R is normal with a mean of δ and variance

$$v^R = \left(\sum_{i=1}^{k} \frac{\rho(Y_i, Y'_i)}{\sigma^2(d_i)}\right)^{-1}, \tag{41}$$

variance so that an approximate $100(1-\alpha)$-percent confidence interval (δ_L, δ_U) for the disattenuated effect size δ is

$$\delta_L = d^R - C_{\alpha/2}\sqrt{v^R}, \qquad \delta_U = d^R + C_{\alpha/2}\sqrt{v^R}, \tag{42}$$

where $C_{\alpha/2}$ is the two-tailed critical value of the standard normal distribution.

Example

Return to the 18 studies of the effects of open versus traditional teaching on student self-concept that were described in Section C of Chapter 2. The sample sizes, effect size estimates, and reliability coefficients for each dependent measure are listed in Table 10 along with $d_i/\sqrt{\rho(Y_i, Y'_i)}$, the estimate corrected for unreliability. Most of the reliabilities are 0.80 or above, so the disattenuated effect size estimates are only slightly larger than the attenuated estimates d_i. The weighted average disattenuated effect size estimate is obtained from (40),

$$d^R = 4.006/455.676 = 0.009,$$

and a 95-percent confidence interval for the common effect size δ^* from (42) is $0.009 \pm 1.96\sqrt{1/455.676}$, so that

$$-0.083 \leq \delta \leq 0.101.$$

G.1.d Testing for Homogeneity of Disattenuated Effect Sizes

If the reliabilities $\rho(Y_i, Y'_i)$ of the measures used in a series of studies differ, then this differential reliability will attenuate effect sizes to a different degree in each study. Thus even if the disattenuated effect sizes are perfectly homogeneous, the attenuated effect sizes will be heterogeneous. If the homogeneity statistic (25) is applied to the estimates $d_1, \ldots, d_k$, the result is a test for the equality of the *attenuated* effect sizes. Since the attenuation due to

TABLE 10

Computations for Studies of TU Effects of Open Education on Student Self-Concept

Study	n^E	n^C	d	$\rho(Y, Y')$	$\frac{d}{\sqrt{\rho(Y, Y')}}$	$\frac{\rho(Y, Y')}{\hat{\sigma}^2(d)}$	$\frac{d\sqrt{\rho(Y, Y')}}{\hat{\sigma}^2(d)}$	$\frac{d^2}{\hat{\sigma}^2(d)}$
1	100	180	0.100	0.80	0.112	51.370	5.743	0.642
2	131	138	−0.162	0.79	−0.182	52.918	−9.645	1.758
3	40	40	−0.090	0.79	−0.101	15.784	−1.598	0.162
4	40	40	−0.049	0.79	−0.055	15.795	−0.871	0.048
5	97	47	−0.046	0.88	−0.049	27.854	−1.366	0.670
6	28	61	−0.010	0.88	−0.011	16.888	−0.180	0.002
7	60	55	−0.431	0.87	−0.462	24.400	−11.275	5.210
8	72	102	−0.261	0.87	−0.280	36.419	−10.191	2.852
9	87	45	0.134	0.87	0.144	25.751	3.700	0.531
10	80	49	0.019	0.76	0.022	23.094	0.503	0.011
11	79	55	0.175	0.76	0.201	24.552	4.929	0.989
12	40	109	0.056	0.89	0.059	26.035	1.545	0.092
13	36	93	0.045	0.89	0.048	23.094	1.102	0.052
14	9	18	0.103	0.85	0.112	5.094	0.569	0.064
15	14	16	0.121	0.85	0.131	6.335	0.831	0.109
16	21	22	−0.482	0.85	−0.523	8.875	−4.640	2.426
17	133	124	0.290	0.78	0.328	49.533	16.265	5.341
18	83	45	0.342	0.76	0.392	21.885	8.585	3.368
						455.676	4.006	24.327

measurement error is an artifact, the investigator may wish to test whether the disattenuated effect sizes are homogeneous. A test for the homogeneity of effect sizes corrected for reliability can be obtained by the statistic

$$Q^R = \sum_{i=1}^{k} \frac{[d_i - d^R\sqrt{\rho_i(Y_i, Y_i')}]^2}{\hat{\sigma}^2(d_i)}, \tag{43}$$

where $\hat{\sigma}^2(d_i)$ is given in (15) of Chapter 5. The statistic Q^R is compared to the chi-square distribution with $k - 1$ degrees of freedom in the same way as is the test using (25).

Example

We now test the homogeneity of the disattenuated effect sizes corresponding to the 18 studies of the effects of open education on self-concept. Using the computational formula given in Section E and the totals 455.676, 4.006, and 24.328 of $\rho(Y, Y')/\hat{\sigma}^2(d)$, $d\sqrt{\rho(Y, Y')}/\hat{\sigma}^2(d)$, and $d^2/\hat{\sigma}^2(d)$, we obtain from (43)

$$Q^R = 24.328 - (4.006)^2/455.676 = 24.293.$$

Comparing 24.293 with the percentage points of the chi-square distribution with 17 degrees of freedom, we see that a value this large would occur between 10 and 20 percent of the time if the disattenuated effect sizes were homogeneous. Hence we do not reject the hypothesis of homogeneity of the disattenuated effect sizes.

G.2 THE EFFECT OF VALIDITY OF RESPONSE MEASURES

The effect of validity of response measures has not been considered in previous sections. The structural models (1) and (36) do not admit the possibility that some response measures have unique factors. For example, some experiments use an expensive standardized test to measure reading achievement, whereas other studies use locally developed tests that are correlated with the standardized test. If the locally developed tests have unique factors, they will not be perfectly valid measures of reading achievement as measured by the standardized test. Sections G.2.a to G.2.d deal with the effect of invalidity of response measures on estimators of effect size.

G.2.a A Model for Invalidity

One model for test invalidity assumes that a collection of tests shares a common factor, but that some tests also have unique factors. If the population of scores on a particular test is generated by a model which includes both the common factor (among all the tests) and a unique factor, then the test is partially invalid. The (conceptual) structural model for the responses in the ith study is given in Table 11. The model of Table 11 for the observations in a series of k studies can be written more compactly as

$$\begin{aligned} Y_{ij}^{E} &= \mu_i^{C} + \delta\sigma_i + \zeta_i + \xi_{ij}^{E} + \theta_{ij}^{E} + \varepsilon_{ij}^{E}, && j = 1, \ldots, n_i^{E}, \quad i = 1, \ldots, k, \\ Y_{ij}^{C} &= \mu_i^{C} + \xi_{ij}^{C} + \theta_{ij}^{C} + \varepsilon_{ij}^{C}, && j = 1, \ldots, n_i^{C}, \quad i = 1, \ldots, k, \end{aligned} \tag{44}$$

where $\delta\sigma_i$ and ζ_i are the contributions of the treatment effect on the common and the ith unique factors respectively; μ_i^{C} is the control group mean, ξ_{ij} and θ_{ij} are the contributions of subject-treatment interaction on the common and unique factors, respectively, and ε_{ij} is an error of measurement. The parameters of the model are not identifiable without additional information or restrictions. The primary concern usually is to estimate δ, the effect of treatment on the common factor, standardized by the common factor standard deviation within groups. Note that when $\zeta_i = \theta_{ij} = \varepsilon_{ij} = 0$, $j = 1, \ldots, n_i$, $i = 1, \ldots, k$, model (44) reduces to (1) and δ_i has the same interpretation as in (2).

TABLE 11

Models for the Experimental and Control Groups in the Case of Invalidity

(a) Experimental group of the ith study

				Treatment effect				Subject-treatment interaction			
Observation		Control group mean		Common factor		Unique factor		Common factor		Unique factor		Measurement error
Y_{i1}^{E}	=	μ_i^{C}	+	$\delta\sigma_i$	+	ζ_i	+	ξ_{i1}^{E}	+	θ_{i1}^{E}	+	ε_{i1}^{E}
	⋮						⋮					
$Y_{in_i^E}^{E}$	=	μ_i^{C}	+	$\delta\sigma_i$	+	ζ_i	+	$\xi_{in_i^E}^{E}$	+	$\theta_{in_i^E}^{E}$	+	$\varepsilon_{in_i^E}^{E}$

(b) Control group of the ith study

				Subject-treatment interaction				
Observation		Control group mean		Common factor		Unique factor		Measurement error
Y_{i1}^{C}	=	μ_i^{C}	+	ξ_{i1}^{C}	+	θ_{i1}^{C}	+	ε_{i1}^{C}
	⋮				⋮			
$Y_{in_i^C}^{C}$	=	μ_i^{C}	+	$\xi_{in_i^C}^{C}$	+	$\theta_{in_i^C}^{C}$	+	$\varepsilon_{in_i^C}^{C}$

G.2.b The Effect of Validity When the Treatment Affects Only the Common Factor

One of the restrictions that can be applied to model (44) is to assume that $\zeta_i = 0$ for each $i = 1, \ldots, k$, that is, that the treatment affects Y only through the common factor. We also assume that ξ_{ij}, θ_{st}, and ε_{uv} have zero means and are independent of one another. These are standard assumptions from classical test theory. Under these assumptions, it is possible to obtain an unbiased estimate of the treatment effect δ if the correlation of each response measure with a valid response measure (that is, a test with no unique factor) of known reliability is available. Suppose that $X_1, \ldots, X_k$ is a series of response measures with reliabilities $\rho(X_1, X_1'), \ldots, \rho(X_k, X_k')$, and suppose that the X_i have no unique factors but share the common factor of another (partially invalid) series of response measures $Y_1, \ldots, Y_k$. If the correlation $\rho(X_i, Y_i)$ of X_i with Y_i is known, then an unbiased estimate of δ in model (44) can be obtained from a series of measurements using the response scales $Y_1, \ldots, Y_k$. Under these assumptions, invalidity biases the estimates d_i downward.

We write the standardized mean difference δ'' as a function of the parameters of model (44), where for simplicity of notation the subscript i referring to the ith experiment is omitted. Recall that $\zeta = 0$, and that ξ, θ, and ε are mutually independent. Consequently, the mean difference is $\mu^{\mathrm{E}} - \mu^{\mathrm{C}}$, the within-group variance is $\sigma_\xi^2 + \sigma_\theta^2 + \sigma_\varepsilon^2$, and

$$\delta'' = \frac{\mu^{\mathrm{E}} - \mu^{\mathrm{C}}}{\sqrt{\sigma_\xi^2 + \sigma_\theta^2 + \sigma_\varepsilon^2}} = \frac{\delta\sigma_\xi}{\sqrt{\sigma_\xi^2 + \sigma_\theta^2 + \sigma_\varepsilon^2}}. \tag{45}$$

The validity coefficient of the response measure can be expressed by its within-group correlation with the common factor. Since the common factor is a linear transformation of ξ, the validity coefficient can be written as

$$\rho(\xi, Y) = \frac{\mathrm{Cov}(\xi, \xi + \theta + \varepsilon)}{\sigma_\xi\sqrt{\sigma_\xi^2 + \sigma_\theta^2 + \sigma_\varepsilon^2}} = \frac{\sigma_\xi}{\sqrt{\sigma_\xi^2 + \sigma_\theta^2 + \sigma_\varepsilon^2}}.$$

Therefore, the population value of the standardized mean difference δ'' can be expressed as $\delta'' = \delta\rho(\xi, Y)$.

It seems unlikely that the population correlation of an invalid measure with the common factor among a series of measures would be known. If X is a measure that is not perfectly reliable, if it shares the Y common factor, and has no unique factor, then the correlation $\rho(\xi, Y)$ can be obtained from $\rho(X, Y)$ by the familiar disattenuation formula (see, e.g., Lord and Novick, 1968):

$$\rho(\xi, Y) = \rho(X, Y)/\sqrt{\rho(X, X')}, \tag{46}$$

where $\rho(X, X')$ is the reliability of the measure X. Thus the population standardized mean difference δ'' can be written in terms of a correlation with a valid but unreliable measure X and the reliability of X:

$$\delta'' = \delta\rho(X, Y)/\sqrt{\rho(X, X')}. \tag{47}$$

Note that $\delta'' \leq \delta$ because $\rho(X, Y) \leq \sqrt{\rho(X, X')}$. This means that invalidity always *reduces* the standardized mean difference when treatment affects only the common factor among the response measures.

G.2.c Correcting Estimators for Validity

Expression (47) for δ'' leads directly to estimators of the effect size that would be obtained from a valid and perfectly reliable test. Suppose that $X_1, \ldots, X_k$ are valid response scales with known reliabilities $\rho(X_1, X'_1), \ldots, \rho(X_k, X'_k)$, $d_1, \ldots, d_k$ are effect size estimates based on partially invalid response scales $Y_1, \ldots, Y_k$, and $\rho(X_1, Y_1), \ldots, \rho(X_k, Y_k)$ are the known correlations between the X_i's and the Y_i's. Then if the treatment affects only the common factor, the estimators

$$d_i\sqrt{\rho(X_1, X'_1)}/\rho(X_1, Y_1), \ldots, d_k\sqrt{\rho(X_k, X'_k)}/\rho(X_k, Y_k) \tag{48}$$

are unbiased estimators of the disattenuated effect size on the common factor. Since each estimator in (48) is a constant times d_i, the variance of $d_i\sqrt{\rho(X_i, X'_i)}/\rho(X_i, Y_i)$ is $\rho(X_i, X'_i)/\rho^2(X_i, Y_i)$ times the variance of d_i. Therefore the weighted estimate of δ corrected for reliability and validity is

$$d^{\mathrm{RV}} = \sum_{i=1}^{k} \frac{d_i \rho(X_i, Y_i)}{\hat{\sigma}^2(d_i)\sqrt{\rho(X_i, X'_i)}} \Bigg/ \sum_{i=1}^{k} \frac{\rho^2(X_i, Y_i)}{\hat{\sigma}^2(d_i)\rho(X_i, X'_i)},$$

where $\hat{\sigma}^2(d_i)$ is given in (15) of Chapter 5.

The large sample distribution of d^{RV} is normal with mean δ and variance

$$v^{\mathrm{RV}} = \left(\sum_{i=1}^{k} \frac{\rho^2(X_i, Y_i)}{\hat{\sigma}^2(d_i)\rho(X_i, X'_i)}\right)^{-1},$$

so that an approximate $100(1-\alpha)$-percent confidence interval $(\delta_{\mathrm{L}}, \delta_{\mathrm{U}})$ for δ is given by

$$\delta_{\mathrm{L}} = d^{\mathrm{RV}} - C_{\alpha/2}\sqrt{v^{\mathrm{RV}}}, \qquad \delta_{\mathrm{U}} = d^{\mathrm{RV}} + C_{\alpha/2}\sqrt{v^{\mathrm{RV}}},$$

where $C_{\alpha/2}$ is the two-tailed critical value of the standard normal distribution.

Example

The results of seven studies of the effects of open versus traditional education on student creativity are summarized in Table 12. Each of the seven studies used a measure that was validated against a divergent production test of creativity with a reliability of $\rho(X, X') = 0.95$. The table lists the validity coefficient $\rho(X, Y_i)$ for each test along with the estimate of effect size corrected for validity, $d\sqrt{\rho(X, X')}/\rho(X, Y_i)$. The weighted mean of the corrected estimates is

$$d^{RV} = 40.856/66.623 = 0.613,$$

and a 95-percent confidence interval (δ_L, δ_U) for δ is given by

$$\delta_L = 0.613 - 1.96/\sqrt{66.623} = 0.373,$$

$$\delta_U = 0.613 + 1.96/\sqrt{66.623} = 0.853.$$

A test for homogeneity of the disattenuated effect sizes on the common factor uses the statistic

$$Q^{RV} = \sum_{i=1}^{k} \frac{[d_i - d^{RV}\rho(X_i, Y_i)/\sqrt{\rho(X_i, X_i')}]^2}{\hat{\sigma}^2(d_i)},$$

which is to be compared with the critical value of the chi-square distribution with $k - 1$ degrees of freedom.

Example

We now test the homogeneity of the corrected effect size estimates from seven studies of the impact of open versus traditional education on student creativity. Using the information given in Table 13 and the computational formula given in Section E with $a_i = 1/\sigma^2(d_i)$ and $x_i = \sqrt{\rho(X, X')}\,d/\sqrt{\rho(X, Y_i)}$, we obtain

$$Q^{RV} = 29.003 - (40.856)^2/66.623 = 3.948.$$

Comparing 3.948 with the percentage points of the chi-square distribution with $7 - 1 = 6$ degrees of freedom, we see that a value this large would occur between 50 and 75 percent of the time if the corrected effect sizes were perfectly homogeneous.

G.2.d The Effect of Validity When the Treatment Affects Both Common and Unique Factors

When the treatment affects the response measure Y through both common and unique factors, the invalidity of the response measure may either increase

TABLE 12

Summary of Effect Sizes from Seven Studies of the Effects of Open versus Traditional Education on Student Creativity[a]

Study	$n^E = n^C = n$	d	$\rho(X, Y)$	$\frac{d\sqrt{\rho(X, X')}}{\rho(X, Y)}$	$\hat{\sigma}^2(d)$	$\frac{\rho^2(X, Y)}{\rho(X, X')\hat{\sigma}^2(d)}$	$\frac{d\rho(X, Y)}{\sqrt{\rho(X, X')\hat{\sigma}^2(d)}}$
1	40	0.524	0.97	0.527	0.0517	19.151	10.084
2	36	0.762	0.97	0.766	0.0596	16.621	12.727
3	22	0.542	0.88	0.600	0.0943	8.649	5.192
4	10	0.278	0.90	0.301	0.2019	4.222	1.271
5	10	0.073	0.97	0.073	0.2001	4.949	0.363
6	10	0.598	0.97	0.543	0.2073	4.778	2.593
7	20	1.052	0.94	1.045	0.1127	8.253	8.626
						66.623	40.856

[a] $\rho(X, X') = 0.95$ for these data.

TABLE 13

Data for Calculation of Homogeneity Statistics for Seven Studies of the Effects of Open versus Traditional Education on Student Creativity[a]

Study	$n^E = n^C = n$	d	$\rho(X, Y)$	$\frac{\sqrt{\rho(X, X')}d}{\rho(X, Y)}$	$\frac{\rho^2(X, Y)}{\rho(X, X')\hat{\sigma}^2(d)}$	$\frac{\rho(X, Y)d}{\sqrt{\rho(X, X')}\hat{\sigma}^2(d)}$	$\frac{d^2}{\hat{\sigma}^2(d)}$
1	40	0.524	0.97	0.527	19.151	10.084	5.309
2	36	0.762	0.97	0.766	16.621	12.727	9.744
3	22	0.542	0.88	0.600	8.649	5.192	3.117
4	10	0.278	0.90	0.301	4.222	1.271	0.383
5	10	0.073	0.97	0.073	4.949	0.363	0.027
6	10	0.598	0.97	0.543	4.778	2.593	1.407
7	20	1.052	0.94	1.045	8.253	8.626	9.016
					66.623	40.856	29.003

[a] $\rho(X, X') = 0.95$ for these data.

or decrease the standardized mean difference. Hence no simple characterization of the effects of invalidity on estimates of δ obtained from the estimator d_i given by (10) of Chapter 5 is possible.

For simplicity of notation we continue to omit the subscripts i and j to denote the ith response measure and jth subject, and continue to assume that $\mathrm{Cov}(\xi, \theta) = \mathrm{Cov}(\xi, \varepsilon) = \mathrm{Cov}(\theta, \varepsilon) = 0$. Now ζ, the effect of treatment via the unique factor, is nonzero. The standardized mean difference under model (44) and these assumptions is denoted by $\tilde{\delta}$. Specifically,

$$\tilde{\delta} = \frac{\mu^{\mathrm{E}} - \mu^{\mathrm{C}}}{\sqrt{\sigma_\xi^2 + \sigma_\theta^2 + \sigma_\varepsilon^2}} = \frac{\delta\sigma_\xi + \zeta}{\sqrt{\sigma_\xi^2 + \sigma_\theta^2 + \sigma_\varepsilon^2}}.$$

Note that if $\zeta = 0$, then $\tilde{\delta}$ reduces to the standardized mean difference δ'' given by (45). If $\zeta = 0$ and $\sigma_\theta^2 = \sigma_\varepsilon^2 = 0$, then $\tilde{\delta} = \delta$. If ζ, the treatment effect via the unique factor, is large enough, namely,

$$\zeta > \left(\sqrt{\sigma_\xi^2 + \sigma_\theta^2 + \sigma_\varepsilon^2} - \sigma_\xi\right)\delta, \tag{49}$$

then $\tilde{\delta} > \delta$. If the inequality in (49) is reversed, then $\tilde{\delta} < \delta$. Because $\tilde{\delta}$ involves the unknown effect ζ of treatment on the unique factor, no correction for the effects of invalidity is possible in this situation. Moreover, invalidity may increase or decrease the observed effect size.

CHAPTER 7

Fitting Parametric Fixed Effect Models to Effect Sizes: Categorical Models

Although there has been an accumulation of a large number of studies in many areas of educational research, this myriad of research studies has not always led to clearer insights about the phenomena under investigation. Some of the difficulties with methods that are sometimes used to synthesize research results have been discussed throughout this book. Usually a first step in the quantitative synthesis of a collection of research studies is to calculate an estimate of effect magnitude for each study. These estimates can then be combined to produce an overall estimate of effect magnitude. Methods for combining estimates of effect magnitude from a collection of studies were discussed in Chapter 6.

If, as is often the case, the underlying (population) effect sizes are not identical in all of the studies, the representation of the results of a set of studies by a single estimate of effect magnitude can be misleading. For example, suppose a treatment produces large positive (population) effects in one-half of a collection of studies, and large negative (population) effects in the other half of a collection of studies. Then the representation of the overall effect of the treatment as zero is obviously misleading, since all the studies actually have underlying effects that are different from zero. The test of homogeneity of effect sizes given in Chapter 6 is designed to detect situations in which underlying (population) effect sizes are not homogeneous.

There is some indication that in many real data sets the assumption of homogeneity of effect sizes is not met. For example, in a collection of studies of the effects of open education on 18 different dependent variables, the hypothesis of homogeneity of effect sizes was rejected for every dependent variable (Giaconia & Hedges, 1982).

Some writers in the area of research synthesis cite substantive reasons for concluding that different studies of the effects of the same treatment yield quite different results. Light and Smith (1971) argue that many contradictions in research evidence may be resolved by grouping studies with similar characteristics. They assert that studies with the same characteristics are more likely to yield similar results, and hence many apparent contradictions among research results arise from differences in the characteristics of studies. Pillemer and Light (1980) argue that grouping of studies according to their characteristics is an essential step in assessing the range of generalizability of a research finding. For example, if a treatment produces essentially the same effect in a wide variety of settings with a variety of people, then more confidence can be placed in the generalizability of the finding of a treatment effect.

Many investigators in quantitative research synthesis (e.g., Kulik, Kulik, & Cohen, 1979) have recognized the potential for heterogeneous effect sizes and have grouped studies that share common characteristics. A common approach is to carry out an analysis of variance (ANOVA) on the effect size estimates to determine if the mean effect sizes differ.

There are two concerns with this procedure. First, the assumptions for the analysis of variance may not be met by the effect size data because effect size estimates probably will not have the same distribution within cells. The variance of an individual observation (effect size estimate) is inversely proportional to the number of subjects in the study. When studies have different sample sizes, the individual "error" variances can differ by a factor of 10 or 20. Second, even if the between-class test is accurate, the use of ANOVA does not provide any indication whether or not studies *within the classes* share a common effect size. Thus, even if an ANOVA correctly detects that two classes of studies have a different average effect size, there is no guarantee that the average effect size within each class is a reflection of a common underlying effect size for that class.

In this chapter, we present a method for fitting models to effect sizes when the independent variables are categorical. This method is essentially an analogue to the analysis of variance for effect sizes. The method is useful when it is possible to group studies that have similar characteristics, such as experimental conditions or stimuli. It provides valid large sample tests for differences in average effect sizes between classes, and also tests for the homogeneity of effect size within classes. The within-class test of homogeneity can be used as a test for the specification of the categorical model. This method is a special case of a method of fitting general linear models to effect sizes given in Chapter 8, but we describe the present method separately because it is so frequently used.

A. AN ANALOGUE TO THE ANALYSIS OF VARIANCE FOR EFFECT SIZES

Assume that an investigator has an a priori grouping of studies, that is, a taxonomy for classifying studies that are likely to produce similar results. This often takes the form of a set of categories into which studies can be placed. Studies may be cross-classified by two or more sets of categories.

The technique presented in this chapter is quite direct. The first step is to determine, by means of a statistical test, whether all studies (regardless of category) share a common effect size. If the hypothesis that all the effect sizes are equal is rejected, then the experimenter breaks the series of studies into groups or a priori classes, in such a way that the effect sizes within each class are approximately equal. It is interesting to note that the fit statistic calculated at the first stage is partitioned into (stochastically) independent parts corresponding to between-class and within-class fits. The between-class fit is an index of the extent to which effect sizes differ across the classes. If the within-class fit (fit to a single effect size within each class) is not rejected, the process stops. If the within-class fit is rejected, then the classes are subdivided further. The process of subdividing and testing for between- and within-class fit continues until an acceptable level of within-class homogeneity is achieved. The procedure provides valid large sample tests for the effects of classifications as well as an indication that the final classes are internally homogeneous with respect to effect size.

An exposition of the specific model and notation used throughout this chapter is given in Section B. Some tests of homogeneity based on weighted estimators are presented in Section C. An explanation of the use of these results for fitting models to a series of studies is then given in Section D. In Section E the use of contrasts among effect sizes for different classes is discussed. Finally, Section F deals with computational aspects.

B. MODEL AND NOTATION

The analogue to the analysis of variance for effect sizes depends on the same assumptions that were used in Section A of Chapter 6. However, it is useful to alter the notation slightly to be consistent with that usually used in the analysis of variance. Assume that the studies are sorted into p disjoint classes that are determined a priori, and that there are $m_1, m_2, \ldots, m_p$ studies in the p classes.

Let $Y^E_{ij\alpha}$ and $Y^C_{ij\alpha}$ be the αth experimental and control group scores in the jth experiment in the ith class. Denote the experimental and control group sample sizes for the jth study in the ith class by n^E_{ij} and n^C_{ij}, respectively. The complete set of observations in all the studies, the parameters and their estimates, are displayed in Table 1.

Assume that, for a fixed class i and for a fixed study j, $Y^E_{ij\alpha}$ and $Y^C_{ij\alpha}$ are normally distributed with means μ^E_{ij} and μ^C_{ij} and common variance σ^2_{ij}:

$$Y^E_{ij\alpha} \sim \mathcal{N}(\mu^E_{ij}, \sigma^2_{ij}), \quad \alpha = 1, \ldots, n^E_{ij}, \quad j = 1, \ldots, m_i, \quad i = 1, \ldots, p,$$

and (1)

$$Y^C_{ij\alpha} \sim \mathcal{N}(\mu^C_{ij}, \sigma^2_{ij}), \quad \alpha = 1, \ldots, n^C_{ij}, \quad j = 1, \ldots, m_i, \quad i = 1, \ldots, p.$$

The *effect size* for the jth experiment in the ith class is the standardized mean difference defined by the parameter

$$\delta_{ij} = (\mu^E_{ij} - \mu^C_{ij})/\sigma_{ij}. \tag{2}$$

When the response scale has unit variance within groups, the effect size parameter δ_{ij} is simply the treatment effect (mean difference) in the jth experiment in the ith class. Note that the effect size remains unaltered under

TABLE 1

Parameters and Estimates for the Control and Experimental Groups

		Experimental				Control			
		Parameters		Estimates		Parameters		Estimates	
Class	Study	Mean	Variance	Mean	Variance	Mean	Variance	Mean	Variance
1	1	μ^E_{11}	σ^2_{11}	$\bar{Y}^E_{11}$	$(s^E_{11})^2$	μ^C_{11}	σ^2_{11}	$\bar{Y}^C_{11}$	$(s^C_{11})^2$
$\vdots$	$\vdots$	$\vdots$	$\vdots$	$\vdots$	$\vdots$	$\vdots$	$\vdots$	$\vdots$	$\vdots$
1	m_1	$\mu^E_{1m_1}$	$\sigma^2_{1m_1}$	$\bar{Y}^E_{1m_1}$	$(s^E_{1m_1})^2$	$\mu^C_{1m_1}$	$\sigma^2_{1m_1}$	$\bar{Y}^C_{1m_1}$	$(s^C_{1m_1})^2$
$\vdots$	$\vdots$	$\vdots$	$\vdots$	$\vdots$	$\vdots$	$\vdots$	$\vdots$	$\vdots$	$\vdots$
p	1	μ^E_{p1}	σ^2_{p1}	$\bar{Y}^E_{p1}$	$(s^E_{p1})^2$	μ^C_{p1}	σ^2_{p1}	$\bar{Y}^C_{p1}$	$(s^C_{p1})^2$
$\vdots$	$\vdots$	$\vdots$	$\vdots$	$\vdots$	$\vdots$	$\vdots$	$\vdots$	$\vdots$	$\vdots$
p	m_p	$\mu^E_{pm_p}$	$\sigma^2_{pm_p}$	$\bar{Y}^E_{pm_p}$	$(s^E_{pm_p})^2$	$\mu^C_{pm_p}$	$\sigma^2_{pm_p}$	$\bar{Y}^C_{pm_p}$	$(s^C_{pm_p})^2$

changes of location or scale. As in Chapter 5, we require the dependent variables to be linearly related so that effect sizes will have the same interpretation across studies.

In subsequent sections we are concerned with testing hypotheses about the δ_{ij}. Three models need to be distinguished. The first is the most homogeneous case in which the studies in *different* classes share a common but unknown effect size δ^*. This is exemplified by the hypothesis

$$\begin{array}{lllllll} H_0: & \text{class } 1: & \delta_{11} & = \delta_{12} & = \cdots = & \delta_{1m_1} & = \delta^*, \\ & \vdots & \vdots & \vdots & & \vdots & \vdots \\ & \text{class } p: & \delta_{p1} & = \delta_{p2} & = \cdots = & \delta_{pm_p} & = \delta^*. \end{array} \tag{3}$$

In the second model the effect sizes *within* a class are equal, but the effect sizes *between* classes differ:

$$\begin{array}{lllllll} H_1: & \text{class } 1: & \delta_{11} & = \delta_{12} & = \cdots = & \delta_{1m_1} & = \delta_1^*, \\ & \vdots & \vdots & \vdots & & \vdots & \vdots \\ & \text{class } p: & \delta_{p1} & = \delta_{p2} & = \cdots = & \delta_{pm_p} & = \delta_p^*. \end{array} \tag{4}$$

The least restrictive case is where the effect sizes may all be different:

$$H_2: \qquad \delta_{ij} \text{ unrestricted.} \tag{5}$$

The test of H_1 versus H_2 is a test of homogeneity of effect size within classes, and the test of H_0 versus H_1 is a test of homogeneity between classes, given that there is homogeneity within classes.

Denote the unbiased estimator of the effect size δ_{ij} given in (10) of Chapter 5 by d_{ij}:

$$d_{ij} = \frac{J(N_{ij} - 2)(\overline{Y}_{ij}^{\mathrm{E}} - \overline{Y}_{ij}^{\mathrm{C}})}{s_{ij}}, \qquad j = 1, \ldots, m_i, \quad i = 1, \ldots, p, \tag{6}$$

where $J(m)$ is given in Table 2 of Chapter 5, $N_{ij} = n_{ij}^{\mathrm{E}} + n_{ij}^{\mathrm{C}}$, $\overline{Y}_{ij}^{\mathrm{E}}$ and $\overline{Y}_{ij}^{\mathrm{C}}$ are the experimental and control group sample means, and s_{ij} is the pooled within-group sample standard deviation for the jth experiment in the ith class.

If all of the n_{ij}^{E} and n_{ij}^{C} become large at the same rate, then the distribution of d_{ij} is approximately normal with a mean δ_{ij} and asymptotic variance

$$\sigma_\infty^2(d_{ij}) = \frac{n_{ij}^{\mathrm{E}} + n_{ij}^{\mathrm{C}}}{n_{ij}^{\mathrm{E}} n_{ij}^{\mathrm{C}}} + \frac{\delta_{ij}^2}{2(n_{ij}^{\mathrm{E}} + n_{ij}^{\mathrm{C}})}, \tag{7}$$

which is estimated by

$$\hat{\sigma}^2(d_{ij}) = \frac{n_{ij}^{\mathrm{E}} + n_{ij}^{\mathrm{C}}}{n_{ij}^{\mathrm{E}} n_{ij}^{\mathrm{C}}} + \frac{d_{ij}^2}{2(n_{ij}^{\mathrm{E}} + n_{ij}^{\mathrm{C}})}. \tag{8}$$

For a series of experiments, define the weighted estimator, d_{i+}, of effect size for the ith class and the weighted grand mean by d_{++}, where

$$d_{i+} = \sum_{j=1}^{m_i} \frac{d_{ij}}{\hat{\sigma}^2(d_{ij})} \bigg/ \sum_{j=1}^{m_i} \frac{1}{\hat{\sigma}^2(d_{ij})} \tag{9}$$

and

$$d_{++} = \sum_{i=1}^{p} \sum_{j=1}^{m_i} \frac{d_{ij}}{\hat{\sigma}^2(d_{ij})} \bigg/ \sum_{i=1}^{p} \sum_{j=1}^{m_i} \frac{1}{\hat{\sigma}^2(d_{ij})}, \tag{10}$$

where $\hat{\sigma}^2(d_{ij})$ is given by (8).

The statistics d_{i+} and d_{++} estimate the weighted mean effect sizes

$$\delta_{i+} = \sum_{i=1}^{m_i} \frac{\delta_{ij}}{\sigma_\infty^2(d_{ij})} \bigg/ \sum_{i=1}^{m_i} \frac{1}{\sigma_\infty^2(d_{ij})}, \tag{11}$$

and

$$\delta_{++} = \sum_{i=1}^{p} \sum_{j=1}^{m_i} \frac{\delta_{ij}}{\sigma_\infty^2(d_{ij})} \bigg/ \sum_{i=1}^{p} \sum_{j=1}^{m_i} \frac{1}{\sigma_\infty^2(d_{ij})}. \tag{12}$$

Note that $\delta_{i+} = \delta_i^*$ under hypothesis H_1 given in (4) and that $\delta_{++} = \delta^*$ under hypothesis H_0 given in (3).

Under the hypothesis of equal effect sizes within classes [hypothesis H_1 given by (4)], the large sample distribution of d_{i+} is normal with mean δ_i^* and asymptotic variance

$$\sigma_\infty^2(d_{i+}) = \left(\sum_{j=1}^{m_i} \frac{1}{\sigma_\infty^2(d_{ij})} \right)^{-1}, \tag{13}$$

which is estimated by

$$\hat{\sigma}^2(d_{i+}) = \left(\sum_{j=1}^{m_i} \frac{1}{\hat{\sigma}^2(d_{ij})} \right)^{-1}. \tag{14}$$

Under the hypothesis of a common effect size between classes [hypothesis H_0 given by (3)], the large sample distribution of d_{++} is normal with a mean δ^* and asymptotic variance

$$\sigma_\infty^2(d_{++}) = \left(\sum_{i=1}^{p} \sum_{j=1}^{m_i} \frac{1}{\sigma_\infty^2(d_{ij})} \right)^{-1}, \tag{15}$$

which is estimated by using the estimate $\hat{\sigma}^2(d_{ij})$ for $\sigma_\infty^2(d_{ij})$.

The estimators $d_{1+}, \ldots, d_{p+}$ are not the maximum likelihood estimators of $\delta_{1+}^*, \ldots, \delta_{p+}^*$ under H_1, but they have the same asymptotic distributions as the maximum likelihood estimators. Similarly, though d_{++} is not the maximum likelihood estimator of δ^* under H_0, it has the same asymptotic distribution as the maximum likelihood estimator. Since the estimators d_{i+} and d_{++} are special cases of the estimator d_+ discussed in Section B.1 of Chapter 6, the comments of Section B.2 of Chapter 6 on the accuracy of the large sample distribution approximation of d_+ apply to d_{i+} and d_{++} as well. In particular, the approximation is quite accurate for effect sizes less than about 1.5 and sample sizes exceeding about 10 per group.

C. SOME TESTS OF HOMOGENEITY

In this section we develop procedures for testing the hypothesis H_0 of a common effect size between classes, as defined in (3), versus the hypothesis H_1 of different effect sizes between classes, as defined in (4), and also for testing H_1 versus the hypothesis H_2 that all effect sizes are different, as defined in (5). The proposed procedures are intuitively appealing and involve easy calculations.

C.1 TESTING WHETHER ALL STUDIES SHARE A COMMON EFFECT SIZE

We start by testing whether the studies, regardless of class, share the same effect size versus the alternative that the studies do not have the same effect size. Although the results of this section can be obtained from the results of Section C.3 below, for the sake of completeness we state the result explicitly. The test proposed in this section is formally a test of H_0 versus H_2, for which the test statistic is based on the total weighted sum of squares:

$$Q_T = \sum_{i=1}^{p} \sum_{j=1}^{m_i} \frac{(d_{ij} - d_{++})^2}{\hat{\sigma}^2(d_{ij})}, \tag{16}$$

where $\hat{\sigma}^2(d_{ij})$ is given by (8). This statistic is the normalized weighted sum of squares of the d_{ij} about the grand mean d_{++}.

If the $m_1 + \cdots + m_p$ studies share a common effect size, that is, if H_0 given in (3) is true, then Q_T has an approximate chi-sqaure distribution with $m_1 + \cdots + m_p - 1$ degrees of freedom, and we compare Q_T with the critical value of the chi-square distribution.

Example

Some data derived from studies examining sex differences in conformity using the fictitious norm group paradigm were described in Section B of Chapter 2. The sample sizes and the estimate d of effect size are given in

Table 5 of that chapter. In Section F we provide the computations for calculating the overall fit statistic $Q_{\mathrm{T}} = 31.799$ for these 10 studies. We see that Q_{T} exceeds 16.9, the 95-percent critical value of the chi-square distribution with nine degrees of freedom. Hence the hypothesis that all studies share a common population effect size is rejected at the $\alpha = 0.05$ significance level.

C.2 TESTING HOMOGENEITY OF EFFECT SIZES ACROSS CLASSES

The test of the hypothesis H_0 that the average effect size does not differ across classes versus the alternative H_1 is analogous to the F-test in the analysis of variance to test that class means are the same. This test is based on the between class goodness-of-fit statistic Q_{B} given by

$$Q_{\mathrm{B}} = \sum_{i=1}^{p} \frac{(d_{i+} - d_{++})^2}{\hat{\sigma}^2(d_{i+})} = \sum_{i=1}^{p} \sum_{j=1}^{m_i} \frac{(d_{i+} - d_{++})^2}{\hat{\sigma}^2(d_{ij})}, \qquad (17)$$

where $\hat{\sigma}^2(d_{ij})$ is given by (8) and $\hat{\sigma}^2(d_{i+})$ is given by (14). Note that this statistic is essentially a weighted sum of squares of the d_{i+} about the grand mean d_{++}.

The approximate distribution of Q_{B} is that of a chi-square statistic with $p - 1$ degrees of freedom. Thus the test of consistency of effect sizes across classes at significance level α consists in comparing the obtained value of Q_{B} with the $100(1 - \alpha)$ percentage point of the chi-square distribution with $p - 1$ degrees of freedom. If Q_{B} exceeds the critical value, the hypothesis that the average effect sizes are equal across classes is rejected; i.e., we reject H_0 in favor of H_1.

Example

In the example of studies concerning sex differences in conformity using the fictitious norm group paradigm as described in Section B of Chapter 2, the sample size, the estimate d of effect size, and the percentage of male authors are given for each study in Table 3. In an earlier analysis of these data Eagly and Carli (1981) contended that the percentage of male authors is related to the magnitude of the sex difference effect size. To test this hypothesis, we group the 10 studies into three classes, namely, studies with 25, 50, and 100 percent male authors, denoted by classes 1, 2, and 3, respectively. In Section F we provide the necessary computations for calculating the between-class fit statistic Q_{B}. Comparing the obtained value 20.706 of Q_{B} with 5.99, the 95 percentage point of the chi-square distribution with two degrees of freedom, we see that the between-class differences are significant at the $\alpha = 0.05$ level. Hence the percentage of male authors is related to the magnitude of the sex difference effect size for these studies.

C.3 TESTING HOMOGENEITY WITHIN CLASSES

To test model specification we test the hypothesis H_1 that effect sizes are homogeneous within classes versus H_2 given in (5). The test uses the within-class goodness-of-fit statistic Q_W given by

$$Q_W = \sum_{i=1}^{p} \sum_{j=1}^{m_i} \frac{(d_{ij} - d_{i+})^2}{\hat{\sigma}^2(d_{ij})}, \tag{18}$$

where $\hat{\sigma}^2(d_{ij})$ is given by (8). Note that Q_W is essentially a weighted sum of squares of the d_{ij} about the d_{i+}.

When the effect sizes are homogeneous within classes, that is, when H_1 is true, the statistic Q_W has an approximate chi-square distribution with $(m_1 - 1) + \cdots + (m_p - 1)$ degrees of freedom, so that we reject H_1 in favor of H_2 if Q_W is greater than the critical chi-square value.

In practice it is often helpful to partition the within-class fit statistic Q_W into p-statistics each of which measures the fit within a class. This helps to isolate those classes for which the fit is particularly poor. The statistic for the goodness of fit within class i is

$$Q_{W_i} = \sum_{j=1}^{m_i} \frac{(d_{ij} - d_{i+})^2}{\hat{\sigma}^2(d_{ij})}, \tag{19}$$

and the overall within-class fit statistic is $Q_W = Q_{W_1} + \cdots + Q_{W_p}$. Note that because all studies are independent, the statistics Q_{W_i} and Q_{W_j} are independent whenever i and j are distinct.

Example

Recall the example of studies of sex differences in conformity using the fictitious norm group paradigm (see Table 3). We group the studies into three classes according to the percentage of male authors (25, 50, and 100 percent, respectively). In Section F we provide the computations for calculating the overall within-class fit statistic, to obtain $Q_W = 11.093$. Comparing Q_W with 14.1, the 95-percent critical value of the chi-square distribution with seven degrees of freedom, we see that Q_W is not significant. In fact, a value of 11.1 would occur between 10 and 25 percent of the time when the effect sizes are homogeneous within classes. Thus the overall fit of the categorical model is not rejected at the $\alpha = 0.05$ significance level. The within-class fit statistics for each class are $Q_{W_1} = 2.565$, $Q_{W_2} = 0.00$, and $Q_{W_3} = 8.528$ for the classes composed of studies with 25, 50, and 100 percent male authors, respectively. Comparing Q_{W_1} and Q_{W_3} with 3.84 and 12.6, (the 95-percent critical values of the chi-square distribution with one and six degrees of freedom, respectively), the within-class fit is not rejected for either of these classes individually.

C.4 AN ANALOGY TO THE ANALYSIS OF VARIANCE

There is a simple relationship among the fit statistics Q_B, Q_W, and Q_T that is analogous to the partitioning of the sum of squares in the analysis of variance, namely, that

$$Q_T = Q_B + Q_W.$$

One interpretation of this formula is that Q_T represents the "total fit" to the model of a single effect size, Q_B represents the "between-class fit," and Q_W represents the "within-class fit." Thus the total fit is partitioned into between- and within-class components. Furthermore, for large samples, the statistics Q_B, Q_W, and Q_T are distributed as chi-square variables. The relationships are summarized in the following table, which is analogous to an analysis of variance source table. We write $k = m_1 + \cdots + m_p$.

Source	Statistic	Degrees of freedom
Between classes	Q_B	$p - 1$
Within classes	Q_W	$k - p$
Total	Q_T	$k - 1$

An important result in the analysis of variance is that the between- and within-group sums of squares are independent. A similar result holds here for the between- and within-class fit statistics. In large samples Q_B and Q_W are independent.

C.5 SMALL SAMPLE ACCURACY OF THE ASYMPTOTIC DISTRIBUTIONS OF THE TEST STATISTICS

The test statistics Q_B, Q_W, and Q_T are equivalent to the statistic Q discussed in Chapter 5, and the comments on the accuracy of Q also apply to these statistics. In particular, if sample sizes are at least 10 per group and the effect sizes are not too large, then the actual significance of the test statistics will not depart too greatly from the nominal significance values obtained from the large sample distribution.

D. FITTING EFFECT SIZE MODELS TO A SERIES OF STUDIES

The statistical results of Section C can be used as part of a general procedure for fitting models to the effect sizes from a series of studies. Our starting point is a series of studies in which each study assesses the effect of a particular treatment through the use of a two-group experimental group/control group design. Suppose further that the dependent variables measure the same construct and are (approximately) linearly equatable, and that the studies are classified according to one of the classification dimensions. The classes obtained by one partitioning may be partitioned again according to a second classification dimension, and in turn partitioned according to other dimensions to yield finer and finer groupings.

In a sense, the procedure for fitting models to effect sizes for each class is analogous to the procedure used to fit hierarchical log linear models to contingency tables. The procedure can be described as follows.

Step 1: Ignore the classification and fit the model of a single effect size to all the studies. Calculate the fit statistic Q_T.

If the value of Q_T is not large or is statistically nonsignificant at some preset level of significance, the process stops, and we conclude that the model of a single effect size fits the data adequately. The estimate of this single effect size is the grand weighted mean d_{++} given by (10). The large sample distribution of d_{++} can be used to determine an approximate confidence interval for δ^* (see Section B).

If the fit statistic Q_T is large or statistically significant, go to Step 2.

Step 2: A large value of the fit statistic Q_T indicates that effect sizes are not homogeneous across all studies, so partition the studies into groups such that the effect sizes are homogeneous within classes. Choose the most important dimension first, that is, the dimension believed to be most related to effect size. Calculate the between-class fit statistic Q_B and the within-class fit statistic Q_W.

If the value of the within-class fit statistic Q_W is small or statistically nonsignificant, the process stops, since the model of a different effect size for each class is consistent with the data. In this case, d_{i+} given in (9) is the estimate of effect size for the ith class and Q_B represents the extent to which the effect sizes differ among classes.

If Q_W is large or statistically significant, go to Step 3.

Step 3: A large value of the fit statistic Q_W indicates that the effect sizes are not homogeneous within classes. At this point it may be useful to partition the within-class fit statistic Q_W into p (if there are p classes) statistics Q_{W_i},

$i = 1, \ldots, p$, where Q_{W_i} indicates the fit statistic within the ith class. Examining the values of Q_{W_i} may help identify classes with especially poor fit, that is, classes in which the effect sizes are heterogeneous. This may suggest whether to exclude some classes or studies from further analysis. Examination of within-class fit may also suggest *which* other classification dimensions are useful. Go to Step 4.

Step 4: Partition the existing classes according to a second classification dimension. Repeat Step 2; that is, calculate the between- and within-class fit statistics Q_B and Q_W for the second dimension. Proceed through Steps 2–4 until an acceptable level of within-class fit is obtained or the classification dimensions are exhausted.

The procedure given is quite practical and involves relatively simple calculations. It has the advantage that the fit to the model can be assessed at each stage, and it also provides a test of the relationship between the classification dimension and effect size.

Example

Recall the example of studies of sex differences using the fictitious norm group paradigm (see Table 5 of Chapter 2). In Section C.1 we calculated Q_T for these studies and obtained the value $Q_T = 31.799$, which led to rejection of the hypothesis that all studies share a common population effect size. On the basis of the hypothesis that male authorship is related to observed sex differences we group the studies into three classes corresponding to studies with 25, 50, and 100 percent male authorship. From Section C.3 we obtained $Q_W = 11.093$, which did not lead to rejection of the homogeneity of effect size within classes. Similarly, the statistics Q_{W_1}, Q_{W_2}, and Q_{W_3} reflecting homogeneity of effect sizes within each class individually were not significant. Therefore we conclude that the model that sex differences in conformity are related to the proportion of male authors accounts for the systematic variation among effect sizes in this data set. The between-class fit statistic Q_B calculated in Section C.2 was $Q_B = 20.706$, which is statistically significant at the $\alpha = 0.05$ level. These values can be put into a summary table similar to that for the analysis of variance as given in Table 2.

The point estimates of effect sizes are $d_{1+} = -0.15$, $d_{2+} = -0.30$, and $d_{3+} = 0.34$ for the classes including studies with 25, 50, and 100 percent male authors, respectively. Ninety-five-percent confidence intervals for δ_1^*, δ_2^*, and δ_3^* are

$$-0.39 \leq \delta_1^* \leq 0.09, \qquad -0.59 \leq \delta_2^* \leq -0.01, \qquad 0.19 \leq \delta_3^* \leq 0.49,$$

which suggests that δ_1^* does not differ from zero, but that δ_2^* and δ_3^* do.

TABLE 2
Summary of Fit Statistics

Source		Q	Degrees of freedom
Between classes	Q_B	20.706	2
Within classes	Q_W	11.093	7
Within class 1	Q_{W_1}	2.565	1
Within class 2	Q_{W_2}	0.000	0
Within class 3	Q_{W_3}	8.528	6
Total	Q_T	31.799	9

E. COMPARISONS AMONG CLASSES

If a priori knowledge or a formal hypothesis test (significant value of Q_B) leads to the conclusion that the effect sizes are not homogeneous across classes, then the investigator can compare the effect sizes of different classes by means of linear combinations or contrasts of the average effect sizes for the classes. Such comparisons are analogous to contrasts in the analysis of variance, and the methods used for choosing contrasts in the analysis of variance (such as orthogonal polynomials to estimate trend coefficients) can be used in the present context.

We consider comparisons of the form

$$\gamma = c_1\delta_{1+} + \cdots + c_p\delta_{p+}, \tag{20}$$

where $c_1, \ldots, c_p$ are known constants and δ_{i+} is the weighted average effect size for the ith class given in (11). If the effect sizes are homogeneous within classes, then $\delta_{i+} = \delta_i^*$, the (single) effect size for the ith class. Most frequently, the comparisons are contrasts in which the c_i are either 0 or ± 1. We use

$$\hat{\gamma} = \sum_{i=1}^{p} c_i d_{i+} \tag{21}$$

as an estimator of γ. Test statistics and confidence intervals for γ are obtained from the large sample normal approximation to the distribution of each d_{i+}.

Recall from Section B that d_{i+} is approximately normally distributed with mean δ_i^* and variance $\sigma_\infty^2(d_{i+})$ given by (13), so that a large sample confidence interval (γ_L, γ_U) for γ is given by

$$\gamma_L = \hat{\gamma} - C_{\alpha/2}\hat{\sigma}_\gamma, \qquad \gamma_U = \hat{\gamma} + C_{\alpha/2}\hat{\sigma}_\gamma, \tag{22}$$

where

$$\hat{\sigma}_{\hat{\gamma}}^2 = c_1^2\hat{\sigma}^2(d_{1+}) + \cdots + c_p^2\hat{\sigma}^2(d_{p+}),$$

$\hat{\sigma}^2(d_{i+})$ is given by (14), and C_α is the two-tailed critical value of the standard normal distribution.

It is important to recognize that contrasts of the d_{i+} are estimates of contrasts among the weighted average population effect sizes δ_{i+}. If the effect sizes within each class are homogeneous, such contrasts are easy to interpret, since all the studies within a class share a common effect size. That is, the contrasts provide a test for relations among the δ_i^*. On the other hand, if effect sizes are heterogeneous within classes, the contrasts are not easily interpreted. It is possible, for example, that the weighted average effect size in one class is larger than the weighted average effect size in a second class, but the majority of the effect sizes in the second class exceed all of those in the first class.

Example

Our example of studies of sex differences in conformity using the fictitious norm group paradigm involved three classes of studies, those in which 25, 50, or 100 percent of the authors were male. Previous analyses have demonstrated that the effect sizes are homogeneous within classes. To contrast the effect size of class 1 with that of class 3, use the contrast coefficients

$$c_1 = -1.0, \qquad c_2 = 0.0, \qquad c_3 = 1.0.$$

The value of the contrast estimate $\hat{\gamma}$ is

$$\hat{\gamma} = -1.0(-0.15) + 0.0(-0.30) + 1.0(0.34) = 0.49,$$

with an estimated variance of

$$\hat{\sigma}_{\hat{\gamma}}^2 = (-1.0)^2(0.015) + (0.0)^2(0.022) + (1.0)^2(0.006) = 0.021.$$

Hence a 95-percent confidence interval for $\gamma = \delta_3^* - \delta_1^*$ is

$$0.21 \le \delta_3^* - \delta_1^* \le 0.77.$$

Because this confidence interval does not include zero, we conclude that the contrast is significant at the $\alpha = 0.05$ level. That is, δ_1^* is significantly different from δ_3^* at the 5-percent level of significance.

E.1 SIMULTANEOUS TESTS FOR MANY COMPARISONS

Frequently a categorical analysis will point out multiple comparisons that may be of interest. If several comparisons among classes or groups are made at significance level α, the chance of incorrectly declaring at least one of the contrasts to be significant will be greater than α. The inflation of the probability of a type I error is particularly severe when the comparisons chosen are determined after inspection of the data. This problem is well known in connection with contrasts in the analysis of variance. One solution to the problem is to use methods for testing contrasts that provide a prespecified probability of making at least one error of type I. Several such simultaneous test procedures are available for the analysis of variance (see Miller, 1981). In this section we adapt two of these simultaneous test procedures for use with effect sizes.

The simplest simultaneous test procedure is the method of Bonferroni inequalities. If l comparisons are to be tested simultaneously, then each comparison is made at the $\alpha/2l$ level of significance. That is, the value $z(\hat{\gamma}) = \hat{\gamma}/\hat{\sigma}_\gamma$ is compared with the $100(1 - \alpha/2l)$ percentage point of the standard normal distribution instead of the $100(1 - \alpha/2)$ percentage point. Using this procedure, the simultaneous significance level of all l comparisons is less than or equal to α.

If the number l of comparisons is large, the Bonferroni method requires large values of $z(\hat{\gamma})$ to obtain significance. In this situation, an alternative procedure based on the Scheffé method in the analysis of variance will probably be more powerful (see Scheffé 1953, 1959). The procedure consists in testing each of l contrasts by rejecting the null hypothesis if

$$[z(\hat{\gamma})]^2 = \hat{\gamma}^2/\hat{\sigma}_\gamma^2$$

exceeds the $100(1 - \alpha)$-percent critical value of the chi-square distribution with l' degrees of freedom. Here l' is the smaller of l, the number of contrasts, and $p - 1$, where p is the number of groups or classes. This procedure gives a simultaneous significance level α for all l contrasts.

When the number of contrasts is small, the procedure based on the Bonferroni method is usually quite powerful. When the number of contrasts is large, the procedure based on the Scheffé method is frequently more powerful. Both procedures provide protection against inflation of significance level by repeated testing.

Example

Our example of studies of sex differences in conformity using the fictitious norm group paradigm involved three classes. The classes were defined by the percentage of male authors (25, 50, or 100 percent) of each study. It might be desirable to compare effect sizes among classes via two contrasts. The first contrast compares the classes involving studies with some female authors (classes 1 and 2) and the class involving studies with all male authors (class 3). The coefficients for the first contrast are $c_1 = -1.0$, $c_2 = -1.0$, and

$c_3 = 2.0$. The second contrast compares the effect sizes of class 1 (studies with 25 percent male coauthors) and class 2 (studies with 50 percent male coauthors). The coefficients for the second contrast are $c_1 = -1.0$, $c_2 = 1.0$, and $c_3 = 0.0$. We test the significance of these contrasts using a simultaneous significance level of $\alpha = 0.05$.

The Bonferroni inequality method involves comparing the standardized contrast values with 2.24, the 100(1 − 0.05/4) percent = 98.75 percent significance level of the standard normal distribution. For the first contrast

$$z_1 = z(\hat{\gamma}_1) = \frac{-1.0(-0.15) - 1.0(-0.30) + 2.0(0.34)}{\sqrt{(-1.0)^2 0.015 + (-1.0)^2 0.022 + (2.0)^2 0.006}} = 4.57,$$

which exceeds the critical value. For the second contrast

$$z_2 = z(\hat{\gamma}_2) = \frac{-1.0(-0.15) + 1.0(-0.30) + 0.0(0.34)}{\sqrt{(-1.0)^2 0.015 + (1.0)^2 0.022 + (0.0)^2 0.006}} = -0.78,$$

which does not exceed the critical value. Hence the simultancous test using the Bonferroni method shows that the first contrast is significant but the second contrast is not.

Using the Scheffé procedure to obtain a simultaneous test at the $\alpha = 0.05$ significance level, we compare the squares of the standardized contrast values with 5.99, the 95-percent critical value of the chi-square distribution with two degrees of freedom. For the first contrast

$$z_1^2 = [z(\hat{\gamma}_1)]^2 = (4.57)^2 = 20.88,$$

which exceeds the critical value. For the second contrast

$$z_2^2 = [z(\hat{\gamma}_2)]^2 = (-0.78)^2 = 0.61,$$

which does not exceed the critical value. Hence the simultaneous test using the Scheffé method also shows that the first contrast is significant but the second contrast is not.

The test using the Bonferroni method requires a smaller value of $z(\hat{\gamma})$ to reject than does the test using the Scheffé method. The Bonferroni method rejects for a standardized contrast value of 2.24 or larger in absolute magnitude, whereas the Scheffé method requires a standardized contrast value of $2.45 = \sqrt{5.99}$ or larger to reject. Thus the Bonferroni inequality method is slightly more sensitive than the Scheffé method in this situation. When the number of contrasts (and the number of classes) is larger, the Scheffé method becomes more sensitive than the Bonferroni inequality method.

F. COMPUTATIONAL FORMULAS FOR WEIGHTED MEANS AND HOMOGENEITY STATISTICS

In practice, computational formulas can simplify calculation of the fit statistics Q_T, Q_B, and Q_W. These formulas are analogous to computational formulas in the analysis of variance and enable the researcher to compute each of the fit statistics in a single pass through the data (e.g., with a packaged computer program). Each of the formulas can be verified by algebraic manipulation. The computational formulas are as follows. Let TW denote the total of the weights (reciprocals of the variances), TWD the total of the weighted d, and TWDS the total of the weighted d-squares. Write

$$\begin{aligned} \mathrm{TW}_i &= \sum_{i=1}^{m_i} \frac{1}{\hat{\sigma}^2(d_{ij})}, & \mathrm{TW} &= \sum_{i=1}^{k} \mathrm{TW}_i, \\ \mathrm{TWD}_i &= \sum_{i=1}^{m_i} \frac{d_{ij}}{\hat{\sigma}^2(d_{ij})}, & \mathrm{TWD} &= \sum_{i=1}^{k} \mathrm{TWD}_i, \\ \mathrm{TWDS}_i &= \sum_{i=1}^{m_i} \frac{d_{ij}^2}{\hat{\sigma}^2(d_{ij})}, & \mathrm{TWDS} &= \sum_{i=1}^{k} \mathrm{TWDS}_i. \end{aligned} \tag{23}$$

These variables are readily computed from n_{ij}^E, n_{ij}^C, d_{ij}, and $\hat{\sigma}^2(d_{ij})$ defined by (8). The homogeneity statistic Q_T is calculated from the total of the three variables across all of the classes:

$$Q_T = \mathrm{TWDS} - (\mathrm{TWD})^2/\mathrm{TW}. \tag{24}$$

The within-class homogeneity statistics for each class are then calculated from totals of the three variables in each class:

$$Q_{W_i} = \mathrm{TWDS}_i - (\mathrm{TWD}_i)^2/\mathrm{TW}_i, \qquad i = 1, \ldots, p. \tag{25}$$

The overall within-class fit statistic is obtained as $Q_W = Q_{W_1} + \cdots + Q_{W_p}$, and the between-class fit statistic is obtained as $Q_B = Q_T - Q_W$. By obtaining the totals of TW, TWD, and TWDS within each class and across classes, Q_T, Q_B, and Q_W can be determined with a minimum of computation.

These totals also permit quick computation of the weighted average effect sizes $d_{1+}, \ldots, d_{p+}$ and d_{++} and of their large sample variances:

$$d_{++} = \mathrm{TWD}/\mathrm{TW}, \tag{26}$$

$$d_{i+} = \mathrm{TWD}_i/\mathrm{TW}_i, \qquad i = 1, \ldots, p. \tag{27}$$

The large sample variance of the weighted grand mean d_{++} is

$$\hat{\sigma}^2(d_{++}) = 1/\text{TW}, \tag{28}$$

and the large sample variance $\hat{\sigma}^2(d_{i+})$ of the class means $d_{i+}, \ldots, d_{p+}$ is given by

$$\hat{\sigma}^2(d_{i+}) = 1/\text{TW}_i. \tag{29}$$

The easiest way to compute these statistics and the weighted means $d_{1+}, \ldots, d_{p+}$ is to create the three new variables $1/\hat{\sigma}^2(d_{ij})$, $d_{ij}/\hat{\sigma}^2(d_{ij})$, and $d_{ij}^2/\hat{\sigma}^2(d_{ij})$ and use a packaged computer program or calculator to obtain sums of these variables.

Computational Examples

For the example of studies of sex differences in conformity that used the fictitious norm group paradigm (Table 5 of Chapter 2), we apply the computational methods suggested in this section. Calculations of $1/\hat{\sigma}^2(d_{ij})$, $d_{ij}/\hat{\sigma}^2(d_{ij})$ and $d_{ij}^2/\hat{\sigma}(d_{ij})$ for each study along with the sums of these variables overall and for each class are given in Table 3.

TABLE 3

Summary of Calculation of Statistics Used in This Chapter

Study	No. of males	No. of females	Class	d	$\hat{\sigma}^2(d)$	$1/\hat{\sigma}^2(d)$	$d/\hat{\sigma}^2(d)$	$d^2/\hat{\sigma}^2(d)$
1	70	71	1	−0.33	0.029	34.775	−11.476	3.787
2	60	59	1	0.07	0.034	29.730	2.081	0.146
3	77	114	2	−0.30	0.022	45.466	−13.640	4.092
4	118	136	3	0.35	0.016	62.233	21.782	7.624
5	32	32	3	0.70	0.066	15.077	10.554	7.388
6	10	10	3	0.85	0.218	4.586	3.898	3.313
7	45	45	3	0.40	0.045	22.059	8.824	3.529
8	30	30	3	0.48	0.069	14.580	6.998	3.359
9	40	40	3	0.37	0.051	19.664	7.275	2.692
10	61	64	3	−0.06	0.032	31.218	−1.873	0.112
Total class 1						64.505	9.395	3.933
Total class 2						45.466	−13.640	4.092
Total class 3						169.417	57.458	28.015
Total all classes						279.388	34.423	36.040

The overall fit statistic Q_T is computed from (24):

$$Q_T = 36.040 - (34.423)^2/279.388 = 31.799.$$

The within-class fit statistics Q_{W_1}, Q_{W_2}, and Q_{W_3} are calculated from (25):

$$Q_{W_1} = 3.933 - (-9.395)^2/64.505 = 2.565,$$
$$Q_{W_2} = 4.092 - (-13.640)^2/45.466 = 0.0,$$
$$Q_{W_3} = 28.015 - (57.458)^2/169.417 = 8.528,$$

and the overall within-class fit statistic is

$$Q_W = 2.565 + 0.0 + 8.528 = 11.093.$$

The between-class fit statistic is calculated as

$$Q_B = Q_T - Q_W = 31.799 - 11.093 = 20.706.$$

The weighted mean effect sizes for each class d_{1+}, d_{2+}, and d_{3+} are calculated from (27):

$$d_{1+} = -9.395/64.505 = -0.15,$$
$$d_{2+} = -13.640/45.466 = -0.30,$$
$$d_{3+} = 57.458/169.417 = 0.34.$$

The approximate variances $\hat{\sigma}^2(d_{1+})$, $\hat{\sigma}^2(d_{2+})$, and $\hat{\sigma}^2(d_{3+})$ of d_{i+}, d_{2+}, and d_{3+} are calculated from (29):

$$\hat{\sigma}^2(d_{1+}) = 1/64.505 = 0.015,$$
$$\hat{\sigma}^2(d_{2+}) = 1/45.466 = 0.022,$$
$$\hat{\sigma}^2(d_{3+}) = 1/169.417 = 0.006.$$

The values calculated in this section were used in the examples throughout Chapter 7.

CHAPTER 8

Fitting Parametric Fixed Effect Models to Effect Sizes: General Linear Models

Practitioners of quantitative research synthesis have been interested in fitting models with continuous independent variables. One procedure often used is to code the characteristics of studies as a vector of predictor variables and then regress the effect size estimates on the predictors to determine the relationship between characteristics of studies and effect sizes (see Glass, 1978). For example, in a meta-analysis of psychotherapy outcome studies, ordinary linear regression was used to determine the relationship between several coded characteristics of studies such as type of therapy, duration of treatment, internal validity of the study, and effect size (Smith & Glass, 1977). The same method has been used in other research syntheses, including the meta-analysis of the effects of class size (Glass & Smith, 1979; Smith & Glass, 1980) and the meta-analysis of the effects of television on school learning (Williams, Haertel, Haertel, & Walberg, 1982).

Although the regression method is appealing, there are some problems with this procedure. First, the assumptions of ordinary regression analysis may not hold, since the variances of the individual effect size estimates are inversely proportional to the sample sizes of the studies. Thus when the studies to be integrated have different sample sizes, the individual "error" variances can be dramatically different. Second, even if the regression coefficients are properly estimated, the regression method will not necessarily give an indication of the goodness of fit or specification of the regression model.

In this chapter we present methods for fitting models to effect size data when the models include either continuous or discrete independent variables. These methods are analogues to multiple regression analyses for effect sizes, and provide estimates of the parameters of the model, large sample tests of significance, and an explicit test of the specification of the model. Thus it is possible to test whether or not a model adequately explains the observed variability in effect size estimates.

Natural tests of model specification are not available in the context of usual normal theory regression analysis. The remarkable fact that model specification can be tested in research synthesis has important advantages for the research reviewer. If tests for model specification fail to reject the specification of an explanatory model for effect sizes, then the reviewer is in a strong position to assert that the results of the series of experiments are explained.

We develop two techniques for fitting general linear models to effect sizes from a series of independent experiments. Section A provides an exposition of the structural model and the notation used, and Section B is concerned with estimation of the model parameters. A large sample test of the specification of the linear model is given in Section C, and computational aspects are discussed in Section D. Results of simulation studies of the small sample behavior of the alternative estimator are presented in Section E. Alternative methods of estimation are the subject of Section F.

A. MODEL AND NOTATION

In this chapter we use the notation of Chapter 6. That is, given a collection of k studies, the population effect sizes are denoted by δ_i and their (unbiased) estimates by d_i given in formula (10) of Chapter 5:

Study	Population effect size	Unbiased estimator of effect size
1	δ_1	d_1
$\vdots$	$\vdots$	$\vdots$
k	δ_k	d_k

We denote the vectors of population and sample effect sizes by $\boldsymbol{\delta} = (\delta_1, \ldots, \delta_k)'$ and $\mathbf{d} = (d_1, \ldots, d_k)'$, respectively.

In the present model we assume that the standardized mean difference δ_i for the ith experiment depends on a vector of p fixed concomitant variables $\mathbf{x}_i = (x_{i1}, \ldots, x_{ip})'$. Specifically, we assume that

$$\begin{aligned} \delta_1 &= x_{11}\beta_1 + \cdots + x_{1p}\beta_p, \\ \delta_2 &= x_{21}\beta_1 + \cdots + x_{2p}\beta_p, \\ &\vdots \\ \delta_k &= x_{k1}\beta_1 + \cdots + x_{kp}\beta_p, \end{aligned} \tag{1}$$

where $\beta_1, \ldots, \beta_p$ are unknown regression coefficients.

The matrix

$$\mathbf{X} = \begin{bmatrix} x_{11} & \cdots & x_{1p} \\ \vdots & & \vdots \\ x_{k1} & \cdots & x_{kp} \end{bmatrix} \tag{2}$$

is called the design matrix in regression analysis, and is assumed to have no linearly dependent columns; that is, $\mathbf{X}$ has rank p. The set of equations (1) can be written succinctly in matrix notation as

$$\boldsymbol{\delta} = \mathbf{X}\boldsymbol{\beta}, \tag{3}$$

where $\boldsymbol{\beta} = (\beta_1, \ldots, \beta_p)'$ is the vector of regression coefficients.

B. A WEIGHTED LEAST SQUARES ESTIMATOR OF THE REGRESSION COEFFICIENTS

The linear model $\boldsymbol{\delta} = \mathbf{X}\boldsymbol{\beta}$ for the effect sizes is analogous to the model that is the basis for ordinary least squares regression analysis. Specifically, we could rewrite our model in matrix notation as

$$\mathbf{d} = \mathbf{X}\boldsymbol{\beta} + \boldsymbol{\varepsilon}, \tag{4}$$

where $\boldsymbol{\varepsilon} = \mathbf{d} - \boldsymbol{\delta}$ is a vector of residuals. Since $\boldsymbol{\varepsilon} = \mathbf{d} - \boldsymbol{\delta}$, it follows that the distribution of $\boldsymbol{\varepsilon}$ is approximately k-variate normal with mean zero and diagonal covariance matrix $\boldsymbol{\Sigma}_d$ given by

$$\boldsymbol{\Sigma}_d = \mathrm{Diag}[\sigma_\infty^2(d_1), \ldots, \sigma_\infty^2(d_k)],$$

where $\sigma_\infty^2(d_i)$ is given by Eq. (14) of Chapter 5, Thus, the elements of $\boldsymbol{\varepsilon}$ are independent but not identically distributed.

If $\boldsymbol{\Sigma}_d$ were known exactly, then we could use the method of generalized least squares to obtain an estimate of $\boldsymbol{\beta}$. Unfortunately, because $\sigma_\infty^2(d_i)$ depends on the unknown parameter δ, the covariance matrix $\boldsymbol{\Sigma}_d$ also depends on the unknown parameter $\boldsymbol{\delta}$. However, it is still possible to estimate $\boldsymbol{\beta}$ by inserting the estimate $\hat{\sigma}^2(d_i)$ for $\sigma_\infty^2(d_i)$ in $\boldsymbol{\Sigma}_d$ and using the estimated covariance matrix $\hat{\boldsymbol{\Sigma}}_d$.

The resulting estimator has the same asymptotic distribution as the maximum likelihood estimator of $\boldsymbol{\beta}$. It is consistent, asymptotically efficient, and much easier to compute than is the maximum likelihood estimator discussed in Section F.1 (see Hedges, 1982c).

The modified generalized least squares estimator $\hat{\boldsymbol{\beta}}$ of $\boldsymbol{\beta}$ under the model (1) is given by

$$\hat{\boldsymbol{\beta}} = (\mathbf{X}'\hat{\boldsymbol{\Sigma}}_d^{-1}\mathbf{X})^{-1}\mathbf{X}'\hat{\boldsymbol{\Sigma}}_d^{-1}\mathbf{d}, \tag{5}$$

which for large samples has a normal distribution with mean $\boldsymbol{\beta}$ and covariance matrix $\boldsymbol{\Sigma}_\beta$ given by

$$\hat{\boldsymbol{\Sigma}}_\beta = (\mathbf{X}'\hat{\boldsymbol{\Sigma}}_d^{-1}\mathbf{X})^{-1}. \tag{6}$$

The large sample normal approximation to the distribution of $\hat{\boldsymbol{\beta}}$ [that is, $\hat{\boldsymbol{\beta}} \sim \mathcal{N}(\boldsymbol{\beta}, \boldsymbol{\Sigma}_\beta)$] can be used to obtain tests of significance or confidence intervals for components of $\boldsymbol{\beta}$. If $\hat{\boldsymbol{\Sigma}}_\beta = (\mathbf{X}'\hat{\boldsymbol{\Sigma}}_d^{-1}\mathbf{X})^{-1}$, σ_{jj} is the jth diagonal element of $\hat{\boldsymbol{\Sigma}}_\beta$, and $\hat{\boldsymbol{\beta}} = (\hat{\beta}_1, \ldots, \hat{\beta}_p)$, then a $100(1-\alpha)$-percent confidence interval for β_j, $1 \le j \le p$, is given by

$$\hat{\beta}_j - C_{\alpha/2}\sqrt{\sigma_{jj}} \le \beta_j \le \hat{\beta}_j + C_{\alpha/2}\sqrt{\sigma_{jj}}, \tag{7}$$

where $C_{\alpha/2}$ is the two-tailed critical value of the standard normal distribution. The usual theory for the normal distribution can be applied if one-tailed or simultaneous confidence intervals are desired.

Sometimes it is useful to test whether some subset $\beta_1, \ldots, \beta_l$ of the regression coefficients are simultaneously zero, that is, $\beta_1 = \beta_2 = \cdots = \beta_l = 0$. To test this hypothesis, compute $\hat{\beta}_1, \ldots, \hat{\beta}_p$ and the statistic

$$Q = (\hat{\beta}_1, \ldots, \hat{\beta}_l)\hat{\boldsymbol{\Sigma}}_{11}^{-1}(\hat{\beta}_1, \ldots, \hat{\beta}_l)', \tag{8}$$

where $\boldsymbol{\Sigma}_{11}$ is the upper $l \times l$ submatrix of

$$\boldsymbol{\Sigma}_\beta = \begin{bmatrix} \boldsymbol{\Sigma}_{11} & \boldsymbol{\Sigma}_{12} \\ \boldsymbol{\Sigma}_{12} & \boldsymbol{\Sigma}_{22} \end{bmatrix}.$$

The test that $\beta_1 = \cdots = \beta_l = 0$ at the 100α-percent significance level consists in comparing Q to the $100(1-\alpha)$ percentage point of the chi-square distribution with l degrees of freedom.

Of course, if $l = p$, then the procedure above yields a test that all the β_j are simultaneously zero. The weighted sum of squares due to regression test statistic becomes

$$Q_{\mathrm{R}} = \hat{\boldsymbol{\beta}}'\hat{\boldsymbol{\Sigma}}_{\beta}^{-1}\hat{\boldsymbol{\beta}}. \tag{9}$$

The test that $\boldsymbol{\beta} = \mathbf{0}$ is simply a test of whether the weighted sum of squares is larger than would be expected if $\boldsymbol{\beta} = \mathbf{0}$.

Example

Data from 10 studies of sex differences in conformity were presented in Table 5 of Chapter 2. Conformity is measured by the extent to which subjects' responses resemble those of a fictitious norm group. Becker (1983) suggested that the number of items on the response measure (and hence its reliability) would influence the magnitude of the sex difference effect size.

This hypothesis was investigated using a linear model to predict effect size δ from the number of items used in the experimental procedure. The regression model is linear with a constant or intercept term and a predictor which is the logarithm of the number of items. Consequently, the design matrix is

$$\mathbf{X} = \begin{bmatrix} 1 & 1 & 1 & 1 & 1 & 1 & 1 & 1 & 1 & 1 \\ \log 2 & \log 2 & \log 2 & \log 38 & \log 30 & \log 45 & \log 45 & \log 45 & \log 5 & \log 5 \end{bmatrix}',$$

and the data vector is

$$\mathbf{d} = (-0.33, 0.07, -0.30, 0.35, 0.69, 0.81, 0.40, 0.47, 0.37, -0.06)'.$$

The covariance matrix is

$$\hat{\boldsymbol{\Sigma}}_d = 0.10 \operatorname{Diag}(0.288, 0.336, 0.220, 0.161, 0.662, 2.166, 0.453, 0.685, 0.508, 0.320).$$

Using SAS Proc GLM as described in Section D, we obtain

$$(\mathbf{X}'\hat{\boldsymbol{\Sigma}}_d^{-1}\mathbf{X}) = 100\begin{bmatrix} 2.795 & 5.930 \\ 5.930 & 17.809 \end{bmatrix}.$$

The estimated regression coefficients are $\hat{\beta}_1 = -0.321$ for the intercept term and $\hat{\beta}_2 = 0.209$ for the effect of the number of items. The covariance matrix of $\hat{\boldsymbol{\beta}}$ is

$$\hat{\boldsymbol{\Sigma}}_{\beta} = (\mathbf{X}'\hat{\boldsymbol{\Sigma}}_d^{-1}\mathbf{X})^{-1} = 10^{-2}\begin{bmatrix} 1.219 & -0.406 \\ -0.406 & 0.191 \end{bmatrix}.$$

Consequently a 95-percent confidence interval for β_2, the regression coefficient for items, is given by $0.209 \pm 1.96\sqrt{0.00191}$ or

$$0.12 \leq \beta_2 \leq 0.29.$$

Because the confidence interval does not contain zero we reject the hypothesis that $\beta_2 = 0$.

Alternatively we could have computed

$$z(\hat{\beta}_2) = \hat{\beta}_2/\sqrt{\sigma_{22}} = 0.209/\sqrt{0.00191} = 4.78,$$

which is to be compared with the critical value 1.96, so that the test leads to rejection of the hypothesis that $\beta_2 = 0$ at the $\alpha = 0.05$ significance level.

The test statistic Q_R for testing that the slope and intercept are simultaneously zero has the value $Q_R = 35.306$, which exceeds 5.99, the 95th percentage point of the chi-square distribution with two degrees of freedom. Hence, we also reject the hypothesis that $\beta_1 = \beta_2 = 0$ at the $\alpha = 0.05$ significance level. This means that the number of items on the response measure is significantly related to effect size.

C. TESTING MODEL SPECIFICATION

It is well known that when the model is correctly specified and the sample size increases, the estimates of regression coefficients approach their respective population values (see, e.g., Goldberger, 1964). That is, the estimates of coefficients in linear models are consistent when the variables that actually determine the dependent variable are included in the model. Estimates from incorrectly specified models may be difficult or impossible to interpret. In data analyses using standard regression methods, it is impossible to test the specification of a regression model in a natural way.

In the case of effect size analyses, tests of model specification do exist whenever the number k of studies exceeds the number p of predictors. These tests of model specification are an important part of the model-building process. If a model for a series of effect sizes is found to be well specified, then the investigator is in a strong position to draw inferences about model parameters. In particular, arguments about bias due to model misspecification are unlikely to be credible.

When the number k of studies is larger than the number p of predictors, the error sum of squares statistic

$$Q_E = \mathbf{d}'\hat{\mathbf{\Sigma}}_d^{-1}\mathbf{d} - Q_R \tag{10}$$

provides a test of model specification. When the model is correctly specified (that is, when $\boldsymbol{\delta} = \mathbf{X}\boldsymbol{\beta}$), Q_E has an approximate chi-square distribution with $k - p$ degrees of freedom. Thus the test for model specification at the 100α-percent level consists in comparing Q_E with the $100(1 - \alpha)$ percentage point of the chi-square distribution with $k - p$ degrees of freedom. If Q_E exceeds this critical value, we reject model specification. Note that Q_E is analogous to the weighted sum of squares about the regression line. Thus the test for model specification corresponds to testing whether the residual variance is larger than expected.

Example

Return to our example of studies of sex differences in conformity using the fictitious norm group paradigm. In the previous section we used a linear model involving a constant term and the logarithm of the number of experimental items to predict effect size. We found that the number of items is a significant predictor of effect size. The question that now arises is whether the model that predicts effect size from the number of items is well specified. Using SAS Proc GLM to compute Q_E as described in Section D, we obtain the value $Q_E = 8.430$, which does not exceed 15.5, the 95 percentage point of the chi-square distribution with eight degrees of freedom. In fact, the obtained value of Q_E is near the mean of the chi-square distribution with eight degrees of freedom. Hence model specification is not rejected, and we conclude that the effect sizes derived from these 10 studies are reasonably consistent with the data analysis model.

D. COMPUTING ESTIMATES AND TEST STATISTICS

The estimate $\hat{\boldsymbol{\beta}}$ of $\boldsymbol{\beta}$ and the test statistics Q_R and Q_E can be calculated using any computer program package that manipulates matrices (such as SAS Proc Matrix). A simpler alternative to the computation of estimates and test statistics is to use a program (such as SPSS or SAS Proc GLM) that can perform weighted least squares analyses.

Weighted least squares analysis involves estimation of linear model parameters by minimizing a weighted sum of squared differences between observations and estimates. Given a design matrix $\mathbf{X}$, a vector $\mathbf{y}$ of observations, and a diagonal weight matrix $\mathbf{W}$, the weighted least squares estimate of $\boldsymbol{\beta}$ in the linear model $\mathbf{y} = \mathbf{X}\boldsymbol{\beta}$ is

$$\hat{\boldsymbol{\beta}}_W = (\mathbf{X}'\mathbf{W}\mathbf{X})^{-1}\mathbf{X}'\mathbf{W}\mathbf{y}. \tag{11}$$

Note that the form of the estimator is the same as that of $\hat{\boldsymbol{\beta}}$ given in (5), so that $\hat{\boldsymbol{\beta}}$ is a special case of $\hat{\boldsymbol{\beta}}_W$ where the weight matrix $\mathbf{W}$ is given by $\hat{\boldsymbol{\Sigma}}_d^{-1}$. The covariance matrix $\hat{\boldsymbol{\Sigma}}_\beta$ of $\hat{\boldsymbol{\beta}}$ is given by $(\mathbf{X}'\mathbf{W}\mathbf{X})^{-1}$, where the weight matrix is $\hat{\boldsymbol{\Sigma}}_d^{-1}$.

Weighted regression programs usually fit an intercept term in the regression equation by default. In this case, when $\mathbf{W} = \mathbf{\Sigma}_d^{-1}$, the weighted sum of squares due to the regression is a statistic Q that is useful in testing that all components of $\boldsymbol{\beta}$ except the intercept are simultaneously zero. In the null case the statistic Q has a chi-square distribution with one degree of freedom for each substantive predictor. The error sum of squares statistic Q_E for testing model specification is the weighted sum of squares about the regression line. When the model is correctly specified, Q_E has a chi-square distribution with $k - p - 1$ degrees of freedom, where p is the number of predictors not including the intercept.

In the unusual case that the regression program fits a "no-intercept" model, then Q_R is the weighted sum of squares due to the regression line. In this case, however, Q_E has a chi-square distribution with $k - p$ degrees of freedom when the model is correctly specified.

The procedure for conducting the analyses described in Sections B and C can be carried out as follows. Set up a weighted regression so that each observation d_i receives the weight

$$w_i = \frac{1}{\hat{\sigma}^2(d_i)} = \frac{2(n_i^E + n_i^C)n_i^E n_i^C}{2(n_i^E + n_i^C)^2 + n_i^E n_i^C d_i^2}.$$

The weighted least squares analysis gives the regression coefficients $\hat{\beta}_1, \ldots, \hat{\beta}_p$ directly. The standard errors for $\hat{\beta}_j$ printed by the program are *incorrect* by a factor of $\sqrt{MS_E}$, where MS_E is the error or residual mean square for the regression. If $\mathrm{SE}(\hat{\beta}_j)$ is the standard error of $\hat{\beta}_j$ printed by the weighted regression program, then the correct standard error [the square root of the jth diagonal element of $(\mathbf{X}'\hat{\mathbf{\Sigma}}_d^{-1}\mathbf{X})^{-1}$] of $\hat{\beta}_j$ is $\mathrm{SE}(\hat{\beta}_j)/\sqrt{MS_E}$. Alternatively, the diagonal elements of the inverse of the sum of squares and cross products matrix $(\mathbf{X}'\mathbf{W}\mathbf{X})^{-1}$ also provide the correct sampling variances for $\hat{\beta}_1, \ldots, \hat{\beta}_p$. The F-tests in the analysis of variance for the regression should be ignored, but the (weighted) sum of squares about the regression is the chi-square statistic Q_E for testing model specification, and the (weighted) sum of squares due to the regression gives the chi-square statistic Q_R for testing that all components of $\boldsymbol{\beta}$ are simultaneously zero (or a related statistic for testing that all components of $\boldsymbol{\beta}$ except the intercept are zero if the program fits an intercept). Therefore all the statistics necessary to compute the analysis can be computed with a single run of a weighted least squares program.

E. THE ACCURACY OF LARGE SAMPLE APPROXIMATIONS

The methods described in this section depend on the asymptotic distributions of $\hat{\boldsymbol{\beta}}$, Q_R, and Q_E. Although large sample approximations are often reasonably accurate in small samples, each approximation must be evaluated individually. The results of the simulations presented in Chapter 5 confirm that the large sample approximations to the distribution of d are quite accurate when $\delta \leq 1.5$ and $n^E + n^C \geq 20$. The asymptotic distribution theory for $\hat{\boldsymbol{\beta}}$, Q_E, and Q_R might also be expected to be reasonably accurate for sample sizes and values of δ in the same ranges.

Simulation studies of the distributions of $\hat{\boldsymbol{\beta}}$, Q_E, and Q_R in some special cases suggest that the theory presented may be useful for the ranges of sample sizes and effect sizes usually encountered in syntheses of educational research. Yet some caution should be used in the application of large sample theory in particular cases without further evidence of the accuracy of that theory in small samples (see Hedges, 1982a, b, c).

The simulations described in this section is based on normal and chi-square random numbers generated by the IMSL (1977) library subroutines GGNML and GGCHS. The range of effect size values was between -1.25 and 1.25 because virtually all meta-analyses have found effect sizes in this range. For each "study," the experimental and control group sample sizes were set equal, i.e., $n_i^E = n_i^C = n_i$. The d values were generated based on the identity

$$d = z/\sqrt{u/m},$$

where z has a normal distribution with mean $\delta = \beta_1 x_1 + \beta_2 x_2$ and variance $2/n_i$, and u is a chi-square random variable with $m = 2n_i - 2$ degrees of freedom.

Two different design matrices were used for $k = 6$ studies in each case. In the first design matrix the predictors have a correlation $\rho = 0.1$, and in the second design matrix the predictors have a correlation $\rho = 0.5$. The design matrices $\mathbf{X}$ are

$$\mathbf{X} = \begin{bmatrix} 2.000 & 0.500 \\ 0.500 & -1.000 \\ 0.000 & -2.150 \\ 0.000 & 2.150 \\ -0.500 & 1.000 \\ -2.000 & -0.500 \end{bmatrix}, \quad \begin{bmatrix} 2.000 & 0.500 \\ 0.500 & 0.500 \\ 0.000 & -0.985 \\ 0.000 & 0.985 \\ -0.500 & -0.500 \\ -2.000 & -0.500 \end{bmatrix},$$

and the regression coefficients $\boldsymbol{\beta} = (\beta_1, \beta_2)'$ were chosen as

$$\boldsymbol{\beta} = \begin{bmatrix} 0.00 \\ 0.00 \end{bmatrix}, \quad \begin{bmatrix} 0.20 \\ 0.20 \end{bmatrix}, \quad \begin{bmatrix} 0.20 \\ 0.50 \end{bmatrix}, \quad \begin{bmatrix} 0.50 \\ 0.50 \end{bmatrix}.$$

For each design matrix $\mathbf{X}$ and each set of regression weights $\boldsymbol{\beta}$, six sets of sample sizes with $n_i^E = n_i^C = 10, 20$, or 50 were used. In three configurations, the sample sizes of all six studies were equal. In the other three sample size

TABLE 1

The Small Sample Behavior of Estimates of Regression Coefficients Based on 2000 Replications for Each Configuration

Correlation between predictors	Common sample sizes: Studies 1–3	Common sample sizes: Studies 4–6	Mean of $\hat{\beta}$	Variance of $\hat{\beta}$	Proportion of confidence intervals including β with nominal significance level of interval: 0.60	0.70	0.80	0.90	0.95	0.99
				$\beta_1 = 0.0, \quad \beta_2 = 0.0$						
0.10	10	10 β_1	−0.004	0.026	0.595	0.692	0.794	0.896	0.943	0.989
0.10	20	20 β_1	−0.001	0.012	0.622	0.717	0.815	0.918	0.955	0.991
0.10	50	50 β_1	−0.000	0.005	0.595	0.704	0.803	0.901	0.956	0.994
0.10	10	20 β_1	0.002	0.016	0.615	0.716	0.808	0.894	0.946	0.990
0.10	10	50 β_1	−0.000	0.008	0.612	0.700	0.814	0.901	0.953	0.989
0.10	20	50 β_1	0.002	0.007	0.615	0.712	0.797	0.904	0.954	0.991
0.10	10	10 β_2	−0.002	0.017	0.634	0.723	0.809	0.908	0.951	0.994
0.10	20	20 β_2	−0.003	0.009	0.603	0.706	0.796	0.894	0.955	0.990
0.10	50	50 β_2	0.000	0.003	0.615	0.709	0.809	0.913	0.957	0.992
0.10	10	20 β_2	−0.003	0.011	0.606	0.710	0.804	0.909	0.950	0.995
0.10	10	50 β_2	0.002	0.006	0.609	0.707	0.809	0.908	0.956	0.990
0.10	20	50 β_2	0.002	0.005	0.589	0.592	0.800	0.895	0.949	0.989
0.50	10	10 β_1	0.001	0.033	0.604	0.696	0.811	0.906	0.942	0.988
0.50	20	20 β_1	−0.001	0.017	0.592	0.686	0.794	0.900	0.943	0.991
0.50	50	50 β_1	0.001	0.006	0.595	0.695	0.792	0.899	0.949	0.989
0.50	10	20 β_1	0.001	0.021	0.600	0.703	0.804	0.903	0.956	0.990
0.50	10	50 β_1	0.003	0.010	0.622	0.727	0.817	0.913	0.959	0.992
0.50	20	50 β_1	−0.002	0.009	0.589	0.703	0.805	0.912	0.952	0.991

0.50	10	10 β_2	−0.008	0.093	0.605	0.698	0.793	0.910	0.955	0.990
0.50	20	20 β_2	0.005	0.047	0.615	0.697	0.799	0.896	0.947	0.989
0.50	50	50 β_2	0.000	0.019	0.598	0.703	0.796	0.890	0.949	0.988
0.50	10	20 β_2	0.001	0.061	0.597	0.696	0.804	0.904	0.949	0.992
0.50	10	50 β_2	−0.005	0.032	0.593	0.697	0.789	0.893	0.945	0.936
0.50	20	50 β_2	0.004	0.026	0.604	0.702	0.801	0.903	0.954	0.994
$\beta_1 = 0.2,\quad \beta_2 = 0.2$										
0.10	10	10 β_1	0.205	0.026	0.591	0.688	0.796	0.894	0.950	0.991
0.10	20	20 β_1	0.200	0.013	0.605	0.695	0.798	0.897	0.950	0.989
0.10	50	50 β_1	0.200	0.005	0.604	0.696	0.795	0.899	0.949	0.990
0.10	10	20 β_1	0.203	0.016	0.609	0.708	0.800	0.915	0.957	0.988
0.10	10	50 β_1	0.204	0.009	0.586	0.691	0.790	0.890	0.941	0.967
0.10	20	50 β_1	0.200	0.007	0.617	0.714	0.810	0.905	0.950	0.990
0.10	10	10 β_2	0.198	0.019	0.602	0.703	0.804	0.894	0.946	0.989
0.10	20	20 β_2	0.202	0.009	0.597	0.702	0.790	0.900	0.950	0.988
0.10	50	50 β_2	0.200	0.003	0.604	0.706	0.815	0.904	0.949	0.989
0.10	10	20 β_2	0.201	0.012	0.610	0.712	0.806	0.899	0.953	0.989
0.10	10	50 β_2	0.200	0.006	0.615	0.715	0.809	0.905	0.954	0.991
0.10	20	50 β_2	0.203	0.005	0.585	0.696	0.802	0.902	0.950	0.993
0.50	10	10 β_1	0.195	0.033	0.615	0.708	0.803	0.908	0.953	0.987
0.50	20	20 β_1	0.199	0.017	0.582	0.691	0.788	0.887	0.943	0.990
0.50	50	50 β_1	0.206	0.007	0.581	0.683	0.798	0.900	0.946	0.989
0.50	10	20 β_1	0.202	0.023	0.600	0.682	0.789	0.890	0.946	0.989
0.50	10	50 β_1	0.202	0.011	0.581	0.687	0.800	0.900	0.949	0.988
0.50	20	50 β_1	0.204	0.009	0.592	0.699	0.802	0.905	0.959	0.992
0.50	10	10 β_2	0.213	0.094	0.602	0.707	0.804	0.901	0.949	0.990
0.50	20	20 β_2	0.208	0.046	0.593	0.688	0.794	0.901	0.950	0.993
0.50	50	50 β_2	0.192	0.018	0.580	0.697	0.794	0.899	0.957	0.995
0.50	10	20 β_2	0.209	0.064	0.580	0.678	0.794	0.891	0.951	0.992
0.50	10	50 β_2	0.210	0.031	0.609	0.705	0.806	0.895	0.949	0.988
0.50	20	50 β_2	0.200	0.027	0.594	0.696	0.792	0.901	0.950	0.989

(Continued)

TABLE 1 (*Continued*)

Correlation between predictors	Common sample sizes: Studies 1–3	Common sample sizes: Studies 4–6	Mean of $\hat{\beta}$	Variance of $\hat{\beta}$	Proportion of confidence intervals including β with nominal significance level of interval: 0.50	0.70	0.80	0.90	0.95	0.99
				$\beta_1 = 0.2,\quad \beta_2 = 0.5$						
0.10	10	10 β_1	0.206	0.028	0.586	0.693	0.796	0.885	0.940	0.987
0.10	20	20 β_1	0.204	0.013	0.504	0.695	0.794	0.893	0.947	0.991
0.10	50	50 β_1	0.199	0.005	0.588	0.688	0.790	0.897	0.950	0.985
0.10	10	20 β_1	0.202	0.017	0.589	0.696	0.792	0.893	0.949	0.990
0.10	10	50 β_1	0.199	0.008	0.580	0.683	0.800	0.907	0.963	0.992
0.10	20	50 β_1	0.204	0.008	0.583	0.677	0.780	0.890	0.947	0.987
0.10	10	10 β_2	0.502	0.020	0.616	0.712	0.806	0.905	0.951	0.989
0.10	20	20 β_2	0.502	0.010	0.601	0.705	0.805	0.890	0.940	0.988
0.10	50	50 β_2	0.502	0.004	0.603	0.695	0.793	0.896	0.945	0.989
0.10	10	20 β_2	0.502	0.014	0.571	0.675	0.783	0.892	0.945	0.986
0.10	10	50 β_2	0.502	0.007	0.597	0.697	0.799	0.899	0.949	0.989
0.10	20	50 β_2	0.501	0.006	0.597	0.689	0.793	0.907	0.955	0.989
0.50	10	10 β_1	0.197	0.034	0.613	0.711	0.801	0.897	0.950	0.991
0.50	20	20 β_1	0.201	0.017	0.592	0.688	0.795	0.899	0.949	0.988
0.50	50	50 β_1	0.202	0.006	0.614	0.721	0.821	0.904	0.953	0.989
0.50	10	20 β_1	0.202	0.021	0.606	0.707	0.809	0.913	0.958	0.994
0.50	10	50 β_1	0.200	0.011	0.599	0.698	0.796	0.900	0.946	0.990
0.50	20	50 β_1	0.202	0.010	0.604	0.696	0.799	0.902	0.950	0.989
0.50	10	10 β_2	0.495	0.095	0.619	0.715	0.809	0.903	0.957	0.989
0.50	20	20 β_2	0.502	0.047	0.610	0.711	0.807	0.904	0.951	0.988
0.50	50	50 β_2	0.500	0.019	0.613	0.696	0.795	0.902	0.947	0.993
0.50	10	20 β_2	0.501	0.061	0.617	0.708	0.814	0.901	0.953	0.990
0.50	10	50 β_2	0.503	0.030	0.606	0.701	0.808	0.908	0.960	0.991
0.50	20	50 β_2	0.498	0.027	0.590	0.704	0.809	0.904	0.951	0.988

0.10	10	10 β_1	0.509	0.031	0.584	0.686	0.784	0.893	0.948	0.991
0.10	20	20 β_1	0.506	0.015	0.594	0.696	0.798	0.697	0.952	0.988
0.10	50	50 β_1	0.501	0.005	0.611	0.717	0.814	0.910	0.959	0.991
0.10	10	20 β_1	0.506	0.021	0.590	0.688	0.789	0.890	0.945	0.985
0.10	10	50 β_1	0.503	0.010	0.577	0.689	0.795	0.909	0.950	0.991
0.10	20	50 β_1	0.504	0.008	0.604	0.702	0.798	0.903	0.951	0.993
0.10	10	10 β_2	0.502	0.019	0.615	0.710	0.807	0.906	0.961	0.990
0.10	20	20 β_2	0.502	0.010	0.603	0.699	0.804	0.899	0.949	0.987
0.10	50	50 β_2	0.502	0.004	0.593	0.695	0.792	0.882	0.943	0.994
0.10	10	20 β_2	0.504	0.014	0.601	0.699	0.797	0.892	0.944	0.987
0.10	10	50 β_2	0.501	0.006	0.610	0.715	0.820	0.910	0.953	0.989
0.10	20	50 β_2	0.502	0.006	0.594	0.691	0.794	0.900	0.947	0.986
0.50	10	10 β_1	0.499	0.039	0.608	0.708	0.801	0.895	0.945	0.987
0.50	20	20 β_1	0.506	0.019	0.598	0.694	0.801	0.901	0.951	0.991
0.50	50	50 β_1	0.500	0.007	0.598	0.704	0.807	0.903	0.953	0.994
0.50	10	20 β_1	0.496	0.025	0.586	0.683	0.788	0.894	0.946	0.992
0.50	10	50 β_1	0.499	0.013	0.594	0.692	0.788	0.892	0.948	0.987
0.50	20	50 β_1	0.507	0.011	0.606	0.701	0.797	0.896	0.945	0.989
0.50	10	10 β_2	0.502	0.108	0.573	0.688	0.786	0.889	0.935	0.986
0.50	20	20 β_2	0.495	0.049	0.584	0.692	0.789	0.896	0.945	0.990
0.50	50	50 β_2	0.497	0.018	0.607	0.706	0.803	0.909	0.953	0.993
0.50	10	20 β_2	0.503	0.064	0.598	0.704	0.796	0.900	0.951	0.989
0.50	10	50 β_2	0.499	0.030	0.620	0.705	0.796	0.903	0.958	0.992
0.50	20	50 β_2	0.489	0.027	0.599	0.698	0.795	0.897	0.949	0.989

TABLE 2

The Small Sample Behavior of the Test Statistic Q_R [a]

Correlation between predictors	Common sample sizes: Studies 1–3	Studies 4–6	Mean of Q_R	Variance of Q_R	Proportion of test statistics exceeding critical chi-squares with nominal significance level of chi-square: 0.40	0.30	0.20	0.10	0.05	0.01
0.10	10	10	1.957	3.887	0.396	0.301	0.198	0.096	0.045	0.010
0.10	20	20	1.950	4.007	0.391	0.287	0.184	0.095	0.045	0.011
0.10	50	50	1.921	3.684	0.393	0.291	0.187	0.090	0.047	0.008
0.10	10	20	1.942	3.734	0.381	0.294	0.198	0.099	0.046	0.008
0.10	10	50	1.945	3.878	0.393	0.284	0.188	0.092	0.046	0.009
0.10	20	50	1.982	3.840	0.395	0.298	0.201	0.098	0.051	0.009
0.50	10	10	1.958	3.823	0.391	0.291	0.190	0.092	0.046	0.011
0.50	20	20	2.034	4.476	0.399	0.299	0.206	0.107	0.055	0.012
0.50	50	50	2.031	4.336	0.405	0.316	0.209	0.102	0.051	0.013
0.50	10	20	1.972	3.710	0.397	0.300	0.194	0.092	0.045	0.009
0.50	10	50	1.953	3.861	0.387	0.283	0.201	0.102	0.043	0.010
0.50	20	50	1.964	3.744	0.406	0.305	0.191	0.092	0.040	0.006

[a] For each sample size configuration and correlation between predictors, $\boldsymbol{\beta} = \mathbf{0}$, and 2000 replications were generated.

TABLE 3

The Small Sample Behavior of the Model Misspecification Test Statistic Q_E

Correlation between predictors	Common sample sizes: Studies 1–3	Common sample sizes: Studies 4–6	Mean of Q_E	Variance of Q_E	Proportion of test statistics exceeding critical chi-squares with nominal significance level of chi-square: 0.40	0.30	0.20	0.10	0.05	0.01
				$\beta_1 = 0.0, \quad \beta_2 = 0.0$						
0.10	10	10	4.180	9.172	0.425	0.323	0.230	0.106	0.058	0.018
0.10	20	20	3.964	8.306	0.392	0.293	0.194	0.107	0.055	0.007
0.10	50	50	3.935	8.114	0.373	0.286	0.197	0.101	0.045	0.010
0.10	10	20	4.102	8.374	0.416	0.327	0.224	0.116	0.057	0.009
0.10	10	50	4.009	8.239	0.400	0.308	0.203	0.102	0.051	0.010
0.10	20	50	4.043	8.179	0.410	0.310	0.206	0.103	0.055	0.010
0.50	10	10	4.153	8.596	0.432	0.316	0.215	0.108	0.063	0.013
0.50	20	20	4.085	7.604	0.417	0.308	0.206	0.102	0.048	0.008
0.50	50	50	3.984	7.753	0.394	0.300	0.198	0.108	0.043	0.008
0.50	10	20	4.250	8.758	0.437	0.333	0.229	0.118	0.060	0.015
0.50	10	50	4.151	8.462	0.420	0.324	0.219	0.108	0.057	0.011
0.50	20	50	4.122	8.395	0.409	0.320	0.221	0.109	0.056	0.011
				$\beta_1 = 0.20, \quad \beta_2 = 0.20$						
0.10	10	10	4.256	8.579	0.430	0.343	0.231	0.116	0.056	0.014
0.10	20	20	4.214	8.767	0.438	0.331	0.231	0.119	0.055	0.010
0.10	50	50	4.077	8.405	0.402	0.311	0.217	0.114	0.054	0.009
0.10	10	20	4.078	8.192	0.408	0.302	0.210	0.103	0.052	0.011
0.10	10	50	4.037	8.055	0.410	0.308	0.203	0.108	0.055	0.010
0.10	20	50	4.014	7.678	0.407	0.310	0.201	0.083	0.046	0.007

(Continued)

TABLE 3 (*Continued*)

Correlation between predictors	Common sample sizes: Studies 1–3	Studies 4–6	Mean of Q_E	Variance of Q_E	Proportion of test statistics exceeding critical chi-squares with nominal significance level of chi-square: 0.40	0.30	0.20	0.10	0.05	0.01
0.50	10	10	4.267	8.993	0.439	0.330	0.235	0.117	0.056	0.014
0.50	20	20	4.079	8.052	0.416	0.321	0.200	0.108	0.049	0.009
0.50	50	50	4.034	8.537	0.393	0.300	0.204	0.112	0.055	0.012
0.50	10	20	4.155	8.473	0.423	0.324	0.219	0.110	0.051	0.013
0.50	10	50	4.020	7.956	0.408	0.308	0.201	0.099	0.054	0.010
0.50	20	50	4.013	7.949	0.393	0.304	0.195	0.094	0.048	0.012
				$\beta_1 = 0.20, \quad \beta_2 = 0.50$						
0.10	10	10	4.279	9.016	0.439	0.339	0.242	0.124	0.062	0.014
0.10	20	20	4.141	8.487	0.422	0.320	0.218	0.115	0.059	0.011
0.10	50	50	4.014	7.505	0.412	0.295	0.196	0.094	0.042	0.010
0.10	10	20	4.114	8.858	0.405	0.305	0.212	0.112	0.061	0.013
0.10	10	50	4.273	9.292	0.424	0.337	0.235	0.124	0.070	0.014
0.10	20	50	4.004	7.696	0.403	0.313	0.210	0.106	0.047	0.008
0.50	10	10	4.226	8.486	0.430	0.321	0.224	0.124	0.060	0.012
0.50	20	20	4.098	8.223	0.423	0.315	0.215	0.111	0.054	0.008
0.50	50	50	4.025	8.264	0.404	0.296	0.214	0.102	0.049	0.009
0.50	10	20	4.169	8.353	0.411	0.319	0.226	0.118	0.058	0.010
0.50	10	50	4.225	9.001	0.424	0.337	0.230	0.119	0.059	0.009
0.50	20	50	4.110	7.931	0.415	0.318	0.214	0.109	0.056	0.010

configurations, one-half of the studies had one sample size and the other half had a different sample size. All of the reported data are based on 2000 replications for each design matrix, $\boldsymbol{\beta}$ value, and sample size configuration.

The results of the empirical sampling investigation of the distribution of $\hat{\boldsymbol{\beta}}$ are given in Table 1. Means and variances of the estimates as well as the empirical proportions of (nonsimultaneous) confidence intervals that included β_1 or β_2 are reported. These results suggest that the bias of $\hat{\boldsymbol{\beta}}$ is small and that confidence intervals (7) are reasonably accurate, even for sample sizes as small as $n_i^E = n_i^C = n_i = 10$.

The results of the empirical sampling study of the distribution of the statistic Q_R are presented in Table 2. The mean and variance of Q_R are tabulated along with the empirical proportions of Q_R values that exceed the (asymptotic) critical values for various significance levels. Note that the empirical distribution of the Q_R is evaluated only under the null hypothesis that $\boldsymbol{\beta} = \mathbf{0}$. The results of this simulation suggest that critical values based on the large sample distribution of Q_R are reasonably accurate, even for sample sizes as small as $n_i^E = n_i^C = n_i = 10$.

The results of the empirical sampling study of the distribution of the specification statistic Q_E are presented in Table 3. The means and variances of Q_E are tabulated along with the empirical proportions of Q_E values that exceed the (asymptotic) critical values for various significance levels. The results of this simulation suggest that critical values based on the large sample distribution of Q_E are reasonably accurate when the sample sizes exceed $n_i^E = n_i^C = n_i = 10$. These critical values seem to be somewhat more accurate when the sample sizes are at least $n_i^E = n_i^C = n_i = 20$.

F. OTHER METHODS OF ESTIMATING THE REGRESSION COEFFICIENTS

In this section we present other methods of fitting general linear models to effect size data. One of the alternative methods is maximum likelihood estimation. Other methods involve transformations of the effect sizes and the use of weighted least squares to fit models to the transformed data. These transformation methods can be used when the studies from which effect sizes are derived are reasonably balanced, that is, when $n_1^E \cong n_1^C, \ldots, n_k^E \cong n_k^C$.

F.1 MAXIMUM LIKELIHOOD ESTIMATORS OF REGRESSION COEFFICIENTS

The method of maximum likelihood can also be used to estimate the regression coefficients $\beta_1, \ldots, \beta_p$ given the design matrix $\mathbf{X}$ and the observed effect sizes $d_1, \ldots, d_k$. The maximum likelihood estimate of $\boldsymbol{\beta}$ is obtained as the solution $\hat{\boldsymbol{\beta}}$ of a system of p equations

$$\sum_{i=1}^{k} \tilde{n}_i x_{ij}\left((L_i - 2)\mathbf{x}_i'\boldsymbol{\beta} \pm \sqrt{(\mathbf{x}_i'\boldsymbol{\beta})^2 + \frac{4L_i(n_i^E + n_i^C)}{\tilde{n}_i}}\right) = 0, \qquad j = 1, \ldots, p, \quad (12)$$

where

$$\tilde{n}_i = n_i^E n_i^C / N_i, \qquad L_i = \tilde{n}_i d_i^2 / \{\tilde{n}_i d_i^2 + (N_i - 2)[J(N_i - 2)]^2\}, \qquad i = 1, \ldots, k,$$

$J(m)$ is given by formula (7) of Chapter 5, and the sign of the ith term is the sign of d_i. The system of equations (12) is rather complicated and cannot be solved algebraically, so that the solution has to be obtained numerically. In any particular instance, standard iterative numerical procedures should yield the solution.

Although the maximum likelihood estimator $\hat{\boldsymbol{\beta}}_M$ of $\boldsymbol{\beta}$ cannot be obtained in closed form, it is possible to obtain an expression for the large sample normal approximation to the distribution of $\hat{\boldsymbol{\beta}}_M$. The large sample joint distribution of $\hat{\boldsymbol{\beta}}_M$ is that of a p-variate normal distribution with mean $\boldsymbol{\beta}$ and covariance matrix $\hat{\boldsymbol{\Sigma}}_\beta$ given by (6). Therefore $\hat{\boldsymbol{\beta}}_M$ has the same large sample distribution as the weighted least squares estimator $\hat{\boldsymbol{\beta}}$ given in (5). The weighted least squares estimator is usually preferred because it is so much easier to compute than is the maximum likelihood estimator. The small sample properties of $\hat{\boldsymbol{\beta}}_M$ have not been investigated (except for $k = 1$), and they may not be the same as those for $\hat{\boldsymbol{\beta}}$. Given the fact that the performance of $\hat{\boldsymbol{\beta}}$ is substantially better than that of $\hat{\boldsymbol{\beta}}_M$ for $k = 1$, there is little reason to expect that $\hat{\boldsymbol{\beta}}_M$ performs better than $\hat{\boldsymbol{\beta}}$ for $k \geq 2$. Consequently there is little reason to recommend $\hat{\boldsymbol{\beta}}_M$ over the weighted least squares estimator.

F.2 ESTIMATORS BASED ON TRANSFORMATIONS OF SAMPLE EFFECT SIZES

When the sample sizes are balanced so that $n_1^E = n_1^C = n_1, \ldots, n_k^E = n_k^C = n_k$, then a variance-stabilizing transformation for d is

$$h = h(d) = \sqrt{2}\,\sinh^{-1}(d/2\sqrt{2}) \qquad (13)$$

(see Section C.2 of Chapter 6). Consequently, the large sample variance will be a constant, and will not depend on the unknown parameters. By transforming d_i to $h_i = h(d_i)$ and δ_i to $\eta_i = h(\delta_i)$, the problem of modeling variability in the δ's is equivalent to modeling the variability in the η's. Define the vectors of transformed parameters and estimates via $\boldsymbol{\eta} = (\eta_1, \ldots, \eta_k)'$ and $\mathbf{h} = (h_1, \ldots, h_k)'$, and use the linear model

$$\boldsymbol{\eta} = \mathbf{X}\boldsymbol{\beta}, \qquad (14)$$

where $\mathbf{X}$ is the design matrix and $\boldsymbol{\beta}$ is the vector of regression coefficients. Note, however, that the numerical values of the regression coefficients in the vector $\boldsymbol{\beta}$ for the transformed parameters will not be the same as those obtained using a linear model for the untransformed parameters.

An estimator of $\boldsymbol{\beta}$ in the present model is

$$\hat{\boldsymbol{\beta}}_h = (\mathbf{X}'\boldsymbol{\Sigma}_h^{-1}\mathbf{X})^{-1}\mathbf{X}'\boldsymbol{\Sigma}_h^{-1}\mathbf{h}, \tag{15}$$

where $\boldsymbol{\Sigma}_h^{-1} = 2\ \mathrm{Diag}(n_1, \ldots, n_k)$. If none of the experimental and control group sample sizes is small, then the estimator $\hat{\boldsymbol{\beta}}_h = (\hat{\beta}_{h,1}, \ldots, \hat{\beta}_{h,p})'$ has approximately a normal distribution with mean $\boldsymbol{\beta}$ and covariance matrix

$$\boldsymbol{\Sigma}_{\beta_h} = (\mathbf{X}'\boldsymbol{\Sigma}_h^{-1}\mathbf{X})^{-1}. \tag{16}$$

Consequently confidence intervals for $\hat{\beta}_j$, the jth component of $\boldsymbol{\beta}$, are given by

$$\hat{\beta}_{h,j} - C_{\alpha/2}\sqrt{\sigma_{h,jj}} \leq \beta_j \leq \hat{\beta}_{h,j} + C_{\alpha/2}\sqrt{\sigma_{h,jj}}, \tag{17}$$

where $\sigma_{h,jj}$ is the jth diagonal element of $\boldsymbol{\Sigma}_{\beta_h}$, and $C_{\alpha/2}$ is the two-tailed critical value of the standard normal distribution (see Hedges & Olkin, 1983b).

To test the hypothesis that $\beta_1 = \cdots = \beta_p = 0$ use the statistic

$$Q_{Rh} = \hat{\boldsymbol{\beta}}'\mathbf{X}'\boldsymbol{\Sigma}_h^{-1}\mathbf{X}\hat{\boldsymbol{\beta}}_h, \tag{18}$$

which has a chi-square distribution with p degrees of freedom when the hypothesis is true. When the number k of studies exceeds the number p of predictors the statistic

$$Q_{Eh} = \mathbf{h}'\boldsymbol{\Sigma}_h^{-1}\mathbf{h} - Q_{Rh} \tag{19}$$

provides a test of model specification. This statistic has a chi-square distribution with $k - p$ degrees of freedom when the model is properly specified (i.e., when $\boldsymbol{\eta} = \mathbf{X}\boldsymbol{\beta}$). The statistics Q_{Rh} and Q_{Eh} are analogous to the statistics Q_R and Q_E described in Sections B and C.

Example

Ten studies of the effects of open versus traditional education on creativity were discussed in Section C of Chapter 2. We now study the effect of grade level on the effect size of open education on creativity. Divide the studies into two groups: those that examined open education in grades 1–3 and those that examined open education in grades 4–8. Because $n_i^E = n_i^C = n_i$ for these studies, we can use the variance-stabilizing transformation (13). The effect size estimate, transformed effect size estimate, and sample size for each study are given in Table 4.

TABLE 4

Effect Size Estimates from 10 Studies of the Effects of Open versus Traditional Education on Student Creativity

Study	Grade level	n_i	d_i	h_i
1	6	90	−0.581	−0.288
2	5	40	0.530	0.263
3	3	36	0.771	0.381
4	3	20	1.031	0.505
5	2	22	0.553	0.275
6	4	10	0.295	0.147
7	8	10	0.078	0.039
8	1	10	0.573	0.284
9	3	39	−0.176	−0.088
10	5	50	−0.232	−0.116

The data vector is

$$\mathbf{h} = (-0.288, 0.263, 0.381, 0.505, 0.275, \\ 0.147, 0.039, 0.284, -0.088, -0.116)',$$

the design matrix is

$$\mathbf{X} = \begin{bmatrix} 1 & 1 & 1 & 1 & 1 & 1 & 1 & 1 & 1 & 1 \\ 2 & 2 & 1 & 1 & 1 & 2 & 2 & 1 & 1 & 2 \end{bmatrix},$$

and the inverse of the covariance matrix of $\mathbf{h}$ is

$$\mathbf{\Sigma}_h^{-1} = \text{diag}\ (180, 80, 72, 40, 44, 20, 20, 20, 78, 100).$$

Using SAS Proc GLM as described in Section D but substituting $\mathbf{h}$ for $\mathbf{d}$ and $\mathbf{\Sigma}_h^{-1}$ for $\mathbf{\Sigma}_d^{-1}$, we obtain

$$\hat{\boldsymbol{\beta}}_h = (0.557, -0.327)'$$

and

$$\mathbf{\Sigma}_{\beta_h} = \begin{bmatrix} 0.01825 & -0.01037 \\ -0.01037 & 0.00644 \end{bmatrix}.$$

Using (17), a 95-percent confidence interval for the regression coefficient corresponding to the effect of grade level on effect size is

$$-0.327 \pm 1.96\sqrt{0.00644}$$

or

$$-0.484 \le \beta_2 \le -0.017.$$

Because this confidence interval does not contain zero, we conclude that the effect size for open education on creativity is significantly smaller for studies using students in grades 4–8.

We also obtain the statistic $Q_{\mathrm{E}h}$ for testing model specification using SAS Proc GLM. The obtained value is $Q_{\mathrm{E}h} = 31.266$, which exceeds the 99-percent critical value of the chi-square distribution with $10 - 2 = 8$ degrees of freedom. Thus the specification of the linear model is rejected. This implies that grade level differences alone do not adequately explain the variability in this group of effect size estimates.

An alternative transformation procedure uses the relationship between the distribution of the effect size and that of the product–moment correlation (see Section B.2 of Chapter 5 and Section C.2 of Chapter 6). If the k experiments are balanced (that is, $n_j^{\mathrm{E}} = n_j^{\mathrm{C}}$, $j = 1, \ldots, k$), we can transform the effect size estimates $d_1, \ldots, d_k$ to the corresponding correlations $r_1, \ldots, r_k$ using the relation

$$r_i = d_i/(d_i^2 + 4)^{1/2}, \qquad i = 1, \ldots, k, \tag{20}$$

and also transform the population effect sizes $\delta_1, \ldots, \delta_k$ to the corresponding population correlations $\rho_1, \ldots, \rho_k$ using the parallel relation

$$\rho_i = \delta_i/(\delta_i^2 + 4)^{1/2}. \tag{21}$$

The linear model then becomes

$$\boldsymbol{\rho} = \mathbf{X}\boldsymbol{\beta}, \tag{22}$$

where $\boldsymbol{\rho} = (\rho_1, \ldots, \rho_k)'$. Methods for the analysis of such models are discussed in Chapter 11.

G. TECHNICAL COMMENTARY

Section B: The estimator (5) can be further refined by iterating in a manner analogous to that used to obtain the estimators $d^{(l)}$ in Section B.1 of Chapter 6. That is, we could use $\hat{\boldsymbol{\beta}}$ to compute an estimate of $\boldsymbol{\delta}$ and use this estimate to refine our estimate of $\hat{\boldsymbol{\Sigma}}_d$. Define $\hat{\boldsymbol{\beta}}^{(0)}$ to be $\hat{\boldsymbol{\beta}}$ and compute

$$\hat{\boldsymbol{\beta}}^{(1)} = (\mathbf{X}'\mathbf{V}_{(0)}^{-1}\mathbf{X})^{-1}\mathbf{X}'\mathbf{V}_{(0)}^{-1}\mathbf{d},$$

where $\mathbf{V}_{(0)}$ is a $k \times k$ diagonal matrix whose ith diagonal element is

$$\frac{N_i}{n_i^{\mathrm{E}} n_i^{\mathrm{C}}} + \frac{(\mathbf{x}_i'\hat{\boldsymbol{\beta}})^2}{2N_i},$$

where $N_i = n_i^{\mathrm{E}} + n_i^{\mathrm{C}}$. Now define

$$\hat{\boldsymbol{\beta}}^{(2)} = (\mathbf{X}'\mathbf{V}_{(1)}^{-1}\mathbf{X})^{-1}\mathbf{X}'\mathbf{V}_{(1)}^{-1}\mathbf{d},$$

where $\mathbf{V}_{(1)}$ is a $k \times k$ diagonal matrix whose ith diagonal element is

$$\frac{N_i}{n_i^{\mathrm{E}} n_i^{\mathrm{C}}} + \frac{(\mathbf{x}_i'\hat{\boldsymbol{\beta}}^{(1)})^2}{2N_i}.$$

This generates an iterative procedure in which

$$\hat{\boldsymbol{\beta}}^{(s)} = (\mathbf{X}'\mathbf{V}_{(s-1)}^{-1}\mathbf{X})^{-1}\mathbf{X}'\mathbf{V}_{(s-1)}^{-1}\mathbf{d}, \qquad s = 1, 2, \ldots,$$

where $\mathbf{V}_{(s)}$ is a $k \times k$ diagonal matrix whose ith diagonal element is

$$\frac{N_i}{n_i^{\mathrm{E}} n_i^{\mathrm{C}}} + \frac{(\mathbf{x}_i'\boldsymbol{\beta}^{(s-1)})^2}{2N_i}.$$

The iterated estimators $\hat{\boldsymbol{\beta}}^{(s)}$ have the same asymptotic distribution as $\hat{\boldsymbol{\beta}}$, but they are likely to be less biased than $\hat{\boldsymbol{\beta}}$ when the model is correctly specified. However, the iteration process does not usually change the estimates very much when the model is correctly specified.

CHAPTER 9

Random Effects Models for Effect Sizes

Previous chapters provided a discussion of statistical methodology for carrying out a quantitative synthesis of research studies under various alternative models. Chapter 6 dealt with the model in which all studies have the same (constant but unknown) effect size. In that model effect size estimates differ only as a result of sampling variability. A test for homogeneity of effect size provides a way of determining whether the observed effect sizes are reasonably consistent with the model of a single underlying effect size.

One immediate extension to the model of a single effect size across studies is to permit effect sizes to be determined by predictor variables whose values for each study are known. The substantive impetus for this model is that studies differ according to many characteristics such as sampling plans, experimental procedures, or stimulus items used. Analyses of effect sizes under the assumption of linear models and tests for the specification (explanatory adequacy) of those models were presented in Chapters 7 and 8.

The idea that characteristics of a study may influence the magnitude of its effect size has been examined in several substantive contexts. For example, Glass and Smith (1979) and Smith and Glass (1980) use ordinary linear regression in their studies of the relationship between class size and student achievement or classroom processes. Although the use of ordinary linear regression on effect sizes cannot be rigorously justified, such analyses may sometimes produce estimates of regression coefficients that are not greatly different from those produced by the methods given in Chapter 8.

A reanalysis of Glass's class size data (Hedges & Stock, 1983) using the methods of Chapter 8 gives results that are somewhat different, but for the most part have the same substantive interpretation. One important conclusion of the reanalysis is that each of the models considered is definitely misspecified. Therefore, differences in class size alone are not sufficient to explain the variability in effect size estimates from the class size data sets.

There are many examples of the use of ordinary linear regression to analyze the relationship of study characteristics and their effect sizes. For example, in their meta-analysis of psychotherapy outcome studies Smith and Glass (1977) used ordinary linear regression to determine the relationship between characteristics of studies (e.g., type of therapy, duration of treatment, internal validity of the study) and effect size. The same methods have been used in quantitative syntheses of gender effects in decoding nonverbal cues (Hall, 1978) and the effects of goal structures on achievement (Johnson, Maruyama, Johnson, Nelson, & Skon, 1981). In general, these analyses have found few consistent relationships between study characteristics and effect size.

One explanation for the elusiveness of such relationships was proposed by Cronbach (1980). He argued that evaluation studies should consider a model in which each treatment site (or study) is a sample realization from a universe of related treatments. Thus "replication" of a treatment across sites may yield many different treatments, each sampled from some universe of possible treatments. If these variations in treatment are more or less effective in producing the outcome, then variations in the true (population) effect of the treatment would be expected. Such variations might be expected to attenuate any relationship between a fixed characteristic (such as age or sex of subjects) and the outcome variable.

According to this formulation there is no single true or population effect of the "treatment" across studies. Rather, there is a *distribution* of true effects; each treatment implementation (site) has its own unique true effect. This leads naturally to the consideration of an average true effect of the treatment as an index of overall efficacy. However, this average true effect will not be very meaningful without some measure of the variation in the true effect of the treatment. For example, it is quite possible for the average true effect to be greater than zero, whereas the true effect of the treatment is negative in nearly half the implementations. The problem of estimating the variability in the true effects is further complicated by the fact that the true effect in any treatment site (or study) is never known. We must estimate that true effect from sample data, and that estimate will itself be subject to sampling fluctuations.

In this chapter we present an analogue to random effects analysis of variance for effect size analyses. This model assumes that the population values of the effect size are samples from a distribution of effect size parameters. Thus the observed variability in sample estimates of effect size is partly due to the variability in the underlying population parameters and partly due to the sampling error of the estimator about the parameter value.

A method for obtaining unbiased estimates of the variance of population effect sizes (the parameter variance component) is given in Section C. Confidence intervals for effect size variance components are obtained in Section D. Section E provides a statistical test of the hypothesis that the variance in population effect sizes is zero. Estimation of the mean of the effect size distribution is discussed in Section F. Empirical Bayes estimation procedures for random effects models are discussed in Section G.

A. MODEL AND NOTATION

Assume that the data arise from a series of k independent studies, where each study is a comparison between an experimental (E) group and a control (C) group. Let Y_{ij}^{E} and Y_{ij}^{C} denote the jth observations on the ith experiment from the experimental and control groups, respectively. The observations can be presented as in Table 1.

Assume that in the ith study each of the n_i^{E} experimental observations is normally distributed with a common mean μ_i^{E} and a common variance σ_i^2. Similarly, the n_i^{C} control observations are normally distributed with common mean μ_i^{C} and variance σ_i^2. Note that variances of the experimental and control groups of the ith experiment are assumed to be identical. More succinctly, we assume that

$$Y_{ij}^{\mathrm{E}} \sim \mathcal{N}(\mu_i^{\mathrm{E}}, \sigma_i^{\mathrm{E}}), \qquad j = 1, \ldots, n_i^{\mathrm{E}}, \quad i = 1, \ldots, k, \tag{1}$$

TABLE 1

Layout of Observations for k Studies

	Observations	
Study	Experimental	Control
1	$Y_{11}^E, \ldots, Y_{1n_1^E}^E$	$Y_{11}^C, \ldots, Y_{1n_1^C}^C$
$\vdots$	$\vdots$	$\vdots$
k	$Y_{k1}^E, \ldots, Y_{kn_k^E}^E$	$Y_{k1}^C, \ldots, Y_{kn_k^C}^C$

and

$$Y_{ij}^{C} \sim \mathcal{N}(\mu_i^{C}, \sigma_i^2), \qquad j = 1, \ldots, n_i^{C}, \quad i = 1, \ldots, k. \tag{1}$$

The effect size for the ith study is defined as in Chapter 5 as

$$\delta_i = (\mu_i^{E} - \mu_i^{C})/\sigma_i, \tag{2}$$

and an unbiased estimator d_i of δ_i is

$$d_i = J(N_i - 2)\,(\bar{Y}_i^{E} - \bar{Y}_i^{C})/s_i, \tag{3}$$

where $N_i = n_i^{E} + n_i^{C}$, $\bar{Y}_i^{E}$, $\bar{Y}_i^{C}$, and s_i are the experimental and control group means and the pooled within-group standard deviation of the ith study, respectively, and $J(m)$ is the correction factor given in Eq. (7) of Chapter 5.

In previous discussions, the effect sizes $\delta_1, \ldots, \delta_k$ were treated as fixed but unknown constants. In this chapter we depart from previous practice and treat the δ_i as if they are chosen from a population of possible values of δ_i. Thus, the ith study provides a realization δ_i from a distribution of true effect sizes. This conceptualization is analogous to random effects analysis of variance, in which treatments, and therefore treatment effect parameters, are sampled from a population of possible treatments.

This random effects model is consistent with Cronbach's idea that treatments (and therefore treatment effects) are sampled from a universe of possible treatments. The treatment is sampled first, and it defines the "population" value δ of the treatment effect. The subjects within the study are then sampled to define the sample estimate of effect size d for the study. Both the variability of the δ's and the variability of the sample estimate about δ within an experiment contribute to the observed variability in the d's across studies. Consequently, the variance of the observed d's across studies is not necessarily a good estimator of the variance of the δ's.

In one sense $\delta_1, \ldots, \delta_k$ are population parameters (the mean of a hypothetical population of d_i values), but in another sense they are sample realizations of the random variable Δ. This may seem to be a confusing duality, but it is much like the situation of the true score in classical test theory. In one sense the true score for an individual is a sample from a population of true scores. On the other hand, the true score for an individual is the mean of a distribution of observed scores for that individual.

The object of the statistical analysis is to estimate the mean effect size $\bar{\Delta}$ (the mean of the population of δ's, estimate the variance $\sigma^2(\Delta)$, and test the hypothesis that $\sigma^2(\Delta)$ is zero. These statistical methods are analogous to those for random effects analysis of variance, in which the object is to estimate the variance components and to test whether variance components are zero. One difference is that in the familiar analysis of variance case the mean of the random effects is usually constrained to be zero, whereas no such constraint is applied to effect sizes.

B. THE VARIANCE OF ESTIMATES OF EFFECT SIZE

The exact variance of the estimator d when the effect size δ is a fixed but unknown constant was presented in Chapter 5. In this chapter we assume that δ is not fixed, but is itself random and has a distribution of its own. Therefore, it is necessary to distinguish between the variance of d assuming a fixed δ and the variance of d incorporating variability of δ. The former is the *conditional sampling variance* of d and the latter is the *unconditional sampling variance* of d. The exact conditional sampling variance $\sigma^2(d_i | \delta_i)$ of d_i is

$$\sigma^2(d_i | \delta_i) = a_i/\tilde{n}_i + (a_i - 1)\delta_i^2, \tag{4}$$

where $\tilde{n}_i = n_i^E n_i^C / N_i$, $N_i = n_i^E + n_i^C$,

$$a_i = (N_i - 2)[J(N_i - 2)]^2/(N_i - 4), \tag{5}$$

and $J(m)$ is given by Eq. (7) in Chapter 5. We emphasize that the conditional sampling variance $\sigma^2(d_i | \delta_i)$ depends on δ_i.

The expression (4) for the conditional sampling variance of d_i shows the dependence on the sample size and on the particular δ_i. This variance can be estimated by using d_i as an estimate of δ_i in (4).

Since the effect size δ_i is the value obtained from a distribution of δ_i, the (unconditional) sampling variance of d_i involves the variance $\sigma^2(\Delta)$ of the distribution of effect sizes. A direct argument shows that the sampling variance is

$$\sigma^2(d_i) = \sigma^2(\Delta) + \sigma^2(d_i | \delta_i) \tag{6}$$

(see Hedges, 1983b). Note that (6) would not be true if a biased estimator (such as Glass's estimator g) were used in place of d. Bias in an estimator can result in a nonzero covariance between sampling error and the parameter values.

C. ESTIMATING THE EFFECT SIZE VARIANCE COMPONENT

In random effects analysis of variance, the expected values of the mean squares are expressed in terms of variance components. These expected values are then replaced by their sample values and the equations are solved for the variance components. This process yields unbiased estimates of the variance components. The rationale for estimating $\sigma^2(\Delta)$ is the same as for the estimation of variance components in random effects analysis of variance.

The formula for the expected value of the sample variance of $d_1, \ldots, d_k$ is first expressed as a function of variance components including the conditional sampling variances as given in (8). Then the expected value of the unconditional variance of the d_i is replaced by the observed variance of the d_i and the equations are solved for $\sigma^2(\Delta)$. This process results in an unbiased estimator of $\sigma^2(\Delta)$. It is worth noting that many other quadratic functions of the observations can also be used to estimate variance components.

We now provide the formulas necessary to obtain an unbiased estimator of $\sigma^2(\Delta)$. Given estimates $d_1, \ldots, d_k$ of the effect sizes $\delta_1, \ldots, \delta_k$, define the usual sample variance of the d_i:

$$s^2(d) = \sum_{i=1}^{k} \frac{(d_i - \bar{d})^2}{k - 1}, \tag{7}$$

where $\bar{d}$ is the unweighted mean of $d_1, \ldots, d_k$. A direct argument shows that the mean (expected value) of $s^2(d)$ is

$$E[s^2(d)] = \sigma^2(\Delta) + \frac{1}{k} \sum_{i=1}^{k} \sigma^2(d_i | \delta_i). \tag{8}$$

On the left-hand side of (8), the observed variance $s^2(d)$ is an unbiased estimator of $E[s^2(d)]$ by definition. On the right-hand side, an unbiased estimator of $\sigma^2(d_i | \delta_i)$ is

$$\hat{\sigma}^2(d_i | \delta_i) = c_i' + c_i'' d_i^2, \tag{9}$$

where $c_i' = 1/\tilde{n}_i$, $c_i'' = (a_i - 1)/a_i$, and a_i is defined in (5).

In (8) we estimate both $E[s^2(d)]$ and $\sigma^2(d_i | \delta_i)$, so that by subtraction we can estimate $\sigma^2(\Delta)$:

$$\hat{\sigma}^2(\Delta) = s^2(d) - \frac{1}{k} \sum_{i=1}^{k} (c_i' + c_i'' d_i^2), \tag{10}$$

which is an unbiased estimator of $\sigma^2(\Delta)$. Because no assumptions have been made about the form of the distribution of Δ, the estimator $\hat{\sigma}^2(\Delta)$ is an unbiased estimator of $\sigma^2(\Delta)$ regardless of the distribution of the random effect sizes.

Example

The results of 11 studies that examined the effects of open education programs on students' attitude toward school were summarized in Table 6 of Chapter 2. If we treat these studies as a sample of possible implementations of the open education treatment, we might wish to estimate the variance $\sigma^2(\Delta)$ of the population of effect sizes corresponding to the universe of all implementations of open education. The calculations necessary to compute

TABLE 2

Studies of the Effects of Open Education on Students' Attitude toward School

Study	n^E	n^C	d	c'	$c'' \times 10^2$	$c' + c''d^2$
1	131	138	0.158	0.0149	0.1886	0.0149
2	40	40	−0.254	0.0500	0.6568	0.0504
3	40	40	0.261	0.0500	0.6568	0.0505
4	90	90	−0.043	0.0222	0.2839	0.0222
5	40	40	0.649	0.0500	0.6568	0.0528
6	79	49	0.503	0.0331	0.4028	0.0341
7	84	45	0.458	0.0341	0.3996	0.0350
8	78	55	0.577	0.0310	0.3872	0.0323
9	38	110	0.588	0.0354	0.3469	0.0366
10	38	93	0.392	0.0371	0.3933	0.0377
11	20	23	−0.055	0.0935	1.2775	0.0935
Total						0.4600
Mean						0.0418

$\hat{\sigma}^2(\Delta)$ are illustrated in Table 2, which lists n^E, n^C, d, and $c' + c''d^2$ for each study. Calculating the variance $s^2(d)$ of the 11 studies using (7) and the average of $c' + c''d^2$, we obtain

$$\hat{\sigma}^2(\Delta) = 0.0928 - 0.0418 = 0.0510.$$

D. THE VARIANCE OF EFFECT SIZE VARIANCE COMPONENTS

In the case of random effects analysis of variance the usual estimates of variance components are unbiased regardless of the distribution of the random effect. However, the variance of a variance component estimate in ANOVA is not easy to estimate without making assumptions about the form of the distribution of the variance components.

Similarly, the estimates of effect size variance components $\sigma^2(\Delta)$ are unbiased regardless of the distribution of the effect sizes. If the variance of an effect size variance component is desired, then assumptions about the form of the distribution of the δ_i must be made. In this section we obtain an expression for the approximate variance of the estimates of the effect size variance components given in Section C under the assumption that the effect sizes have a normal distribution.

Define the $k \times k$ matrix $\mathbf{B}$ by

$$\mathbf{B} = \frac{1}{k}\begin{bmatrix} ka_1 - 1 & -1 & \cdots & -1 \\ -1 & ka_2 - 1 & \cdots & -1 \\ \vdots & \vdots & \vdots & \vdots \\ -1 & -1 & \cdots & ka_k - 1 \end{bmatrix}, \tag{11}$$

where the a_i are defined in (5), and the $k \times k$ diagonal matrix $\mathbf{D}_v^2 = \text{Diag}(v_1^2, \ldots, v_k^2)$ has diagonal elements

$$v_i^2 = \sigma^2(\Delta) + \sigma^2(d_i \mid \delta_i), \qquad i = 1, \ldots, k, \tag{12}$$

which are estimated by

$$\hat{v}_i^2 = \hat{\sigma}^2(\Delta) + \hat{\sigma}^2(d_i \mid \delta_i), \qquad i = 1, \ldots, k, \tag{13}$$

where $\hat{\sigma}^2(\Delta)$ is given by (10) and $\hat{\sigma}^2(d_i \mid \delta_i)$ is given by (9). Then the variance of $\hat{\sigma}^2(\Delta)$ is approximated by

$$\text{Var}[\hat{\sigma}^2(\Delta)] = 2\left(\sum_{i=1}^{k} b_{ii}\hat{v}_i^2\right)^2 + 4(\mathbf{e}'\mathbf{B}\mathbf{D}_v\mathbf{B}\mathbf{e})\bar{\Delta}^2, \tag{14}$$

where $\mathbf{e}$ is a k-dimensional vector of ones. The variance in (14) can be estimated by

$$\text{Est Var}[\hat{\sigma}^2(\Delta)] = 2\left(\sum_{i=1}^{k} b_{ii}\hat{v}_i^2\right)^2 + 4(\mathbf{e}'\mathbf{B}\mathbf{D}_v\mathbf{B}\mathbf{e})[\text{Est}(\bar{\Delta})]^2,$$

where $\mathbf{D}_{\hat{v}} = \text{Diag}(\hat{v}_1^2, \ldots, \hat{v}_k^2)$, and Est $(\bar{\Delta})$ is given later in (17).

Example

Return to the data from 11 studies of the effects of open education on student attitude toward school discussed in the example in Section C. We use the estimate $\hat{\sigma}^2(\Delta) = 0.0510$ for $\sigma^2(\Delta)$ and the values $\hat{\sigma}^2(d_i \mid \delta_i) = c_i + c_i' d_i^2$ from Table 2 for $\sigma^2(d_i \mid \delta_i)$ in (13) to calculate.

$$\mathbf{D}_{\hat{v}} = \text{Diag}(0.0659, 0.1014, 0.1015, 0.0732, 0.1038, 0.0851, \\ 0.0860, 0.0833, 0.0876, 0.0887, 0.1445).$$

We use $\text{Est}(\bar{\Delta}) = 0.301$ calculated in Section F as an estimate of the common value of $\bar{\Delta}$. The matrix $\mathbf{B}$ can be written as $\mathbf{A} - \mathbf{C}$, where $\mathbf{C}$ is a matrix whose elements are $1/k = 1/11 = 0.0909$, and $\mathbf{A} = \text{Diag}(a_1, \ldots, a_k)$, where the a_i are defined by (5):

$$\mathbf{A} = \text{Diag}(1.002, 1.007, 1.007, 1.003, 1.007, 1.004, \\ 1.004, 1.004, 1.003, 1.004, 1.0013).$$

Therefore

$$\text{Est Var}[\hat{\sigma}^2(\Delta)] = 0.0165.$$

If the effect sizes are normally distributed within each class, the estimate $\hat{\sigma}^2(\Delta)$ has an asymptotic normal distribution. Therefore, the approximate variance (14) can be used to obtain approximate confidence intervals for $\sigma^2(\Delta)$. The large sample normal approximation to the distribution of $\hat{\sigma}^2(\Delta)$ is probably not very good unless the number of studies k is quite large. For example, the analogous approximation for the distribution of variance component estimates in the random effects analysis of variance is very poor. Therefore we recommend exercising caution in the use of such confidence intervals until more is known about the accuracy of the large sample approximation to the distribution of the $\hat{\sigma}^2(\Delta)$.

E. TESTING THAT THE EFFECT SIZE VARIANCE COMPONENT IS ZERO

It is sometimes useful to test the hypothesis that the effect size variance is zero in addition to estimating this quantity. This test is analogous to the F-test in random effects analysis of variance. Note that if $\sigma^2(\Delta) = 0$, then

$$\delta_1 = \delta_2 = \cdots = \delta_k = \bar{\Delta};$$

that is, the effect sizes are fixed but unknown parameters. Therefore, a test that $\sigma^2(\Delta) = 0$ in the current model corresponds to a test for homogeneity of effect sizes in models with fixed parameters. Such a test was given in Section D of Chapter 6. When the hypothesis is true, the homogeneity test statistic is

$$Q = \sum_{i=1}^{k} \frac{(d_i - d_+)^2}{\hat{\sigma}^2(d_i \mid \delta_i)}, \tag{15}$$

where

$$d_+ = \sum_{i=1}^{k} \frac{d_i}{\hat{\sigma}^2(d_i \mid \delta_i)} \Bigg/ \sum_{i=1}^{k} \frac{1}{\hat{\sigma}^2(d_i \mid \delta_i)}. \tag{16}$$

The statistic Q has an approximate chi-square distribution with $k - 1$ degrees of freedom. If the obtained value of Q exceeds the $100(1 - \alpha)$ percentile point of the distribution, then we reject the hypothesis that $\sigma^2(\Delta) = 0$ at the 100α-percent significance level.

Example

In our example of the 11 studies of the effects of open education on student attitude toward school, we test that $\sigma^2(\Delta) = 0$ by computing the test statistic (15). Calculating Q using the method given in (29) of Chapter 6 and comparing the obtained value $Q = 23.159$ with 18.31, the 95-percentile point of the chi-square distribution with 10 degrees of freedom, we reject the hypothesis that $\sigma^2(\Delta) = 0$.

Although the test proposed is a large sample test, the simulations reported in Chapter 6 suggest that the nominal significance levels are quite accurate when the experimental and control group sample sizes in each study exceed 10 per group and δ is between -1.5 and 1.5. These sample sizes and effect sizes are typical of many bodies of educational and psychological research.

Note that the significance test that $\sigma^2(\Delta) = 0$ in the random effect size model is identical to the test of homogeneity of effect sizes in the fixed effect size model. This is analogous to the situation with F-tests in one-way random effects analysis of variance. In the analysis of variance case, the null distributions of the test statistics are identical in the two models, but the nonnull distributions of the F-ratios differ. Similarly, although the null distributions of Q are identical in fixed and random effect size models, the nonnull distributions of Q differ under the two models. We do not explore the nonnull distribution of Q in the random effects model except to note that it is not a chi-square distribution, and it is stochastically larger when $\sigma^2(\Delta) > 0$ than when $\sigma^2(\Delta) = 0$.

F. ESTIMATING THE MEAN EFFECT SIZE

In Chapter 6 we discussed how to estimate a common underlying effect size from a series of studies. In this section we describe methods for estimating the mean $\bar{\Delta}$ of the effect size distribution from the results of a series of studies. It is important to note that the mean of the effect size distribution does not have the same interpretation as a single underlying effect size in fixed effect size models. In the case of random effect size models, some individual studies may have negative population effect sizes even though $\bar{\Delta} > 0$. This corresponds to the substantive idea that some realizations of a treatment may actually be harmful even if the average effect of the treatment $\bar{\Delta}$ is beneficial.

Each of the estimators d_i is an unbiased estimator of δ_i, and therefore the unweighted average effect size $\bar{d}$ is also an unbiased estimator of $\bar{\Delta}$, the average effect size. However, the unweighted mean $\bar{d}$ is not the most precise estimator of $\bar{\Delta}$, because studies that have larger sample sizes yield more precise estimators of the corresponding δ_i than studies based on smaller sample sizes. Thus estimators from studies with larger sample sizes should be weighted more. It can be shown that the "most precise" weighted estimate

$$w_1 d_1 + \cdots + w_k d_k \tag{17}$$

of $\bar{\Delta}$ has weights

$$w_i(\Delta, \delta) = \frac{1}{v_i^2} \Bigg/ \sum_{j=1}^{k} \frac{1}{v_j^2}, \tag{18}$$

where the v_i are defined in (12) (see Hedges, 1983b). Because the weights in (18) involve the unknown parameters $\sigma^2(\Delta)$ and $\delta_1, \ldots, \delta_k$, we estimate the unknown parameters and use those estimates to compute the weights

$$\hat{w}_i(\Delta, \delta) = \frac{1}{\hat{v}_i^2} \Bigg/ \sum_{j=1}^{k} \frac{1}{\hat{v}_j^2}, \tag{19}$$

where the $\hat{v}_i$ are defined in (13). Although it is possible to obtain very precise estimates of the δ_i if the sample sizes are reasonably large in each study, the precision of the estimate of $\sigma^2(\Delta)$ depends primarily on the number k of studies. Therefore, if the number of studies is small, the estimates of the weights may still be fairly imprecise even though the sample size in each study is quite large.

Note that the most precise estimate of $\bar{\Delta}$ does not depend on the distribution of the random effects δ_i, but that the distribution of the estimate does depend on the distribution of the δ_i. If we assume that the random effect sizes δ_i have a normal distribution, and that the number k of studies is large, then $\hat{w}_1 d_1 + \cdots + \hat{w}_k d_k$ has an approximately normal distribution with a mean of $\bar{\Delta}$ and approximate variance $[(\Sigma_1^k\ (1/v_j^2)]^{-1}$. This fact can be employed in obtaining an approximate confidence interval for $\bar{\Delta}$.

Example

Returning to our example of 11 studies of the effects of open education on attitude toward school, we previously computed $\hat{\sigma}^2(\Delta) = 0.0510$ in Section C. Here we estimate the mean $\bar{\Delta}$ using our estimate of $\sigma^2(\Delta)$. Table 3 contains values for $\hat{v}_i^{-2}$ and $d_i/\hat{v}_i^2$. Using the totals of the entries in these columns, we obtain the estimate of $\bar{\Delta}$

$$\text{Est}(\bar{\Delta}) = 37.0512/123.1815 = 0.301$$

and the variance of the estimate

$$1/123.1815 = 0.0081.$$

If we assume that the random effect sizes δ_i have a normal distribution, an approximate 95-percent confidence interval for $\bar{\Delta}$ is

$$0.12 = 0.301 - 1.96(0.090) \leq \bar{\Delta} \leq 0.301 + 1.96(0.090) = 0.48.$$

Because the confidence interval does not contain zero we reject the hypothesis that $\bar{\Delta}$ is zero at the $\alpha = 0.05$ level of significance.

TABLE 3

Studies of the Effects of Open Education on Students' Attitude toward School

Study	d	$\hat{\sigma}^2(d_i \mid \delta_i)$	$1/\hat{v}_i^2$	$d_i/\hat{v}_i^2$
1	0.158	0.0149	15.1683	2.3966
2	−0.254	0.0504	9.8596	−2.5043
3	0.261	0.0505	9.8573	2.5728
4	−0.043	0.0222	13.6561	−0.5872
5	0.649	0.0528	9.6370	6.2544
6	0.503	0.0341	11.7529	5.9117
7	0.458	0.0350	11.6326	5.3277
8	0.577	0.0323	12.0060	6.9275
9	0.588	0.0366	11.4147	6.7119
10	0.392	0.0377	11.2774	4.4207
11	−0.055	0.0935	6.9196	−0.3806
			123.1815	37.0512

G. EMPIRICAL BAYES ESTIMATION FOR RANDOM EFFECTS MODELS

We have treated random effects models for effect sizes as if they had a sampling distribution. Another conceptualization is to view the parameters $\delta_1, \ldots, \delta_k$ as having a prior distribution. In this formulation the observed effect sizes are used to estimate the mean and variance (the so-called hyperparameters) of the prior distribution. It is usually necessary to assume a particular functional form of the prior distribution. The assumption of a normal prior distribution for δ is convenient and often realistic. If we also treat the large sample normal approximation to the distribution of the effect size estimate as the exact distribution, then empirical Bayes estimation can be carried out in a straightforward manner (see Champney, 1983). We also assume a normal prior distribution for $\delta_1, \ldots, \delta_k$ with common mean δ_* and common variance σ_*^2. The problem is to estimate δ_* and σ_*^2 from $d_1, \ldots, d_k$.

One general approach to this type of problem is to use an iterative procedure involving the EM algorithm (Dempster, Laird, and Rubin, 1977). The problem of estimating δ_* and σ_*^2 is treated as a missing data problem where $d_1, \ldots, d_k$ are observed but $\delta_1, \ldots, \delta_k$ are "missing." The basic idea is to start with initial estimates of δ_* and σ_*^2 and to use the observed values of $d_1, \ldots, d_k$ to estimate $\delta_1, \ldots, \delta_k$. This is called the E step (for estimation or expectation) of the EM algorithm. Then the estimates of $\delta_1, \ldots, \delta_k$ can be used to calculate improved estimates of δ_* and σ_*^2. This is called the M step of the EM algorithm because the improved estimates of δ_* and σ_*^2 are maximum likelihood estimates derived from the E-step estimates of $\delta_1, \ldots, \delta_k$. The estimates of δ_* and σ_*^2 derived in the M step of the algorithm are then used as initial estimates for another E step, which is followed by another M step, and so on until the estimates (or the associated likelihood function) no longer change substantially with further iterations.

We use the notation $\hat{\delta}_{*(j)}$, $\hat{\sigma}_{*(j)}^2$, and $\hat{\delta}_{i(j)}$ for the estimates of δ_*, σ_*^2, and δ_i, respectively, at the jth iteration of the EM algorithm. Using Est($\bar{\Delta}$) given in Section F and $\hat{\sigma}^2(\Delta)$ given in Section C as initial estimates of δ_* and σ_*^2, that is,

$$\hat{\delta}_{*(0)} = \hat{\bar{\Delta}}, \qquad \hat{\sigma}_{*(0)}^2 = \hat{\sigma}^2(\Delta),$$

a direct Bayesian argument (see, e.g., Box and Tiao, 1973) shows that

$$\hat{\delta}_{i(1)} = \frac{d_i + \hat{\sigma}^2(d_i|\delta_i)\hat{\delta}_{*(0)}/\hat{\sigma}_{*(0)}^2}{1 + \hat{\sigma}^2(d_i|\delta_i)/\hat{\sigma}_{*(0)}^2},$$

$$\hat{\sigma}^2(\hat{\delta}_{i(1)}) = [\hat{\sigma}^2(d_i|\delta_i) + \hat{\sigma}_{*(0)}^2]^{-1}, \tag{20}$$

where $\hat{\sigma}^2(d_i|\delta_i)$ is given in (9). This completes the first E step of the algorithm. In the M step of the algorithm, the maximum likelihood estimates of δ_* and σ_*^2 are calculated by treating $\hat{\delta}_{1(1)}, \ldots, \hat{\delta}_{k(1)}$ and $\hat{\sigma}^2(\hat{\delta}_{1(1)}), \ldots, \hat{\sigma}^2(\hat{\delta}_{k(1)})$ as observed data. This leads to the estimates

$$\hat{\delta}_{*(1)} = \sum_{i=1}^{k} \frac{\hat{\delta}_{i(1)}}{k} \tag{21}$$

and

$$\hat{\sigma}_{*(1)}^2 = \frac{1}{k}\left[\sum_{i=1}^{k} [\hat{\delta}_{i(1)}^2 + \hat{\sigma}^2(\hat{\delta}_{i(1)})] - \left(\sum_{i=1}^{k} \hat{\delta}_{i(1)}\right)^2 \Big/ k\right]. \tag{22}$$

The iterative process continues with another E step using $\hat{\delta}_{*(1)}$ and $\hat{\sigma}_{*(1)}^2$ in place of $\hat{\delta}_{*(0)}$ and $\hat{\sigma}_{*(0)}^2$ in (20). In general, the estimates $\hat{\delta}_{1(j)}, \ldots, \hat{\delta}_{k(j)}$ and $\delta_1, \ldots, \delta_k$ in the jth E step are obtained from $d_1, \ldots, d_k$ and the estimates $\hat{\delta}_{*(j-1)}$ and $\hat{\sigma}_{*(j-1)}^2$ of δ_* and σ_*^2 at the $(j - 1)$th step via

$$\hat{\delta}_{i(j)} = \frac{d_i + \hat{\sigma}^2(d_i|\delta_i)\hat{\delta}_{*(j-1)}/\hat{\sigma}_{*(j-1)}^2}{1 + \hat{\sigma}^2(d_i|\delta_i)/\hat{\sigma}_{*(j-1)}^2},$$

$$\hat{\sigma}^2(\delta_{i(j)}) = [\hat{\sigma}^2(d_i|\delta_i) + \hat{\sigma}_{*(j-1)}^2]^{-1}. \tag{23}$$

The estimates $\hat{\delta}_{*(j)}$ and $\hat{\sigma}^2_{*(j)}$ in the jth M step are obtained from $\hat{\delta}_{1(j)}, \ldots, \hat{\delta}_{k(j)}$ and $\hat{\sigma}^2(\delta_{1(j)}), \ldots, \hat{\sigma}^2(\delta_{k(j)})$ via

$$\hat{\delta}_{*(j)} = \sum_{i=1}^{k} \frac{\hat{\delta}_{i(j)}}{k},$$
$$\hat{\sigma}^2_{*(j)} = \frac{1}{k}\left[\sum_{i=1}^{k} [\hat{\delta}^2_{i(j)} + \hat{\sigma}^2(\delta_{i(j)})] - \left(\sum_{i=1}^{k} \hat{\delta}_{i(j)}\right)^2 \Big/ k\right]. \tag{24}$$

The iterative process continues until the change in the estimates (or the associated likelihood function) becomes negligible.

The EM algorithm usually requires only a small number of iterations to converge to estimates of δ_* and σ^2_*, but it is usually not suitable for hand calculation. Hence it is necessary to write a simple computer program to carry out the computations. In our experience the estimates of δ_* and σ^2_* produced by this method are very close to the estimates $\text{Est}(\bar{\Delta})$ and $\hat{\sigma}^2(\Delta)$ derived in Sections F and D, respectively.

Remark: Champney (1983) studied empirical Bayes estimation of δ_* and σ^2_* using a slightly different procedure. We have treated $\hat{\sigma}^2(d_i|\delta_i)$ in (20) as if it were a constant independent of d_i, when in fact it depends on δ_i and hence on d_i to a small extent. To avoid the slight dependence of $\hat{\sigma}^2(d_i|\delta_i)$ on δ_i, Champney used the variance-stabilizing transformation (17) of Chapter 6. When $n_1^{\mathrm{E}} = n_1^{\mathrm{C}}, \ldots, n_k^{\mathrm{E}} = n_k^{\mathrm{C}}$, Champney computes each $\hat{\delta}_{i(j)}$ by computing the corresponding quantity in the transformed scale of the $h_i = h(d_i)$, $\eta_i = h(\delta_i)$, $\eta_* = h(\delta_*)$, and $\tilde{\sigma}^2_*$, the variance of $\eta_1, \ldots, \eta_k$, where $h(x)$ is given in (17) of Chapter 6. Thus the computation of the estimates $\hat{\delta}_{i(j)}$ at the jth stage involves first computing

$$\hat{\eta}_{i(j)} = \frac{\hat{\tilde{\sigma}}^2_{*(j-1)} h_i + 2n_i \hat{\eta}_{*(j-1)}}{\hat{\tilde{\sigma}}^2_{*(j-1)} + 2n_i}$$

and

$$\hat{\sigma}^2(\hat{\eta}_{i(j)}) = (2n_i + \hat{\tilde{\sigma}}^2_{*(j-1)})^{-1}.$$

These estimates are then transformed to estimates of $\hat{\delta}_{i(j)}$ and $\hat{\sigma}^2(\delta_{i(j)})$ via

$$\hat{\delta}_{i(j)} = h^{-1}(\hat{\eta}_{i(j)})$$

and

$$\hat{\sigma}^2(\hat{\delta}_{i(j)}) = h^{-1}(2n_i + \tilde{\sigma}^2_{*(j-1)} + \hat{\eta}^2_{*(j-1)}) - [h^{-1}(\hat{\eta}_{i(j-1)})]^2.$$

Champney also evaluated several variations of this procedure, including the use of a different variance-stabilizing transformation for each study to handle the case in which the experimental and control group sample sizes are not equal. In addition, he studied the behavior of the empirical Bayes estimates by simulation studies (see Champney, 1983). Finally, he examined extensions of his random effects models to the problem of estimating δ_* and σ_*^2 in the case in which only effect size estimates corresponding to statistically mean differences can be observed (see Chapter 14).

CHAPTER 10

Multivariate Models for Effect Sizes

An explicit assumption that we have made is that effect sizes arise from different experiments, and consequently are stochastically independent. This assumption of independence will be met if indeed the estimators are derived from different experiments, or if the estimates are derived from independent samples of subjects within the same study. However, not all collections of effect size estimates can easily be reduced to collections of independent estimators. For example, a study may report data on subscales of a mathematics achievement test, but not for overall mathematics achievement. If the scores on overall mathematics achievement were available, one might use these data to determine an effect size estimate. But when the data on overall achievement are unavailable, it it not clear how to derive a single effect size estimate from the study.

In another context, suppose several treatments are to be compared to a control. One may wish to estimate an effect size for each treatment, but the comparison with a common control makes such estimates stochastically dependent. One approach to obtaining a single estimate is to discard effect size estimates for all but one treatment in a study, but this approach will be excessively wasteful. The situation in which there are many treatments per study is characteristic of the data used in Glass's class size meta-analyses (Glass & Smith, 1979, Smith & Glass, 1980).

The use of nonindependent effect size estimates is important in the analysis of the joint behavior of several dependent variables. For example, outcome studies of a teaching method may involve measures of both reading achievement and children's attitude toward school. Another case in which the joint distribution of effect sizes is relevant is that of studies where both pretest and posttest scores are available. The pretest effect size is a natural measure of the degree of preexisting difference between the experimental and control groups of an experiment, and the relationship between the pretest and posttest effect size is of interest as a measure of the extent to which preexisting differences are reflected in posttest effect sizes.

Although the multivariate situations described can be dealt with by discarding data, such a procedure is obviously wasteful. Another possibility is to treat correlated effect sizes as independent and then to adjust the significance level. Sometimes this procedure works reasonably well. However, the most natural method is to treat correlated data by employing multivariate techniques.

A key feature of multivariate procedures is that they deal with all the correlated effect sizes simultaneously. A disadvantage of these multivariate techniques is that they usually require knowledge of the correlations between variables—information that is not always available. In some instances, such correlations may actually be available. For example, test-norming studies may provide very good estimates of correlations between subscales of psychological tests. Such estimates can be treated as "known values" to provide the correlations necessary to use the methods given in this chapter.

The notation that is used is given first in Section A. A crucial result needed is the multivariate distribution of a vector of effect sizes derived from correlated observations, and this is given in Section B. Section C provides a discussion of the estimation of a common effect size from a vector of correlated estimates. Methods for estimating effect sizes from a series of studies in which some studies provide several correlated estimates and other studies provide a single estimate are discussed in Section D. The estimation of a vector of correlated effect sizes from several experiments is discussed in Section E, a test for homogeneity of vectors of effect sizes across studies is given in Section F, and a discussion of pooling is provided in Section G.

A. MODEL AND NOTATION

Suppose that in each of k studies there are measures on p response variables for each subject. The notation requires one subscript for the study, one for the subject, and one for the response variable. Let Y_{ilj}^{E} and Y_{ilj}^{C} denote the observations on the lth variable for the jth subject in the ith study in the experimental and control groups, respectively. Suppose, further, that there are p measures for each individual. Denote the vectors of responses for the jth subjects in the experimental and control groups of the ith study by

$$\mathbf{Y}_{ij}^{\mathrm{E}} = (Y_{i1j}^{\mathrm{E}}, \ldots, Y_{ipj}^{\mathrm{E}})', \qquad \mathbf{Y}_{ij}^{\mathrm{C}} = (Y_{i1j}^{\mathrm{C}}, \ldots, Y_{ipj}^{\mathrm{C}})'.$$

The sample sizes for the ith study are n_i^{E} for the experimental group and n_i^{C} for the control group, and we write $N_i = n_i^{\mathrm{E}} + n_i^{\mathrm{C}}$. The data obtained can be dispayed as in Table 1.

Assume that for the ith study the observations of the experimental and control groups, $\mathbf{Y}_{i\alpha}^{\mathrm{E}}$ and $\mathbf{Y}_{i\alpha}^{\mathrm{C}}$, are independently distributed with the p-variate normal distribution with mean vectors

$$\boldsymbol{\mu}_i^{\mathrm{E}} = (\mu_{i1}^{\mathrm{E}}, \ldots, \mu_{ip}^{\mathrm{E}})', \qquad \boldsymbol{\mu}_i^{\mathrm{C}} = (\mu_{i1}^{\mathrm{C}}, \ldots, \mu_{ip}^{\mathrm{C}})'$$

and with common covariance matrix $\boldsymbol{\Psi}^{(i)}$. That is,

$$\mathbf{Y}_{i\alpha}^{\mathrm{E}} \sim \mathcal{N}(\boldsymbol{\mu}_i^{\mathrm{E}}, \boldsymbol{\Psi}^{(i)}), \qquad \alpha = 1, \ldots, n_i^{\mathrm{E}}, \qquad i = 1, \ldots, k,$$

$$\mathbf{Y}_{i\alpha}^{\mathrm{C}} \sim \mathcal{N}(\boldsymbol{\mu}_i^{\mathrm{C}}, \boldsymbol{\Psi}^{(i)}), \qquad \alpha = 1, \ldots, n_i^{\mathrm{C}}, \qquad i = 1, \ldots, k.$$

The parameters and their sample estimates are displayed as in Table 2.

The vector $\boldsymbol{\delta}_i$ of effect sizes represents the population effect size for each component separately. In particular, the vector of effect sizes is

$$\boldsymbol{\delta}_i = (\delta_{i1}, \ldots, \delta_{ip})',$$

TABLE 1
Data Layout

	Experimental group		Control group	
		Measure		Measure
Study	Person	$1 \quad \cdots \quad p$	Person	$1 \quad \cdots \quad p$
1	1	$Y_{111}^{\mathrm{E}}, \ldots, Y_{1p1}^{\mathrm{E}}$	1	$Y_{111}^{\mathrm{C}}, \ldots, Y_{1p1}^{\mathrm{C}}$
⋮	⋮	⋮ ⋮	⋮	⋮ ⋮
1	n_1^{E}	$Y_{11n_1^{\mathrm{E}}}^{\mathrm{E}}, \ldots, Y_{1pn_1^{\mathrm{E}}}^{\mathrm{E}}$	n_1^{C}	$Y_{11n_1^{\mathrm{C}}}^{\mathrm{C}}, \ldots, Y_{1pn_1^{\mathrm{C}}}^{\mathrm{C}}$
⋮	⋮	⋮ ⋮	⋮	⋮ ⋮
k	1	$Y_{k11}^{\mathrm{E}}, \ldots, Y_{kp1}^{\mathrm{E}}$	1	$Y_{k11}^{\mathrm{C}}, \ldots, Y_{kp1}^{\mathrm{C}}$
⋮	⋮	⋮ ⋮	⋮	⋮ ⋮
k	n_k^{E}	$Y_{k1n_k^{\mathrm{E}}}^{\mathrm{E}}, \ldots, Y_{kpn_k^{\mathrm{E}}}^{\mathrm{E}}$	n_k^{C}	$Y_{k1n_k^{\mathrm{C}}}^{\mathrm{C}}, \ldots, Y_{kpn_k^{\mathrm{C}}}^{\mathrm{C}}$

TABLE 2

(a) Parameters and (b) Estimates

Study	Means		Covariance matrices	Correlation matrices
	Experimental	Control		
	(a) Parameters			
1	$\boldsymbol{\mu}_1^E = (\mu_{11}^E, \ldots, \mu_{1p}^E)'$	$\boldsymbol{\mu}_1^C = (\mu_{11}^C, \ldots, \mu_{1p}^C)'$	$\boldsymbol{\Psi}^{(1)}$	$\mathbf{P}^{(1)}$
$\vdots$	$\vdots$	$\vdots$	$\vdots$	$\vdots$
k	$\boldsymbol{\mu}_k^E = (\mu_{k1}^E, \ldots, \mu_{kp}^E)'$	$\boldsymbol{\mu}_k^C = (\mu_{k1}^C, \ldots, \mu_{kp}^C)'$	$\boldsymbol{\Psi}^{(k)}$	$\mathbf{P}^{(k)}$
	(b) Estimates			
1	$\overline{\mathbf{Y}}_1^E = (\overline{Y}_{11}^E, \ldots, \overline{Y}_{1p}^E)'$	$\overline{\mathbf{Y}}_1^C = (\overline{Y}_{11}^C, \ldots, \overline{Y}_{1p}^C)'$	$\mathbf{S}^{(1)}$	$\mathbf{R}^{(1)}$
$\vdots$	$\vdots$	$\vdots$	$\vdots$	$\vdots$
k	$\overline{\mathbf{Y}}_k^E = (\overline{Y}_{k1}^E, \ldots, \overline{Y}_{kp}^E)'$	$\overline{\mathbf{Y}}_k^C = (\overline{Y}_{k1}^C, \ldots, \overline{Y}_{kp}^C)'$	$\mathbf{S}^{(k)}$	$\mathbf{R}^{(k)}$

where the components δ_{il} are defined by

$$\delta_{il} = \frac{\mu_{il}^E - \mu_{il}^C}{\sqrt{\psi_{ll}^{(i)}}}, \qquad l = 1, \ldots, p, \quad i = 1, \ldots, k, \tag{1}$$

and $\psi_{ll}^{(i)}$ is the lth diagonal element of the covariance matrix $\boldsymbol{\Psi}^{(i)}$.

The vector d_i of sample effect size estimates for the ith study is defined by

$$\mathbf{d}_i = (d_{i1}, \ldots, d_{ip})', \tag{2}$$

where

$$d_{il} = \frac{J(N_i - 2)\,(\overline{Y}_{il}^E - \overline{Y}_{il}^C)}{s_{ll}^{(i)}}, \qquad l = 1, \ldots, p, \quad i = 1, \ldots, k,$$

$\overline{Y}_{il}^E$ and $\overline{Y}_{il}^C$ are the experimental and control group means on the lth measure for the ith experiment, $s_{ll}^{(i)}$ is the lth diagonal element of the pooled within-group covariance matrix $\mathbf{S}^{(i)}$ for the ith study, and $J(m)$ is a constant given by Eq. (7) of Chapter 5. The population effect sizes and their unbiased estimators are displayed below.

Study	Effect sizes	
	Parameters	Unbiased estimates
1	$\delta_1 = (\delta_{11}, \ldots, \delta_{1p})'$	$d_1 = (d_{11}, \ldots, d_{1p})'$
$\vdots$	$\vdots$	$\vdots$
k	$\delta_k = (\delta_{k1}, \ldots, \delta_{kp})'$	$d_k = (d_{k1}, \ldots, d_{kp})'$

Remark: Sometimes it is useful to restrict the form of $\boldsymbol{\Psi}^{(i)}$ so that

$$\boldsymbol{\Psi}^{(i)} = \mathbf{D}^{(i)}\mathbf{P}\mathbf{D}^{(i)},$$

where $\mathbf{D}^{(i)}$ is a $p \times p$ diagonal matrix of the standard deviations in the ith study, and $\mathbf{P}$ is a matrix of correlations that does not depend on the particular study i. This restriction is equivalent to requiring that the p measures in each study have the same correlational structure.

B. THE MULTIVARIATE DISTRIBUTION OF EFFECT SIZES

The vector $\mathbf{d}_i$ of effect sizes has a very complicated distribution in small samples. The exact distribution is known and is related to certain multivariate noncentral t-distributions (Krishnan, 1972). Current representations of this exact distribution are sufficiently complicated that exact results are too cumbersome to use in practical situations. However, because the asymptotic distribution of the vector $\mathbf{d}_i$ is quite simple, we can use this approximate result.

The large sample distribution of $\mathbf{d}_i$ as n_i^E and n_i^C increase with n_i^E/N_i and n_i^C/N_i fixed is normal with a mean $\boldsymbol{\delta}_i$ and asymptotic covariance matrix $\boldsymbol{\Phi}^{(i)} = (\phi_{st}^{(i)})$ with elements

$$\phi_{st}^{(i)} = \rho_{st}^{(i)}\sigma_\infty(d_{is})\sigma_\infty(d_{it}), \tag{3}$$

where $\sigma_\infty^2(d)$ is the asymptotic variance of d given in (14) of Chapter 5, and $\rho_{st}^{(i)}$ is the common correlation between $Y_{is\alpha}^E$ and $Y_{it\alpha}^E$ (or $Y_{is\alpha}^C$ and $Y_{it\alpha}^C$) given in Table 1.

One important conclusion derived from this result is that the effect sizes derived from components s and t have the same asymptotic correlation as the original observations. It is a surprising fact that the asymptotic distributions of the $\mathbf{d}_i$ involve only the correlations among the original observations. This is an important point because the correlations are more likely to be known than are the covariances. For example, if the components are subscales of a psychological test, the correlation matrix is often included in the technical manual for the test.

The large sample approximation to the multivariate distribution of effect sizes is that each p-dimensional vector $\mathbf{d}_i$ is normally distributed with mean vector $\boldsymbol{\delta}_i$ and covariance matrix $\boldsymbol{\Sigma}^{(i)} = (\sigma_{st}^{(i)})$, with elements

$$\sigma_{ss}^{(i)} = \sigma_\infty^2(d_{is}) = \frac{n_i^E + n_i^C}{n_i^E n_i^C} + \frac{\delta_{is}^2}{2(n_i^E + n_i^C)}, \tag{4}$$

$$\sigma_{st}^{(i)} = \rho_{st}^{(i)}\sqrt{\sigma_{ss}^{(i)}\sigma_{tt}^{(i)}}, \tag{5}$$

where $\rho_{st}^{(i)}$ is the correlation between the sth and tth measures in the ith study. In large samples, the population correlations $\rho_{st}^{(i)}$ can be replaced by the sample estimate $r_{st}^{(i)}$ of the correlation, and this approximation can be used in exactly the same way as if the correlations were known.

One application of this large sample result is to the case in which studies provide both pretest and posttest data. In particular, pretest effect sizes indicate the magnitude of preexisting differences between experimental and control groups. Because pretest and posttest scores are usually correlated, the effect sizes will also be correlated. Consequently, in such cases pretest effect sizes will be good predictors of posttest effect sizes, and this permits the use of small pretest effect sizes as indicators of studies with small bias due to preexisting differences.

C. ESTIMATING A COMMON EFFECT SIZE FROM A VECTOR OF CORRELATED ESTIMATES

Some studies report data in a manner that makes it difficult to extract a single effect size estimate from the study, as in the case of studies that report data for subscales of a psychological test, but not the summary statistics for the total test. If the subscales do not seem to measure different constructs, the investigator may believe that the effect sizes on some or all of the subscales will be the same. It is sometimes convenient to combine estimates across studies in two steps. First the correlated estimators within each study are pooled, and then the independent estimators are, in turn, pooled across studies. Because it is poor practice to pool estimators that estimate different parameter values, we may wish to use the vector of (correlated) effect size estimators to test the hypothesis that the population effect sizes from the different subscales are the same. A statistical test of this hypothesis is given in Section C.1. If the subscales are found to be homogeneous, then we can pool estimates from the different subscales. Methods for obtaining optimal weights for pooling are given in Section C.2.

C.1 TESTING HOMOGENEITY OF CORRELATED EFFECT SIZES

Suppose we wish to test, for the ith study, whether the component estimators share a common effect size, that is, whether

$$\delta_{i1} = \delta_{i2} = \cdots = \delta_{ip} = \delta_i^*. \tag{6}$$

To test this hypothesis compute the statistic

$$Q_i = \mathbf{d}_i'\hat{\mathbf{M}}_i\mathbf{d}_i, \tag{7}$$

where $\mathbf{d}_i = (d_{il}, \ldots, d_{ip})'$ is the vector of estimates of $\boldsymbol{\delta}_i$, $\mathbf{M}_i$ is the matrix

$$\mathbf{M}_i = \boldsymbol{\Lambda}^{(i)} - \boldsymbol{\Lambda}^{(i)}\mathbf{ee}'\boldsymbol{\Lambda}^{(i)}/\mathbf{e}'\boldsymbol{\Lambda}^{(i)}\mathbf{e}, \tag{8}$$

$\mathbf{e}$ is a p-dimensional column vector consisting of ones, $\boldsymbol{\Lambda}^{(i)}$ is the inverse of the covariance matrix $\boldsymbol{\Sigma}^{(i)} = (\sigma_{st}^{(i)})$ given by (4) and (5), and $\hat{\mathbf{M}}_i$ is the estimate of $\mathbf{M}_i$ obtained by using sample estimates, given by (15) of Chapter 5, of the variances of the effect size estimates and of the correlations between observations.

When each N_i increases with n_i^{E}/N_i and n_i^{C}/N_i fixed, and (6) holds, the statistic Q_i has an approximate chi-square distribution with $p - 1$ degrees of freedom. Thus the test that $\delta_{i1} = \cdots = \delta_{ip}$ consists in comparing the statistic Q_i with the critical values of the chi-square distribution with $p - 1$ degrees of freedom. If the value of Q_i is small or statistically nonsignificant, the investigator should probably pool the estimates from all components of $\mathbf{d}_i$ to obtain the most precise estimate of $\boldsymbol{\delta}_i^*$.

We emphasize that the statistical test may help to determine whether to pool similar measures, but it should not be the sole basis for decisions about pooling estimates of δ_i^*. Components of $\mathbf{d}_i$ that are derived from dissimilar measures should rarely, if ever, be pooled.

Example

A study with $n_i^{\mathrm{E}} = n_i^{\mathrm{C}} = 40$ provides data on three subscales of a reading test that were administered to the same students and are therefore correlated. We want to test whether the effect sizes δ_{i1}, δ_{i2}, and δ_{i3} on the three subscales are equal in order to determine if it is reasonable to pool the effect size data across the subscales. The vector of effect size estimates is

$$\mathbf{d}_i = (-0.254,\ -0.158,\ -0.261)',$$

and the correlation matrix of the observations is known (from a norming study) to be

$$R = \begin{bmatrix} 1.00 & 0.83 & 0.87 \\ 0.83 & 1.00 & 0.85 \\ 0.87 & 0.85 & 1.00 \end{bmatrix}.$$

The covariance matrix $\hat{\boldsymbol{\Sigma}}^{(i)}$ of $\mathbf{d}_i$ is

$$\hat{\boldsymbol{\Sigma}}^{(i)} = \mathbf{D}_i\mathbf{R}\mathbf{D}_i = \begin{bmatrix} 0.0504 & 0.0417 & 0.0439 \\ 0.0417 & 0.0502 & 0.0427 \\ 0.0439 & 0.0427 & 0.0504 \end{bmatrix},$$

where $\mathbf{D}_i = \text{Diag}(\sqrt{0.0504}, \sqrt{0.0502}, \sqrt{0.0504})$ is a diagonal matrix of standard deviation of elements of $\mathbf{d}_i$. Calculating the inverse $\hat{\mathbf{\Lambda}}^{(i)}$ of $\hat{\mathbf{\Sigma}}^{(i)}$, we obtain

$$\hat{\mathbf{\Lambda}}^{(i)} = \begin{bmatrix} 92.896 & -30.356 & -55.068 \\ -30.356 & 81.705 & -42.901 \\ -55.068 & -42.901 & 104.144 \end{bmatrix}$$

and

$$\hat{\mathbf{M}}_i = \hat{\mathbf{\Lambda}}^{(i)} - \frac{\hat{\mathbf{\Lambda}}^{(i)}\mathbf{ee}'\hat{\mathbf{\Lambda}}^{(i)}}{\mathbf{e}'\hat{\mathbf{\Lambda}}^{(i)}\mathbf{e}} = \begin{bmatrix} 90.369 & -33.213 & -57.156 \\ -33.213 & 78.475 & -45.262 \\ -57.156 & -45.262 & 102.418 \end{bmatrix}.$$

Finally, $Q = \mathbf{d}_i'\hat{\mathbf{M}}_i\mathbf{d}_i = 0.79$, which is compared to a critical value 5.99 of the chi-square distribution with two degrees of freedom. Consequently we do not reject the hypothesis that $\delta_{i1} = \delta_{i2} = \delta_{i3}$, and the effect size estimates from the three subtests can be pooled.

C.2 ESTIMATION OF EFFECT SIZE FROM CORRELATED ESTIMATES

If the estimates $d_{i1}, \ldots, d_{ip}$ are reasonably consistent with the model of the same effect size for each component, then the investigator may wish to compute an estimate of this common effect size. The most precise estimate of the common effect size δ_i^* is given by

$$\hat{\delta}_i = d_{i1}w_{i1} + \cdots + d_{ip}w_{ip} = \mathbf{w}'\mathbf{d}_i, \tag{9}$$

with weight vector

$$\mathbf{w}_i = (w_{i1}, \ldots, w_{ip})' = \mathbf{\Lambda}^{(i)}\mathbf{e}/\mathbf{e}'\mathbf{\Lambda}^{(i)}\mathbf{e}. \tag{10}$$

The weights $\mathbf{w}_i$ are estimated by inserting estimates of $\mathbf{\Lambda}^{(i)}$ in (10). In large samples, $\hat{\delta}_i$ is approximately normally distributed with mean δ_i and estimated asymptotic variance

$$\hat{\sigma}^2(\hat{\delta}_i) = 1/\mathbf{e}'\hat{\mathbf{\Lambda}}^{(i)}\mathbf{e}. \tag{11}$$

If some of the k studies yield a vector of p effect sizes, the formula (9) can be used to obtain a single estimate of effect size δ_i for each of those studies. Then the large sample approximation can be used to obtain the variance of $\hat{\delta}_i$. The result is a series of k independent estimators of effect size, one from each study. These independent effect size estimators can then be combined using the methods given in the next section. Note that, in the special case $p = 1$,

$$\hat{\delta}_i = d_{i1} = d_i,$$

and the large sample variance of $\hat{\delta}_i$ reduces to $\hat{\sigma}^2(d)$, as defined in (15) of Chapter 5.

Example

Return to the study yielding effect size estimates on three reading test subscales. The results of the test of homogeneity of effect sizes suggest that the effect sizes calculated for the three subscales can be pooled. Calculating

$$\mathbf{w}_i = \hat{\mathbf{\Lambda}}^{(i)}\mathbf{e}/\mathbf{e}'\hat{\mathbf{\Lambda}}^{(i)}\mathbf{e} = (0.338, 0.382, 0.279)',$$

we obtain the weighted estimator of effect size,

$$\hat{\boldsymbol{\delta}}_i = \mathbf{w}_i'\mathbf{d}_i = -0.220,$$

with an estimated variance of

$$\hat{\sigma}^2(\hat{\delta}_i) = 1/\mathbf{e}'\hat{\mathbf{\Lambda}}^{(i)}\mathbf{e} = 0.0453.$$

Note that the variance of the pooled estimator $\hat{\delta}_i$ is only about 10 percent smaller (0.045 versus 0.050) than the variance of the three estimators d_{i1}, d_{i2}, and d_{i3} that were pooled to obtain $\hat{\delta}_i$. Because the estimators are highly correlated, they contain little independent information, and hence the best pooled estimate is only slightly superior to any single estimate.

D. ESTIMATING A COMMON EFFECT SIZE AND TESTING FOR HOMOGENEITY OF EFFECT SIZES

In this section we present the details for combining independent estimators of effect size derived from a series of k studies using the methods of Section C. Since the estimator $\hat{\delta}_i$ given in (9) includes the special case of a single effect size estimate from a study, the methods presented here can be used to combine estimates from a series of k independent experiments each of which yields one or more effect size estimates.

D.1 AN ESTIMATOR OF EFFECT SIZE

Suppose $\delta_1, \ldots, \delta_k$ are the effect sizes from k independent studies, and $\hat{\delta}_1, \hat{\delta}_2, \ldots, \hat{\delta}_k$ are the estimates obtained from (9) when the effect sizes are equal to δ^*. Then an optimal weighted mean is

$$\hat{\delta}_+ = w_1\hat{\delta}_1 + \cdots + w_k\hat{\delta}_k, \tag{12}$$

where the estimated weights are

$$\hat{w}_i = \frac{1}{\hat{\sigma}^2(\hat{\delta}_i)} \bigg/ \sum_{j=1}^{k} \frac{1}{\hat{\sigma}^2(\hat{\delta}_j)},$$

and $\hat{\sigma}^2(\hat{\delta}_i) = 1/\mathbf{e}'\hat{\mathbf{\Lambda}}^{(i)}\mathbf{e}$ is the large sample estimated variance of $\hat{\delta}_i$.

In large samples, the distribution of $\hat{\delta}_+$ is approximately normal with a mean of δ^* and a variance given by

$$\sigma^2(\hat{\delta}_+) = \left(\sum_{i=1}^{k} \frac{1}{\sigma_\infty^2(\hat{\delta}_i)}\right)^{-1}, \tag{13}$$

which is estimated by

$$\hat{\sigma}^2(\hat{\delta}_+) = \left(\sum_{i=1}^{k} \frac{1}{\hat{\sigma}^2(\hat{\delta}_i)}\right)^{-1}. \tag{14}$$

The large sample approximation to the distribution of $\hat{\delta}_+$ can be used to construct confidence intervals for δ^* or to test hypotheses about δ^* (such as $\delta^* = 0$) in the same way that the large sample normal approximation to the distribution of the weighted mean d_+ was used in Chapter 6.

Example

Return to the effect size estimate calculated in the example of Section C.2. We now use the pooled estimate derived in that example and combine it with the pooled estimates of effect size derived from two other experiments. The pooled estimates $\hat{\delta}_i$ from each experiment are listed in Table 3 along with their variances $\hat{\sigma}^2(\hat{\delta}_i)$, $1/\hat{\sigma}^2(\hat{\delta}_i)$, $\hat{w}_i$, and $\hat{w}_i\hat{\delta}_i$. The pooled estimate of δ is

$$\hat{\delta}_+ = (0.299)(-0.220) + (0.329)(-0.162) + (0.372)(0.103) = -0.081.$$

The estimated variance of $\hat{\delta}_+$ is

$$\hat{\sigma}^2(\hat{\delta}_+) = (22.173 + 24.331 + 27.472)^{-1} = 0.0135,$$

TABLE 3

Pooled Effect Size Estimates from Three Studies of Reading Achievement

Study	n^E	n^C	$\hat{\delta}_i$	$\hat{\sigma}^2(\hat{\delta}_i)$	$1/\hat{\sigma}^2(\hat{\delta}_i)$	$\check{w}_i$	$\check{w}_i\hat{\delta}_i$
1	40	40	−0.220	0.0453	22.075	0.299	−0.066
2	45	45	−0.162	0.0411	24.331	0.329	−0.053
3	55	55	0.103	0.0364	27.472	0.372	0.038
					73.878	1.000	−0.081

and a 95-percent confidence interval for δ^* is $-0.081 \pm 1.96\sqrt{0.0135}$ or

$$-0.309 \leq \delta^* \leq 0.147.$$

Because this confidence interval contains zero we cannot reject the hypothesis that $\delta = 0$ at the $\alpha = 0.05$ level of significance.

D.2 TESTING HOMOGENEITY OF EFFECT SIZES

Before pooling estimates of effect sizes, across experiments, it is wise to investigate whether one underlying parameter is being estimated. In Chapter 5 we provided a test of homogeneity of effect sizes for *independent* experiments, and we now give an analogous test for the multivariate version when there are *correlated* effect sizes.

The test of homogeneity is based on the test statistic

$$Q_\delta = \sum_{i=1}^{k} \frac{(\hat{\delta}_i - \hat{\delta}_+)^2}{\hat{\sigma}^2(\hat{\delta}_i)}, \tag{15}$$

where $\hat{\delta}_+$ is the weighted mean effect size estimate given in (12), and $\hat{\sigma}^2(\hat{\delta}_i)$ is the estimated large sample variance of $\hat{\delta}_i$ given in (11).

If $\delta_1 = \cdots = \delta_k$ and each of the studies has a large sample size, then Q_δ has a chi-square distribution with $k - 1$ degrees of freedom. The test of homogeneity of effect sizes proceeds in the same way as the test of homogeneity of effect size given in Chapter 5; that is, large values of Q_δ lead to rejection of the hypothesis that $\delta_1 = \cdots = \delta_k$.

Example

In the example of Section D.1 we calculated a weighted average of pooled effect size estimates from three studies of reading achievement given in Table 3. In that section we calculated the weighted mean effect size as $\hat{\delta}_+ = -0.081$. To test whether $\hat{\delta}_1 = \hat{\delta}_2 = \hat{\delta}_3$ we calculate the test statistic

$$Q_\delta = \frac{(-0.220 + 0.081)^2}{0.0453} + \frac{(-0.162 + 0.081)^2}{0.0411} + \frac{(0.103 + 0.081)^2}{0.0364} = 1.52.$$

Comparing the obtained value 1.52 with the percentage point of the chi-square distribution with two degrees of freedom, we see that a value as large or larger would occur nearly 50 percent of the time if $\hat{\delta}_1 = \hat{\delta}_2 = \hat{\delta}_3$. Consequently we conclude that the estimates from the three experiments are estimating a common effect size.

E. ESTIMATING A VECTOR OF EFFECT SIZES

Sometimes studies measure several outcomes reflecting different constructs. For example, studies of the effectiveness of a teaching method examine academic achievement as well as attitude toward school. In such a situation there is little reason to believe that effect sizes on the different constructs are identical, so we need to estimate a vector of effect sizes (one for each construct) from the series of studies. In this section we show how to estimate a vector of correlated effect sizes from a series of independent experiments when each experiment produces a vector of estimates of effect size for each construct.

Let $\mathbf{d}_1, \ldots, \mathbf{d}_k$ denote k vectors of effect size estimates (2), and let $\boldsymbol{\delta}_i$ denote the vector of effect sizes. We assume that $\boldsymbol{\delta}_1 = \boldsymbol{\delta}_2 = \cdots = \boldsymbol{\delta}_k = \boldsymbol{\delta}^*$. That is, the effect size vectors are homogeneous across experiments. For large samples the approximate distribution of $\mathbf{d}_i$ is multivariate normal with a mean vector $\boldsymbol{\delta}_i$ and covariance matrix $\boldsymbol{\Sigma}^{(i)}$ given by (4) and (5), $i = 1, \ldots, k$.

A reasonable assumption for this model is that the difference between the covariance matrices arises only from differences in the variances, so that the correlations are constant from study to study. More specifically, the covariance matrix for the ith study is

$$\boldsymbol{\Sigma}^{(i)} = \mathbf{D}_i'\mathbf{P}\mathbf{D}_i,$$

where

$$\mathbf{D}_i = \operatorname{Diag}[\sigma_\infty(d_{i1}), \ldots, \sigma_\infty(d_{ip})]$$

is a diagonal matrix of standard deviations, and $\mathbf{P}$ is the matrix of correlations of the original observations. That is, we require that the correlations among components of the vector of estimates of effect size be the same for each study.

This requirement is equivalent to the requirement that the correlation matrices among the observations on each construct be the same across studies. This is quite plausible because the correlations among components of the observation vectors should be determined by the relationship among the underlying constructs. Although different studies may use different operationalizations of those constructs (psychological tests), the particular operationalization should not affect the correlation structure. Hence, if the same constructs are being measured across studies, we would expect the correlations of observations among effect sizes to be the same across studies.

We construct an estimator $\mathbf{d}_+$ of the effect size vector $\boldsymbol{\delta}$ from the weighted average components of effect size estimates, namely,

$$\mathbf{d}_+ = (d_{+1}, \ldots, d_{+p})',$$

where d_{+m} is the weighted average effect size over all estimates of the mth component:

$$d_{+m} = d_{1m}w_{1m} + \cdots + d_{km}w_{km}, \tag{16}$$

the weights are

$$w_{im} = \frac{1}{\hat{\sigma}^2(d_{im})} \Bigg/ \sum_{j=1}^{k} \frac{1}{\hat{\sigma}^2(d_{jm})}, \tag{17}$$

and $\hat{\sigma}^2(d_{im})$ is given by (4), with d_{im} replacing δ_{im}.

The asymptotic distribution of $\mathbf{d}_+$ leads to a simple large sample approximation to the distribution of $\mathbf{d}_+$. When the hypothesis

$$\boldsymbol{\delta}_1 = \cdots = \boldsymbol{\delta}_k = \boldsymbol{\delta}^*$$

holds, and when the sample sizes of all of the studies are large, then $\mathbf{d}_+$ is approximately multivariate normally distributed with mean vector $\boldsymbol{\delta}^*$ and asymptotic covariance matrix $\boldsymbol{\Sigma} = (\sigma_{lm})$ given by

$$\sigma_{ll} = \left(\sum_{i=1}^{k} \frac{1}{\sigma_\infty^2(d_{il})} \right)^{-1}, \qquad \sigma_{lm} = \rho_{lm}\sqrt{\sigma_{ll}\sigma_{mm}}, \tag{18}$$

where ρ_{lm} is an element of $\mathbf{P}$, the common correlation matrix of the effect sizes.

This large sample normal approximation can be used [by inserting into (18) estimates $\hat{\sigma}^2(d_{il})$ for $\sigma_\infty^2(d_{il})$] to obtain confidence intervals for components of the vector $\boldsymbol{\delta}^*$ or for significance tests of the components of $\boldsymbol{\delta}^*$. Making use of the Bonferroni inequality, we can also obtain $100(1 - \alpha)$-percent simultaneous confidence intervals for q of the p components. This is done by using the $100(1 - \alpha/2q)$ percentage point of the standard normal distribution in place of the $100(1 - \alpha/2)$ percentage point.

Example

Four studies examine the effects of open education on attitude toward school and on reading achievement. The attitude and achievement measures are derived from the same subjects and the correlation between the particular measures used is $\rho = 0.63$. The vectors of attitude and achievement effect size estimates for the four studies are

$$\mathbf{d}_1 = \begin{bmatrix} 0.458 \\ 0.100 \end{bmatrix}, \quad \mathbf{d}_2 = \begin{bmatrix} 0.363 \\ 0.241 \end{bmatrix}, \quad \mathbf{d}_3 = \begin{bmatrix} 0.162 \\ -0.121 \end{bmatrix}, \quad \mathbf{d}_4 = \begin{bmatrix} 0.294 \\ 0.037 \end{bmatrix},$$

and the estimated covariance matrices for the four studies are

$$\hat{\boldsymbol{\Sigma}}^{(1)} = \begin{bmatrix} 0.0513 & 0.0319 \\ 0.0319 & 0.0501 \end{bmatrix}, \qquad \hat{\boldsymbol{\Sigma}}^{(2)} = \begin{bmatrix} 0.0354 & 0.0222 \\ 0.0222 & 0.0351 \end{bmatrix},$$

$$\hat{\boldsymbol{\Sigma}}^{(3)} = \begin{bmatrix} 0.0546 & 0.0344 \\ 0.0344 & 0.0545 \end{bmatrix}, \qquad \hat{\boldsymbol{\Sigma}}^{(4)} = \begin{bmatrix} 0.0286 & 0.0179 \\ 0.0179 & 0.0286 \end{bmatrix}.$$

The estimate $\mathbf{d}_+$ of the effect size vector $\boldsymbol{\delta}^*$ is $\mathbf{d}_+ = (d_{+1}, d_{+2})$, where d_{+1} and d_{+2} are the weighted average effect sizes calculated on each variable separately. The values of d_{ij}, $\hat{\sigma}^2(d_{ij})$, $1/\hat{\sigma}^2(d_{ij})$, w_{ij}, and $w_{ij}d_{ij}$ used to calculate d_{+1} and d_{+2} are given in Table 4, from which we obtain the estimate

$$\mathbf{d}_+ = (0.321, 0.078)',$$

and the variances $\hat{\sigma}^2(d_{+1}) = 1/101.022 = 0.0099$ and $\hat{\sigma}(d_{+2}) = 1/102.160 = 0.0098$. Using the Bonferroni method, we obtain 95-percent simultaneous confidence intervals for δ_1^* and δ_2^* as $0.321 \pm 2.24\sqrt{0.0099}$ and $0.078 \pm 2.24\sqrt{0.0098}$ or

$$0.098 \leq \delta_1^* \leq 0.544, \qquad -0.144 \leq \delta_2^* \leq 0.300.$$

Since the confidence interval for δ_1^* does not contain zero, we reject the hypothesis that $\delta_1^* = 0$ at the $\alpha = 0.05$ level of signficance. We do not reject the

TABLE 4

Effect Size Estimates from Four Studies of (a) Attitude toward School and (b) Reading Achievement

Study	n^E	n^C	d	$\hat{\sigma}^2(d)$	$1/\hat{\sigma}^2(d)$	w	wd
			(a) Attitude toward school				
1	40	40	0.458	0.0513	19.489	0.193	0.088
2	60	55	0.363	0.0354	28.231	0.280	0.102
3	34	40	0.162	0.0546	18.319	0.181	0.029
4	79	64	0.294	0.0286	34.983	0.346	0.102
					101.022	1.000	0.321
			(b) Reading achievement				
1	40	40	0.100	0.0501	19.975	0.195	0.020
2	60	55	0.241	0.0351	28.489	0.279	0.067
3	34	40	−0.121	0.0545	18.345	0.180	−0.022
4	79	64	0.037	0.0283	35.351	0.346	0.013
					102.160	1.000	0.078

hypothesis that $\delta_{2\cdot}^{*} = 0$, since the corresponding confidence interval contains zero.

F. TESTING HOMOGENEITY OF VECTORS OF EFFECT SIZES

In Section E we presented methods for estimating a vector $\boldsymbol{\delta}$ of effect sizes from k studies each of which produces an estimate $\mathbf{d}_i$ of $\boldsymbol{\delta}$. However, before estimating the vector $\boldsymbol{\delta}$ it is important to determine whether the estimates $\mathbf{d}_1, \ldots, \mathbf{d}_k$ are reasonably consistent with the model that all studies share a common underlying vector of effect sizes.

A simple large sample test for the equality of $\boldsymbol{\delta}_1, \boldsymbol{\delta}_2, \ldots, \boldsymbol{\delta}_k$ can be derived from the asymptotic distributions of the estimators $\mathbf{d}_1, \ldots, \mathbf{d}_k$. To carry out the test define the column vector of dimension kp by

$$\mathbf{d}_* = (\mathbf{d}_1, \ldots, \mathbf{d}_k)'$$

and the $kp \times kp$ large sample estimate $\hat{\boldsymbol{\Sigma}}^*$ of the covariance matrix of $\mathbf{d}_*$ by

$$\hat{\boldsymbol{\Sigma}}^* = \mathrm{Diag}(\hat{\boldsymbol{\Sigma}}^{(1)}, \ldots, \hat{\boldsymbol{\Sigma}}^{(k)}),$$

where $\hat{\boldsymbol{\Sigma}}^{(1)}, \ldots, \hat{\boldsymbol{\Sigma}}^{(k)}$ are the large sample estimates of the covariance matrices of $\mathbf{d}_1, \ldots, \mathbf{d}_k$ defined by (4) and (5).

A large sample test of the hypothesis that

$$\boldsymbol{\delta}_1 = \cdots = \boldsymbol{\delta}_k$$

is based on the statistic

$$Q_* = \mathbf{d}_*'\mathbf{C}\mathbf{d}_*, \tag{19}$$

where

$$\mathbf{C} = \hat{\boldsymbol{\Lambda}}^* - \hat{\boldsymbol{\Lambda}}^*\mathbf{e}\mathbf{e}'\hat{\boldsymbol{\Lambda}}^*/\mathbf{e}'\hat{\boldsymbol{\Lambda}}^*\mathbf{e}, \tag{20}$$

$\hat{\boldsymbol{\Lambda}}^*$ is the inverse of $\hat{\boldsymbol{\Sigma}}^*$, and $\mathbf{e}$ is a column vector of kp ones.

If $\boldsymbol{\delta}_1 = \cdots = \boldsymbol{\delta}_k$ and the sample sizes in all of the studies are reasonably large, then Q_* has a chi-square distribution with $p(k - 1)$ degrees of freedom. The test of homogeneity of the $\boldsymbol{\delta}_i$ using Q_* is completely analogous to the test of homogeneity of univariate effect sizes given in Chapter 5.

Example

Return to the four studies of the effects of open education on attitude toward school and reading achievement described in Section E and summarized in Table 4. In these data $\mathbf{d}_*$ is given by

$$\mathbf{d}_* = (0.458, 0.100, 0.363, 0.241, 0.162, -0.121, 0.294, 0.037)',$$

and the estimated covariance matrices $\hat{\mathbf{\Sigma}}^{(i)}$ are $\mathbf{D}_i\mathbf{R}\mathbf{D}_i$, where

$$\mathbf{R} = \begin{bmatrix} 1.00 & 0.63 \\ 0.63 & 1.00 \end{bmatrix}$$

and

$$\mathbf{D}_1 = \text{Diag}(\sqrt{0.0513}, \sqrt{0.0501}), \qquad \mathbf{D}_2 = \text{Diag}(\sqrt{0.0354}, \sqrt{0.0351}),$$
$$\mathbf{D}_3 = \text{Diag}(\sqrt{0.0546}, \sqrt{0.0545}), \qquad \mathbf{D}_4 = \text{Diag}(\sqrt{0.0286}, \sqrt{0.0283}),$$

which gives

$$\hat{\mathbf{\Sigma}}^{(1)} = \begin{bmatrix} 0.0513 & 0.0319 \\ 0.0319 & 0.0501 \end{bmatrix}, \qquad \hat{\mathbf{\Sigma}}^{(2)} = \begin{bmatrix} 0.0354 & 0.0222 \\ 0.0222 & 0.0351 \end{bmatrix},$$

$$\hat{\mathbf{\Sigma}}^{(3)} = \begin{bmatrix} 0.0546 & 0.0344 \\ 0.0344 & 0.0545 \end{bmatrix}, \qquad \hat{\mathbf{\Sigma}}^{(4)} = \begin{bmatrix} 0.0286 & 0.0179 \\ 0.0179 & 0.0283 \end{bmatrix}.$$

We calculate $\hat{\mathbf{\Sigma}}^* = \text{Diag}(\hat{\mathbf{\Sigma}}^{(1)}, \hat{\mathbf{\Sigma}}^{(2)}, \hat{\mathbf{\Sigma}}^{(3)}, \hat{\mathbf{\Sigma}}^{(4)})$, $\hat{\mathbf{\Lambda}}^*$ is the inverse of $\hat{\mathbf{\Sigma}}^*$, and the value of $\mathbf{C} = \hat{\mathbf{\Lambda}}^* - \hat{\mathbf{\Lambda}}^*\mathbf{e}\mathbf{e}'\hat{\mathbf{\Lambda}}^*/\mathbf{e}'\hat{\mathbf{\Lambda}}^*\mathbf{e}$ is

$$\mathbf{C} = \begin{bmatrix} 31.219 & -21.781 & -1.616 & -1.666 & -1.056 & -1.061 & -2.019 & -2.019 \\ -21.781 & 31.842 & -1.722 & -1.776 & -1.256 & -1.131 & -2.153 & -2.153 \\ -1.616 & -1.722 & 44.472 & -32.117 & -1.547 & -1.555 & -2.958 & -2.958 \\ -1.666 & -1.776 & -32.117 & 44.858 & -1.595 & -1.603 & -3.050 & -3.050 \\ -1.056 & -1.126 & -1.550 & -1.595 & 29.357 & -20.166 & -1.934 & -1.934 \\ -1.061 & -1.131 & -1.555 & -1.603 & -20.166 & 29.403 & -1.943 & -1.943 \\ -2.019 & -2.153 & -2.958 & -3.050 & -1.934 & -1.943 & 54.278 & -40.222 \\ -2.019 & -2.153 & -2.958 & -3.050 & -1.934 & -1.943 & -40.222 & 54.278 \end{bmatrix}$$

Comparing $Q_* = \mathbf{d}_*'\mathbf{C}\mathbf{d}_* = 7.827$ with the percentage points of the chi-square distribution with $(2)(4)-2 = 6$ degrees of freedom, we see that values as large as this would occur between 25 and 50 percent of the time when $\boldsymbol{\delta}_1 = \boldsymbol{\delta}_2 = \boldsymbol{\delta}_3 = \boldsymbol{\delta}_4$. Thus we conclude that the effect sizes on attitude toward school and on reading achievement are homogeneous for the four studies in this example.

G. IS POOLING OF CORRELATED ESTIMATORS NECESSARY?

In Sections C and D of this chapter methods were presented for utilizing correlated effect sizes that arise when a study uses several measures of the same underlying dependent variable. The rationale for these methods is that effect size estimates calculated using each of the measures contain some information about a putative population effect size. We combine all of the available effect size estimates using the methods of Section C to extract all of the information (about the common effect size) that is provided by the effect size estimates. By pooling the (correlated) effect size estimates we obtain a more precise estimate of the common underlying (population) effect size.

It is important to recognize that pooling of correlated estimates does not increase the precision of the combined estimates as dramatically as does pooling of independent estimates. Correlated estimates contain "overlapping information." The higher the correlation, the less "independent information" is contained in each of the estimates. Thus if the correlations are high, the pooled estimate may be a little more precise than any one of the estimators before pooling. The example given in Section C.2 illustrates this point. Although three correlated effect size estimates were pooled, the variance of the pooled estimate was only about 10 percent less than that of the individual estimates (0.045 versus 0.050).

The exact relationship between the variance of the pooled estimator and the correlation structure of the estimates was described in Section C.2, but the formulas given in that section provide little direct insight about the variance of the pooled estimate. One simple special case that does provide some insight about the situation is when two correlated estimates are pooled. In this case the formula for the large sample variance of $\hat{\delta}_i$ reduces to

$$\sigma^2(\hat{\delta}_i) = \frac{1 + \rho}{2} \sigma^2(d_{i1}),$$

where ρ is the correlation between the observations on the two measures of the dependent variable and $\sigma^2(d_{i1})$ is the sampling variance of d_{i1}. [Note that $\delta_{i1} = \delta_{i2}$ implies that $\sigma^2(d_{i1}) = \sigma^2(d_{i2})$.] Consequently the relative efficiency of the pooled estimator $\hat{\delta}_i$ versus d_{i1} is

$$\sigma^2(\hat{\delta}_i)/\sigma^2(d_{i1}) = (1 + \rho)/2.$$

Pooling of effect size estimates is sensible only when the outcome measures clearly reflect the same construct. Alternative measures of the same construct will tend to be highly correlated. Thus pooling of correlated effect size estimates is reasonable only when the estimates are highly correlated. But this is exactly the situation in which pooling results is only a small gain in efficiency. Thus pooling of correlated effect size estimates will produce only modest gains in efficiency. For example, two psychological tests that measure the same cognitive construct will often have a correlation greater than 0.70 and often the correlation will be 0.80 or greater. In such cases pooling of two estimators will increase the efficiency of estimation by no more than 10–15 percent.

In most cases, the gain in efficiency resulting from pooling of correlated estimates does not justify the effort required. Little information is lost by choosing a particular measure a priori and using only effect size estimates calculated from observations on that measure. The statistical analysis can then proceed using only independent estimators of effect size. This suggested procedure obviates the need for information on the correlation structure of the effect size estimates. In rare cases when the small gains in efficiency that can be achieved by pooling correlated estimators are necessary, the methods suggested in Sections C and D of this chapter provide valid albeit more complicated analyses.

CHAPTER 11

Combining Estimates of Correlation Coefficients

In many areas of educational and psychological research, it is useful to assess the relationship between continuous variables. Correlation coefficients have been used extensively as an index of the relationship between two normally distributed variables. Since the correlation coefficient is a scale-free measure of the relationship between variables, it is invariant under substitution of different but linearly equatable measures of the same construct. The correlation coefficient is therefore a natural candidate as an index of effect magnitude suitable for cumulation across studies.

Practitioners of research synthesis have often used the correlation coefficient as an index of effect magnitude. In some cases the researchers sought an estimate of the average correlation and calculated the average of the correlations from each study (e.g., Bloom, 1964; Cohen, 1981). In other cases investigators were interested in examining the relationship between coded characteristics of studies and the magnitude of the correlation. For example, Uguroglu and Walberg (1979) used ordinary multiple regression to study the effects of grade level, average self-concept, and average IQ on the correlation of motivation and achievement in 40 studies.

Although the regression method is appealing, there are some problems with the procedure. First, the assumptions of regression analysis are not met, since the variance of a sample correlation is inversely proportional to the sample size of the study. Moreover, the variance of a sample correlation is strongly dependent on the population correlation. Thus unless the studies to be integrated all have the same sample size and the same population correlation, the individual "error" variances can be dramatically different. Second, even if the regression coefficients are correctly estimated, the conventional regression method will not give an indication of the goodness of fit or specification of the regression model. Tests of goodness of fit are crucial to the interpretation of even simple models, such as the model that uses an average of the sample correlations to estimate a common population correlation.

Methods for estimating a correlation coefficient from a single sample are reviewed in Section A, and the effects of measurement error are discussed in Section B. The formal model and notation used in subsequent sections are included in Section C along with methods for estimating a common correlation from several studies. Tests for homogeneity of correlations across studies are given in Section D. A method developed by Hedges and Olkin (1983b) for fitting linear models to correlations from a series of studies is given in Section E together with a test for model specification. An analogue to the random effects analysis of variance for correlations is given in Section F.

A. ESTIMATING A CORRELATION FROM A SINGLE STUDY

In the present section we review the standard theory for point and interval estimation of correlation coefficients. The correlation coefficient has been extensively studied in the last 75 years, and this review is not an exhaustive coverage of all the work on correlations. Instead we attempt to present the material that is most relevant to the problem of combining estimates of correlations. Section A.1 treats the problem of point estimation of a correlation. The asymptotic distributions of estimators of the correlation are given in Section A.2 along with some transformations that are useful in obtaining confidence intervals. A nomograph for obtaining exact confidence intervals for ρ is discussed in Section A.3.

A.1 POINT ESTIMATION OF A CORRELATION FROM A SINGLE STUDY

We first set forth some notation. Suppose that a study contains n pairs of measurements $(u_1, v_1), \ldots, (u_n, v_n)$ on variables U and V that have a bivariate normal distribution with correlation

$$\rho = \rho(U, V) = \mathrm{Cov}(U, V)/\sqrt{\mathrm{Var}(U)\,\mathrm{Var}(V)}. \tag{1}$$

The sample product–moment correlation is

$$r = \sum_{i=1}^{n} (u_i - \bar{u})(v_i - \bar{v}) \Big/ \sqrt{\sum_{i=1}^{n} (u_i - \bar{u})^2 \sum_{i=1}^{n} (v_i - \bar{v})^2}, \quad (2)$$

where $\bar{u}$ and $\bar{v}$ are the means of $u_1, \ldots, u_n$ and $v_1, \ldots, v_n$, respectively.

The sample product-moment correlation r was proposed by Karl Pearson (1896) as an estimator of the population correlation ρ. The exact distribution of the sample correlation under the assumption of bivariate normality was first derived by R. A. Fisher (1915), who obtained the distribution in a rather complicated form. The distribution was later studied and tabulated (David, 1938). Simpler forms of the distribution that are more suitable for computation were given by Hotelling (1953). Even the simplest forms of the distribution are complicated, and we do not give them here. Examination of the sampling distribution of the correlation reveals that r is the maximum likelihood estimator of ρ, but r is not an unbiased estimator of ρ. The exact mean and bias of r are obtained only as infinite series, but an approximate mean value of r (to order $1/n$) is given by

$$E(r) \cong \rho - \rho(1 - \rho^2)/2n,$$

so that the bias of r as an estimator of ρ is approximately

$$\text{Bias}(r) \cong -\rho(1 - \rho^2)/2n.$$

Thus the sample correlation r tends to underestimate the absolute magnitude of the population correlation ρ. The sampling variance of r is also obtained as an infinite series, but the approximate variance of r (to order $1/n$) is given by

$$\text{Var}(r) \cong (1 - \rho^2)^2/n.$$

If the sample size n is moderately large (say, over 15), the bias of the sample correlation r is seldom of practical concern. Even for smaller samples, if the true correlation is close to 0 or ± 1, the bias is negligible. If a study with a small sample size is used to estimate a correlation, and the true correlation is in the range of 0.4 to 0.6, then the bias can be a serious concern. It would therefore be desirable to have an unbiased estimator $G(r)$ of ρ. Such an unbiased estimator exists in the form of an infinite series. An approximation $\tilde{G}(r)$ to the exact unbiased estimator $G(r)$ is

$$\tilde{G}(r) \cong r + r(1 - r^2)/2(n - 3), \quad (3)$$

which is accurate to within 0.01 if $n \geq 8$ and to within 0.001 if $n \geq 18$. Exact values of the unbiased estimator $G(r)$ as a function of r for various sample sizes n are given in Table 1. From this we see that when $r = 0.500$ and $n = 10$, the unbiased estimator is 0.525. (The derivation of the unbiased estimator is given in Olkin and Pratt, 1958.)

TABLE 1

Table for Converting r to the Minimum Variance Unbiased Estimator $G(r)$ of the Ordinary Bivariate Correlation Coefficient, for n Degrees of Freedom

	r										
n	0	0.1	0.2	0.3	0.4	0.5	0.6	0.7	0.8	0.9	1.0
4	0	0.148	0.280	0.398	0.506	0.605	0.695	0.780	0.858	0.931	1
6	0	0.117	0.232	0.343	0.450	0.552	0.650	0.744	0.833	0.918	1
8	0	0.110	0.220	0.327	0.432	0.534	0.633	0.730	0.823	0.913	1
10	0	0.107	0.214	0.319	0.423	0.525	0.625	0.722	0.817	0.910	1
12	0	0.106	0.211	0.315	0.418	0.520	0.620	0.718	0.814	0.908	1
14	0	0.105	0.209	0.312	0.415	0.516	0.616	0.715	0.812	0.907	1
16	0	0.104	0.207	0.311	0.413	0.514	0.614	0.713	0.810	0.906	1
18	0	0.103	0.206	0.309	0.411	0.512	0.612	0.711	0.809	0.905	1
20	0	0.103	0.206	0.308	0.410	0.511	0.611	0.710	0.808	0.905	1
22	0	0.103	0.205	0.307	0.409	0.510	0.610	0.709	0.807	0.904	1
24	0	0.102	0.205	0.307	0.408	0.509	0.609	0.708	0.806	0.904	1
26	0	0.102	0.204	0.306	0.407	0.508	0.608	0.707	0.806	0.903	1
28	0	0.102	0.204	0.305	0.407	0.507	0.607	0.707	0.805	0.903	1
30	0	0.102	0.204	0.305	0.406	0.507	0.607	0.706	0.805	0.903	1
∞	0	0.1	0.2	0.3	0.4	0.5	0.6	0.7	0.8	0.9	1

The unbiased estimator $G(r)$ has the same range as r of -1 to $+1$, and differs from r by terms of order $1/n$. Consequently, it will have the same asymptotic distribution as r. The unbiased estimator $G(r)$ has the form $G(r) = r[1 + C(r)]$, where $C(r)$ is positive unless $r = 0$. Simulation studies indicate that the unbiased estimator has a larger variance than r in general, but a smaller mean-squared error for most practical values of ρ and n. Consequently, the unbiased estimator of the correlation coefficient, unlike the unbiased estimator of effect size, may be *less* precise (in terms of variance) than the usual biased estimator. However, when the bias is also taken into account, it is generally more precise (in terms of mean-squared error).

A.2 APPROXIMATIONS TO THE DISTRIBUTION OF THE SAMPLE CORRELATION COEFFICIENT

The large sample distribution of a sample correlation r is normal with mean ρ and variance $(1 - \rho^2)^2/n$, or symbolically,

$$\sqrt{n}(r - \rho) \sim \mathcal{N}(0, (1 - \rho^2)).$$

Unfortunately, the variance of r in the large sample approximation depends strongly on ρ, the unknown true value of the correlation. Even if the population correlation ρ is known, the approximate distribution is not very accurate. Hence it is not very useful unless sample sizes are quite large (e.g., n is several hundred).

In order to normalize the distribution of r and to make the variance independent of ρ, Fisher (1921) proposed the z-transformation

$$z \equiv z(r) = \frac{1}{2} \log \frac{1 + r}{1 - r}. \tag{4}$$

Values of z for given r are given in Appendix C. The corresponding transformation for ρ is

$$\zeta = \frac{1}{2} \log \frac{1 + \rho}{1 - \rho}. \tag{5}$$

The z-transformation stabilizes the variance in the sense that when n is large, z is approximately normally distributed with mean ζ and variance $1/n$. A more accurate approximation (for moderate values of n) to the distribution of z is obtained by using the asymptotic variance $1/(n - 3)$ instead of $1/n$. Consequently $\sqrt{n - 3}\, z$ has, approximately, the standard normal distribution:

$$\sqrt{n - 3}(z - \zeta) \sim \mathcal{N}(0, 1). \tag{6}$$

It is equally important to note that the large sample normal approximation to the distribution of z is quite accurate even for relatively modest sample sizes.

The large sample normal approximation of z can be used to obtain an approximate confidence interval for ρ by first obtaining a confidence interval for ζ and then transforming back. An approximate $100(1 - \alpha)$-percent confidence interval (ζ_L, ζ_U) for ζ is given by

$$\zeta_L = z - C_{\alpha/2}/\sqrt{n - 3}, \qquad \zeta_U = z + C_{\alpha/2}/\sqrt{n - 3},$$

where $C_{\alpha/2}$ is the two-tailed critical value of the standard normal distribution. The corresponding confidence interval (ρ_L, ρ_U) for ρ is therefore given by

$$\rho_L = z^{-1}(\zeta_L), \qquad \rho_U = z^{-1}(\zeta_U), \tag{7}$$

where $z^{-1}(x)$ is the inverse function

$$z^{-1}(x) = (e^{2x} - 1)/(e^{2x} + 1). \tag{8}$$

Approximate confidence intervals given by (7) are quite accurate unless the sample size is very small. For ease of computation, tables for transforming r to z and z to r are given in Appendix C, and are also available in most books of statistical tables. An alternative approximation that is even more accurate in very small samples was introduced by Kraemer (1974), (1975):

$$t = \sqrt{n-2}(r-\rho)/\sqrt{(1-r^2)(1-\rho^2)}, \tag{9}$$

which has an approximate Student's t-distribution with $n-2$ degrees of freedom.

Example

Suppose that a study with a sample size of $n = 50$ yields a correlation coefficient of $r = 0.400$. To obtain a 95-percent confidence interval via Fisher's z-transform, use (4) or Appendix C to transform r to $z = 0.424$. A 95-percent confidence interval for ζ is given by $0.424 \pm 1.96/\sqrt{47}$, so that

$$\zeta_{\mathrm{L}} = 0.138, \qquad \zeta_{\mathrm{U}} = 0.710.$$

Then obtain the confidence limits for ρ via (7) or Appendix C:

$$\rho_{\mathrm{L}} = 0.137, \qquad \rho_{\mathrm{U}} = 0.611.$$

A.3 EXACT CONFIDENCE INTERVALS FOR CORRELATIONS

In the case of bivariate normality exact two-sided confidence intervals for ρ can be obtained from nomographs provided by David (1938). The exact confidence limits are charted for sample sizes 3–8, 10, 12, 15, 20, 25, 50, 100, 200, and 400, for confidence coefficients of 0.90, 0.95, 0.98, and 0.99. Appendix G gives nomographs for these confidence coefficients. The value of the sample correlation is read off the abscissa, and this value is followed vertically until the bands labeled with the appropriate sample size are intersected. The ordinate values of the intersections are the confidence limits of an appropriate confidence interval for ρ given r. For example, if $r = 0.65$ and $n = 40$, we find by interpolation in Appendix G that the 95-percent confidence interval is (0.43, 0.80).

B. THE EFFECTS OF MEASUREMENT ERROR

It is well known that measurement error has the effect of attenuating correlations. Specifically, suppose that, instead of measuring the variables U and V, we measure

$$X = U + \eta, \qquad Y = V + \xi,$$

where η and ξ are measurement errors. If we assume (as in classical test theory) that η and ξ are independently normally distributed with zero mean and that η and ξ are independent of U and V, then it is easy to calculate the relationship between the correlation $\rho(U, V)$ of U and V and the correlation $\rho(X, Y)$ of X and Y.

A straightforward computation shows that

$$\rho(X, Y) = \rho(U, V)\sqrt{\rho(X, X')}\sqrt{\rho(Y, Y')},$$

where

$$\rho(X, X') = \sigma_u^2/(\sigma_u^2 + \sigma_\eta^2), \qquad \rho(Y, Y') = \sigma_v^2/(\sigma_v^2 + \sigma_\xi^2)$$

are the *reliabilities* of the measures X and Y, respectively (see, e.g., Lord & Novick, 1968).

If the reliabilities of X and Y are known, then the disattenuated correlation

$$\hat{\rho}(U, V) = r_{xy}/\sqrt{\rho(X, X')\rho(Y, Y')}$$

is an estimator of $\rho(U, V)$. Another effect of this correction is that it increases the variance of the estimator r_{xy} by a factor of $1/\rho(X, X')\rho(Y, Y')$, so that the methods described in this chapter can be used with disattenuated correlations provided the variance of each sample correlation is increased by a factor of $1/\sqrt{\rho(X, X')\rho(Y, Y')}$.

C. ESTIMATING A COMMON CORRELATION FROM SEVERAL STUDIES

Suppose that we have a series of k studies each of which produces a product-moment correlation coefficient. Specifically, suppose that the ith study consists of pairs of measurements U_i, V_i that have a bivariate normal distribution with correlations

$$\rho_i = \mathrm{Cov}(U_i, V_i)/\sqrt{\mathrm{Var}(U_i)\,\mathrm{Var}(V_i)}, \qquad i = 1, \ldots, k. \tag{10}$$

Samples of size $n_1, \ldots, n_k$ are taken from the k studies. If u_{ij} and v_{ij} are the jth observations in the ith study, then the observations for the k studies can be arranged in the following display.

	Study			
	1	2	$\cdots$	k
Observations	(u_{11}, v_{11})	(u_{21}, v_{21})	$\cdots$	(u_{k1}, v_{k1})
	$\vdots$	$\vdots$		$\vdots$
	(u_{1n_1}, v_{1n_1})	(u_{2n_2}, v_{2n_2})	$\cdots$	(u_{kn_k}, v_{kn_k})
Means	$(\bar{u}_1, \bar{v}_1)$	$(\bar{u}_2, \bar{v}_2)$	$\cdots$	$(\bar{u}_k, \bar{v}_k)$
Correlations	r_1	r_2	$\cdots$	r_k

The sample product–moment correlation r_i for the ith study is

$$r_i = \sum_{j=1}^{n_i} (u_{ij} - \bar{u}_i)(v_{ij} - \bar{v}_i) \Bigg/ \sqrt{\sum_{j=1}^{n_i} (v_{ij} - \bar{v}_i)^2 \sum_{j=1}^{n_i} (u_{ij} - \bar{u}_i)^2}, \qquad i = 1, \ldots, k. \tag{11}$$

The statistical literature contains results for estimating correlations under various assumptions about the parameters. For example, the means of (U_i, V_i) might be known or the variances of U_i and V_i might be assumed to be equal or known. Such assumptions would change the results presented below. We choose not to study such restrictions in the case of meta-analysis because they occur rarely in practice. (For a discussion of classical procedures for estimating and testing correlations under restrictions see Olkin & Siotani, 1965, Olkin, 1967, and Olkin & Sylvan, 1977.)

We now consider estimating a common underlying population correlation ρ based on independent estimates from several samples. Suppose that

$$\rho_1 = \rho_2 = \cdots = \rho_k = \rho$$

holds. How should we estimate the common ρ based on k sample correlations $r_1, r_2, \ldots, r_k$? One simple method for obtaining an estimate of ρ is to use a weighted combination of the estimators from each study. Variations of this method are presented in Section C.1. The method of maximum likelihood provides another method for obtaining an estimate of ρ, and this method is discussed in Section C.2.

C.1 WEIGHTED ESTIMATORS OF A COMMON CORRELATION

Several approaches can be used to estimate ρ via weighted linear combinations of estimators. A central issue is whether we should weight the product-moment correlations $r_1, \ldots, r_k$, the unbiased estimators $G(r_1), \ldots, G(r_k)$, the z-transformed versions $z_1, \ldots, z_k$, Kraemer's t-transforms $t_1, \ldots, t_k$, or perhaps some other functions of the r's.

The most direct approach, that of calculating a linear combination of $r_1, \ldots, r_k$, has little to recommend it unless the sample sizes of all k studies are very, very large. An alternative is to estimate ρ from a linear combination of the unbiased estimators $G(r_1), \ldots, G(r_k)$. Some simulation studies suggest that this approach is superior to the approach of using linear combinations of $r_1, \ldots, r_k$, but differs very little from the more standard approach based on linear combinations of z-transformed correlations (see Viana, 1980).

The method usually used to estimate the common ρ is to transform each r by a z-transform to yield $z_1, \ldots, z_k$ given in (4) or Appendix C and then calculate the weighted average

$$z_+ = w_1 z_1 + \cdots + w_k z_k, \tag{12}$$

where the weights are $w_i = (n_i - 3)/\Sigma_{j=1}^k (n_j - 3)$.

It can be shown that when $\rho_1 = \cdots = \rho_k = \rho$ and $n_1, \ldots, n_k$ become large at the same rate, that is, $n_1/N, \ldots, n_k/N$ remain fixed, where $N = \Sigma\, n_j$, then z_+ is approximately normally distributed with a mean of $\zeta = z(\rho)$ and variance $1/(N - 3k)$.

The large sample normal approximation to the distribution of z_+ can be used to test hypotheses about ρ by transforming to ζ. For example, the hypothesis

$$\rho = \rho_0$$

corresponds to the hypothesis $\zeta = \zeta_0 = z(\rho_0)$. Hence we can test that $\rho = \rho_0$ at significance level α by using the test statistic

$$(z_+ - \zeta_0)\sqrt{N - 3k},$$

which is to be compared to the 100α percent two-tailed critical value of the standard normal distribution. Obviously, the case $\rho_0 = 0$ (or, equivalently, $\zeta_0 = 0$) also provides a test that the common correlation is zero.

In a similar manner, we can construct confidence intervals for ρ by first constructing a confidence interval (ζ_L, ζ_U) for ζ,

$$\zeta_L = z_+ - C_{\alpha/2}/\sqrt{N - 3k}, \qquad \zeta_U = z_+ + C_{\alpha/2}/\sqrt{N - 3k},$$

and then using the transformation (8) to obtain a confidence interval for ρ.

Example

Suppose that 10 studies yield the data given in Table 2. Using the sample sizes, z-transform values z_i, and $(n_i - 3)z_i$ values given in the table, we obtain

$$z_+ = 131.24/280 = 0.469,$$

which is transformed by (8) to yield the estimate 0.437 of ρ. To test the hypothesis that $\rho = 0$ we test the hypothesis that $\zeta = 0$ using the statistic

$$z_+\sqrt{N - 3k} = 0.469\sqrt{310 - 3(10)} = 7.85,$$

which exceeds 1.96, the 95-percent two-tailed critical value of the standard normal distribution. A 95-percent confidence interval for ζ is given by (ζ_L, ζ_U), where

$$\zeta_L = 0.469 - 1.96/\sqrt{310 - 3(10)} = 0.352,$$
$$\zeta_U = 0.469 + 1.96/\sqrt{310 - 3(10)} = 0.586.$$

TABLE 2

Correlation Coefficients from 10 Studies

Study	Sample size	r	z	$n-3$	$(n-3)z$
1	20	0.41	0.44	17	7.41
2	30	0.53	0.59	27	15.93
3	27	0.51	0.56	24	13.51
4	42	0.43	0.46	39	17.94
5	49	0.37	0.39	46	17.87
6	12	0.39	0.41	9	3.71
7	17	0.45	0.49	14	6.79
8	35	0.40	0.42	32	13.56
9	38	0.36	0.38	35	13.19
10	40	0.52	0.58	37	21.33
	310			280	131.24

Using the transformation (8), a confidence interval for ρ is

$$0.338 \leq \rho \leq 0.527.$$

C.2 THE MAXIMUM LIKELIHOOD ESTIMATOR OF A COMMON CORRELATION

The method of maximum likelihood can also be used to obtain estimates of ρ given the sample correlations $r_1, \ldots, r_k$ from k independent studies. In general, the maximum likelihood estimator cannot be expressed in closed form. However, it can readily be obtained numerically as the solution of

$$g(\hat{\rho}) = \frac{N\hat{\rho}}{1 - \hat{\rho}^2} - \sum_{i=1}^{k} \frac{n_i r_i}{1 - \hat{\rho} r_i} = 0. \tag{13}$$

For example, if the data for 10 studies are as in Table 2, then the graph of the function $g(\rho)$ is as in Fig. 1. However, from the data we see that the maximum likelihood estimator must lie between 0.36 and 0.53, so that we really need only graph the function in this range, as in Fig. 2, from which we obtain the value 0.437.

Alternatively, we can obtain the result numerically. A very simple iterative procedure is to use as a first approximation to $\hat{\rho}$ the values

$$r_{\text{Min}} = \text{Min}\{r_1, \ldots, r_k\},$$

$$r_{\text{Med}} = \text{Med}\{r_1, \ldots, r_k\},$$

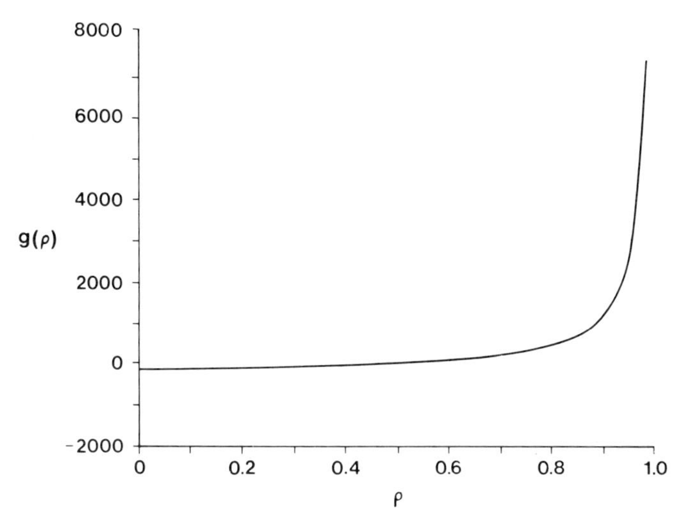

Fig. 1 Graphical solution for maximum likelihood estimator of a common correlation coefficient.

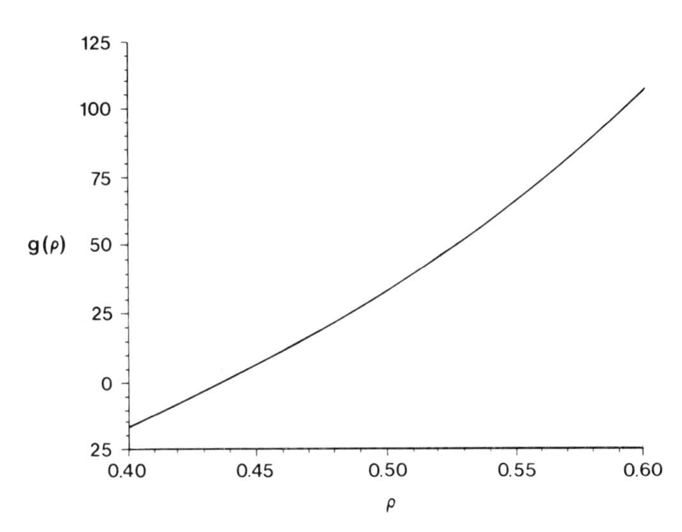

Fig. 2 Selected range for graphical solution for maximum likelihood estimator of a common correlation cofficient.

or

$$r_{\text{Max}} = \text{Max}\{r_1, \ldots, r_k\}.$$

These values serve to further pinpoint the solution. For example,

$$g(r_{\text{Min}}) = g(0.36) < 0, \qquad g(r_{\text{Med}}) = g(0.42) < 0, \qquad g(r_{\text{Max}}) = g(0.53) > 0,$$

so that the solution lies between 0.42 and 0.53. By choosing the midvalue, 0.47, we obtain $g(0.47) > 0$, so that the answer lies between 0.42 and 0.47. Again, the midvalue 0.45 yields $g(0.45) > 0$, so that now we have narrowed the solution to lie in the interval 0.42, 0.45. Several more iterations will yield the exact solution.

In the special case of $k = 2$, the maximum likelihood estimator of ρ can be obtained in closed form:

$$\hat{\rho} = \frac{N(1 + r_1 r_2) - \sqrt{N^2(1 - r_1 r_2)^2 - 4n_1 n_2 (r_1 - r_2)^2}}{2(n_1 r_2 + n_2 r_1)} \tag{14}$$

(see Olkin and Siotani, 1965).

The asymptotic distribution of the maximum likelihood estimator of ρ for any k can also be obtained. When the population correlations are the same, i.e., $\rho_1 = \cdots = \rho_k = \rho$, then as $n_1, \ldots, n_k$ become large at the same rate (that is, $n_i/N, \ldots, n_k/N$ are fixed) the maximum likelihood estimator $\hat{\rho}$ of ρ has a large sample distribution that is normal with mean ρ and variance $(1 - \rho^2)^2/N$. In this case the large sample approximation is accurate only when each of the sample sizes is quite large.

Remark: An alternative representation of the maximum likelihood equation is

$$\sum_{i=1}^{k} \frac{n_i(r_i - \hat{\rho})}{1 - r_i \hat{\rho}} = 0. \tag{15}$$

Although Eqs. (13) and (15) do not appear to be equivalent, they actually are. Either form will readily lead to a numerical solution, though the simplest procedure is probably the graphic one. [For a discussion of (15) and an iterative numerical solution see Kraemer, 1975.]

D. TESTING HOMOGENEITY OF CORRELATIONS ACROSS STUDIES

Before pooling the estimates of ρ from a series of studies, it is important to determine whether the estimates $r_1, \ldots, r_k$ are reasonably consistent with the model of a single underlying population correlation. If the sample correlations are not consistent with the model of a single common population correlation, then it can be misleading to pool the estimates using any of the procedures discussed previously. Statistical tests of homogeneity of correlations provide a method to determine whether a series of sample correlations is more varied than would be expected on the basis of sampling variability if $\rho_1 = \cdots = \rho_k$. We discuss two procedures for testing for homogeneity of correlations. One method is based on the use of z-transforms of the sample correlations (Section D.1), and the other is based on the likelihood ratio test (Section D.2). Tests for the homogeneity of correlations are discussed by Olkin and Siotani (1965), Olkin (1967), and Kraemer (1979).

D.1 A TEST OF HOMOGENEITY BASED ON FISHER'S z-TRANSFORM

Suppose that $r_1, \ldots, r_k$ are a series of independent sample correlations and $z_1, \ldots, z_k$ are the z-transforms. Then a test of the hypothesis of homogeneity of the population correlations is to reject the hypothesis if the statistic

$$Q = \sum_{i=1}^{k} (n_i - 3)(z_i - z_+)^2, \tag{16}$$

(where z_+ is the weighted average correlation given in (12)), exceeds a critical value from the chi-square distribution with $k - 1$ degrees of freedom.

Remark: When $\rho_1 = \cdots = \rho_k$ and $n_1/N, \ldots, n_k/N$ remain fixed as $N \to \infty$, then Q has an asymptotic chi-square distribution with $k - 1$ degrees of freedom. Note that the statistic Q is essentially the weighted sum of squares of the z_i about the weighted mean z_+, where the weights are the reciprocals of the variances. This statistic is analogous to the statistic (25) of Chapter 6 for testing for the homogeneity of effect sizes.

If the homogeneity of the correlations is rejected for a series of studies, the investigator may elect not to obtain a common estimate of ρ from those studies. However, if the sample sizes are very large, it is wise to examine the actual value of the estimates, since relatively small variation in the values of the estimates can lead to significant values of a homogeneity test statistic. If sample sizes are very large and the actual values of the correlations are not too different, the investigator may decide to pool the estimates even though there is reason to suspect some variation among the ρ_i.

Remark: It is sometimes computationally advantageous to use the relation

$$\sum a_i(z_i - z_+)^2 = \sum a_i z_i^2 - \left(\sum a_i z_i\right)^2 / \sum a_i, \tag{17}$$

where

$$z_+ = \sum a_i z_i / \sum a_i,$$

as discussed in Section E of Chapter 6.

Example

For the 10 studies in the example in Section C.1, we calculate the value of the homogeneity statistic Q via the computational formula (17) with $a_i = n_i - 3$. Using the totals of $n_i - 3$, $(n_i - 3)z_i$, and $(n_i - 3)z_i^2$ summarized in Table 3, we obtain

$$Q = 63.24 - (131.24)^2/280 = 1.73.$$

Comparing 1.73 with 16.9, the 95 percent point of the chi-square distribution, we see that the correlations are homogeneous at the 5-percent level.

D.2 THE LIKELIHOOD RATIO TEST OF HOMOGENEITY OF CORRELATIONS

An alternative test for homogeneity of correlation coefficients is the likelihood ratio test, which is based on the statistic

$$\text{LRS} = -2\left(\sum_{i=1}^{k} n_i \log \frac{1 - r_i^2}{(1 - r_i \hat{\rho})^2} + N \log(1 - \hat{\rho}^2)\right), \tag{18}$$

where $\hat{\rho}$ is the maximum likelihood estimator of ρ discussed in Section C.2.

TABLE 3

Computations for Testing for Homogeneity for Correlation Coefficients

Study	Sample size	r	z	$n - 3$	$(n - 3)z$	$(n - 3)z^2$	$n \log \frac{1 - r^2}{(1 - r\hat{\rho})^2}$
1	20	0.41	0.44	17	7.41	3.23	4.22
2	30	0.53	0.59	27	15.93	9.40	5.91
3	27	0.51	0.56	24	13.51	7.60	5.48
4	42	0.43	0.46	39	17.94	8.25	8.90
5	49	0.37	0.39	46	17.87	6.94	10.07
6	12	0.39	0.41	9	3.71	1.53	2.50
7	17	0.45	0.49	14	6.79	3.29	3.60
8	35	0.40	0.42	32	13.56	5.74	7.35
9	38	0.36	0.38	35	13.19	4.97	7.73
10	40	0.52	0.58	37	21.33	12.29	8.01
	310	4.37	4.72	280	131.24	63.24	63.77

The value of the LRS is compared to the chi-square distribution with $k - 1$ degrees of freedom, and large values of the LRS lead to rejection of the hypothesis of homogeneity of the correlations.

Example

We now compute the likelihood ratio test statistic for the data given in Table 3. Using the value $\hat{\rho} = 0.437$ obtained in Section C.2, we obtain

$$\text{LRS} = -2\{63.77 + 310 \log[1 - (0.437)^2]\} = 3.85.$$

Comparing 3.85 with 16.9, the 95 percentage point of the chi-square distribution with nine degrees of freedom, we see that the correlations are not significantly heterogeneous at the $\alpha = 0.05$ significance level. Note that the value 3.85 of the likelihood ratio chi-square statistic differs from the value 1.73 obtained from the statistic Q given in Section D.1.

E. FITTING GENERAL LINEAR MODELS TO CORRELATIONS

In this section we examine models in which correlations depend on a linear combination of known predictor variables. These predictor variables can be either discrete or continuous, and represent study characteristics that are likely to be related to the correlation. Often the predictor variables arise from discrete categorizations of studies such as sex or socioeconomic status of subjects. Other times the predictors are continuous variables, such as the time elapsed between certain measurements. The methods presented here are analogous to multiple linear regression, and any coding scheme used in ordinary linear regression can be used in this context. Thus this section provides a method for fitting general linear models to correlations that is analogous to the methods given in Section B of Chapter 8 for fitting general linear models to effect sizes.

The explicit model (and notation), the estimator of the model parameters, and a test for model specification are given in Sections E.1, E.2, and E.3, respectively. Statistical packages for computing estimates and test statistics are discussed in Section E.4.

E.1 MODEL AND NOTATION

Suppose that we have a series of k studies with sample sizes $n_1, \ldots, n_k$, and that the ith study provides a correlation coefficient r_i as an estimator of the parameter ρ_i. Denote the z-transforms of the sample estimates and population parameters by $z_1, \ldots, z_k$ and $\zeta_1, \ldots, \zeta_k$, respectively. Suppose further, that the z-transformed correlation ζ_i for the ith study depends on a linear combination of p known predictor variables:

$$\begin{aligned} \zeta_1 &= x_{11}\beta_1 + \cdots + x_{1p}\beta_p, \\ \zeta_2 &= x_{21}\beta_1 + \cdots + x_{2p}\beta_p, \\ &\vdots \\ \zeta_k &= x_{k1}\beta_1 + \cdots + x_{kp}\beta_p, \end{aligned} \tag{19}$$

where the vector of predictor variables for the ith study is $\mathbf{x}_i = (x_{i1}, \ldots, x_{ip})'$, and $\boldsymbol{\beta} = (\beta_1, \ldots, \beta_p)'$ is a vector of regression coefficients.

A compact representation of the model can be given in matrix form, by defining the design matrix

$$\mathbf{X} = \begin{bmatrix} x_{11} & \cdots & x_{1p} \\ x_{21} & \cdots & x_{2p} \\ \vdots & & \vdots \\ x_{k1} & \cdots & x_{kp} \end{bmatrix}$$

of predictor values and the vector $\boldsymbol{\zeta} = (\zeta_1, \ldots, \zeta_k)'$. Then the model (19) for $\zeta_1, \ldots, \zeta_k$ is

$$\boldsymbol{\zeta} = \mathbf{X}\boldsymbol{\beta}, \tag{20}$$

which is a familiar form in regression analysis.

E.2 ESTIMATING REGRESSION COEFFICIENTS

The usual least squares estimator of the regression coefficients when all the sample sizes are equal is

$$\hat{\boldsymbol{\beta}}_{\text{LS}} = (\mathbf{X}'\mathbf{X})^{-1}\mathbf{X}'\mathbf{z},$$

where $\mathbf{z} = (z_1, \ldots, z_k)'$ is the vector of transformed correlations. But because the sample sizes are unequal, we need to modify these estimates. Note that the covariance matrix of $\mathbf{z}$ is defined by

$$\boldsymbol{\Sigma}_z^{-1} = \text{Diag}(n_1 - 3, \ldots, n_k - 3).$$

The weighted estimator of the regression coefficients is now

$$\hat{\boldsymbol{\beta}} = (\mathbf{X}'\boldsymbol{\Sigma}_z^{-1}\mathbf{X})^{-1}\mathbf{X}'\boldsymbol{\Sigma}_z^{-1}\mathbf{z}. \tag{21}$$

This is a generalized least squares procedure and can be obtained using standard computer packages (see Section E.4).

When $\boldsymbol{\zeta} = \mathbf{X}\boldsymbol{\beta}$, and $n_1, \ldots, n_k$ become large at the same rate, the large sample distribution of $\hat{\boldsymbol{\beta}}$ is multivariate normal with mean $\boldsymbol{\beta}$ and covariance matrix $(\mathbf{X}'\boldsymbol{\Sigma}_z^{-1}\mathbf{X})^{-1}$ (see Hedges & Olkin, 1983b). As a consequence, we can use this result to provide tests of significance or confidence intervals for the β_i's.

To test an hypothesis about any particular β_j, say that $\beta_j = 0$, we reject the hypothesis at a level of significance α if

$$\hat{\beta}_j^2/\sigma_{jj}$$

exceeds the $100(1 - \alpha)$ percentage point of the chi-square distribution with one degree of freedom. The variances σ_{jj} are obtained as the diagonal elements of the (known) matrix

$$\boldsymbol{\Sigma}_\beta = (\mathbf{X}'\boldsymbol{\Sigma}_z^{-1}\mathbf{X})^{-1},$$

which can be obtained from standard computer packages (see Section E.4).

Similarly, we can use the large sample normal approximation to the distribution of $\hat{\boldsymbol{\beta}}$ to obtain confidence intervals for β_j. A $100(1 - \alpha)$-percent nonsimultaneous confidence interval for β_j is given by

$$\hat{\beta}_j - C_{\alpha/2}\sqrt{\sigma_{jj}} \le \beta_j \le \hat{\beta}_j + C_{\alpha/2}\sqrt{\sigma_{jj}}, \tag{22}$$

where $C_{\alpha/2}$ is the two-tailed critical value of the standard normal distribution.

Simultaneous $100(1 - \alpha)$-percent confidence intervals for $\beta_1, \ldots, \beta_p$ can be obtained by using the Bonferroni method and substituting $C_{\alpha/2p}$ for $C_{\alpha/2}$ in (22) (see, e.g., Miller, 1981).

Sometimes it is useful to test whether some subset $\beta_1, \ldots, \beta_l$ of the regression coefficients are simultaneously zero. To test the hypothesis $\beta_1 = \beta_2 = \cdots = \beta_l = 0$, first compute $\hat{\beta}_1, \ldots, \hat{\beta}_p$ and then

$$Q = (\hat{\beta}_1, \ldots, \hat{\beta}_l)\boldsymbol{\Sigma}_{11}^{-1}(\hat{\beta}_1, \ldots, \hat{\beta}_l)', \tag{23}$$

where $\boldsymbol{\Sigma}_{11}$ is the upper $l \times l$ submatrix of

$$\boldsymbol{\Sigma}_\beta = \begin{bmatrix} \boldsymbol{\Sigma}_{11} & \boldsymbol{\Sigma}_{12} \\ \boldsymbol{\Sigma}_{12} & \boldsymbol{\Sigma}_{22} \end{bmatrix}.$$

The hypothesis that $\beta_1 = \cdots = \beta_l = 0$ is rejected at significance level α if Q exceeds the $100(1 - \alpha)$ percentage point of the chi-square distribution with l degrees of freedom.

Of course, when $l = p$, we have a test that all the β's are zero. The test statistic now becomes $Q_R = \hat{\boldsymbol{\beta}}'\boldsymbol{\Sigma}_\beta^{-1}\hat{\boldsymbol{\beta}}$ (the weighted sum of squares due to regression).

Example

A linear model analysis was used with the data on teacher indirectness given in Table 10 of Chapter 2 to determine whether grade level was related to the correlation between teacher indirectness and achievement. Grade level of the students was scored dichotomously, with a 1 assigned for grades 1–6 and a 2 assigned for grades 7–12. The vector of z-transformed correlations is

$$\mathbf{z} = (-0.07, 0.32, 0.52, 0.45, 0.18, 0.30, 0.42)',$$

the design matrix (with intercept) is

$$\mathbf{X} = \begin{bmatrix} 1 & 1 & 1 & 1 & 1 & 1 & 1 \\ 1 & 1 & 2 & 2 & 2 & 2 & 1 \end{bmatrix}',$$

and the diagonal matrix $\mathbf{\Sigma}_z^{-1}$ is

$$\mathbf{\Sigma}_z^{-1} = \text{Diag}(12, 13, 12, 13, 12, 14, 12).$$

Using SAS Proc GLM as described in Section E.4, the estimated regression coefficients are

$$\hat{\boldsymbol{\beta}} = (0.086, 0.139)'.$$

The matrix $\mathbf{\Sigma}_\beta$ of covariances of the elements of $\hat{\boldsymbol{\beta}}$ is

$$\mathbf{\Sigma}_\beta = \begin{bmatrix} 0.128 & -0.074 \\ -0.074 & 0.047 \end{bmatrix}.$$

To test that $\hat{\beta}_2$ is different from zero we calculate the test statistic

$$Q = \hat{\beta}_2 \Sigma_{22}^{-1} \hat{\beta}_2 = (0.139)^2/0.047 = 0.41.$$

This value is compared to 3.84 and so is not significant. Hence the correlation between teacher indirectness and achievement does not differ between the two grade level categories.

E.3 TESTING MODEL SPECIFICATION

Since the large sample variances of the z_i are known values, it is possible to test the specification (goodness of fit) of the linear model whenever the number of studies exceeds the number of predictors (that is, when $k > p$). This can be important because the estimates of $\boldsymbol{\beta}$ in models that are misspecified do not necessarily converge to the true values for large samples. Therefore parameter estimates in misspecified models are difficult or impossible to interpret. The ability to test model specification directly is an important advantage of the data analysis of correlations in the same way that it was for effect sizes (see Chapter 8).

If $k > p$ the test for model specification is based on the test statistic Q_E (the weighted error sum of squares about the regression line) given by

$$Q_E = \mathbf{z}'\boldsymbol{\Sigma}_z^{-1}\mathbf{z} - \hat{\boldsymbol{\beta}}'(\mathbf{X}\boldsymbol{\Sigma}_z^{-1}\mathbf{X})\hat{\boldsymbol{\beta}}. \tag{24}$$

When the model is correctly specified, i.e., when $\zeta = \mathbf{X}\boldsymbol{\beta}$, and the sample size is large, Q_E has an approximate chi-square distribution with $k - p$ degrees of freedom. If the test leads to rejection (Q_E is large compared to the critical value) the investigator should try to identify reasons for the poor fit. Methods for doing this are given in Chapter 12.

Example

In the analysis of the correlations between teacher indirectness and achievement discussed in Section E.2, the SAS Proc GLM as described in Section E.4 gives the test statistic Q_E:

$$Q_E = 2.53.$$

Comparing 2.53 with the percentage points of the chi-square distribution with $7 - 2 = 5$ degrees of freedom, we see that a chi-square as large as 2.53 would occur between 75 and 90 percent of the time if the model were correctly specified. Consequently, we do not reject the specification of the linear model for the correlations of teacher indirectness and achievement.

E.4 COMPUTATION OF ESTIMATES AND TEST STATISTICS

An easy way to obtain the regression coefficients and test statistics is to use a computer package (such as SAS Proc GLM) that has a program for weighted regression analysis. The weighted regression involves the same procedures as an ordinary (unweighted) regression analysis except that a variable

$$w_i = n_i - 3$$

is specified as the "weight" for each case. The estimates $\hat{\beta}_1, \ldots, \hat{\beta}_p$ are obtained directly from the computer printout. The matrix $\mathbf{V} = (\mathbf{X}'\boldsymbol{\Sigma}_z^{-1}\mathbf{X})^{-1}$ of covariances of the $\hat{\beta}_i$ is usually called the $\mathbf{X}'\mathbf{W}\mathbf{X}$ inverse matrix on computer printouts. The statistic Q_E for testing model specification is the same as the (weighted) "error sum of squares" or the (weighted) sum of squares about the regression line. When the model $\zeta = \mathbf{X}\boldsymbol{\beta}$ is correctly specified, then Q_E has a chi-square distribution with $k - p$ degrees of freedom, where p is the number of predictors including the intercept. The statistic Q_R is the (weighted) "sum of squares due to the regression."

F. RANDOM EFFECTS MODELS FOR CORRELATIONS

In the previous sections of this chapter we have dealt with research synthesis models in which the population correlations $\rho_1, \ldots, \rho_k$ were fixed, but unknown, constants. For example, the population correlation (or its z-transform) is determined by some predictor variables whose values are known for each study. The substantive rationale for this model is that studies differ according to many characteristics, such as sampling schemes, duration between measurements, or intensity of treatment.

In Chapter 9 we developed the idea of random effects models for effect sizes as an alternative to models of fixed effect sizes that were determined by known predictors. Random effects models differ from fixed effects models in that the underlying population parameters (in this case the population correlations) vary from study to study in a manner that is not entirely explained by known variables. That is, the parameters behave as if they were sampled from a distribution of possible parameter values. Thus the observed estimates vary not only because of sampling error but also because of variation in the true parameter values. Recall that random effects models for research synthesis are analogous to random effects analysis of variance models, where treatments (and treatment effects) are sampled from a universe of possible treatments.

Random effects models for correlations have been discussed in at least one important substantive context, that of validity generalization. In this context validity coefficients are calculated in each of several studies, and the object of the research synthesis is to determine the variance of the "population" correlations that are the validity coefficients (see, e.g., Hunter, Schmidt, & Jackson, 1982). If this variance is small, then the validity coefficient is reasonably generalizable across situations. If not, then the validity must be situation-specific.

Below we present a random effects model for correlations. In Section F.1 we define the model in which population correlations are assumed to be sampled from an unknown distribution of population correlations. In Section F.2 a statistical test is given that the variance in the population correlations is zero. In Section F.3 an unbiased estimator of the variance of the population correlations is given.

F.1 MODEL AND NOTATION

Suppose $\rho_1, \ldots, \rho_k$ denote k population effect magnitudes. From the ith population (study) we take a sample of size n_i and compute the sample correlation r_i, $i = 1, \ldots, k$. We depart from previous practice, however, by treating the population correlations $\rho_1, \ldots, \rho_k$ as realizations of a random variable P with an unknown distribution and variance $\sigma^2(P)$. Thus the ρ_i are not fixed constants but the result of sampling from a universe of possible population correlations. The object of the statistical analysis is to estimate $\sigma^2(P)$ from $r_1, \ldots, r_k$, to test whether $\sigma^2(P) = 0$, and perhaps to estimate the mean $\bar{P}$ of the distribution of population correlation coefficients.

The sampling distribution of the correlation r for fixed effects models (that is, when ρ is constant) is known, and is usually represented as an infinite series. It leads to exact expressions for the mean and variance of r also in the form of infinite series. Because ρ is not fixed but has a sampling distribution of its own, it is necessary to distinguish between the variance of r assuming a fixed ρ and the variance of r incorporating the variability in ρ.

We do not need an exact expression for the conditional or unconditional sampling variance of r. Instead we work with the sampling distributions of the unbiased estimator $G(r)$ given in Section A.1.

Denote the unbiased estimator of ρ_i in the ith study by $\hat{\rho}_i$, so that we have

$$\hat{\rho}_1 = G(r_1), \ldots, \hat{\rho}_k = G(r_k). \tag{25}$$

However, for most purposes the approximation (3) to $G(r)$ can be treated as unbiased.

Note that $G(r)$ is conditionally unbiased (unbiased for fixed ρ); but $G(r)$ is also an unbiased estimator of the mean $\bar{P}$ of the distribution of ρ values in the random effects model. The unconditional sampling variance $\sigma^2(\hat{\rho}_i)$ is the sum of the variance of the population correlations $\sigma^2(P)$ and the conditional sampling variance $\sigma^2(\hat{\rho}_i | P)$ of $\hat{\rho}_i$ given P. That is,

$$\sigma^2(\hat{\rho}_i) = \sigma^2(P) + \sigma^2(\hat{\rho}_i | P). \tag{26}$$

Equation (26) would *not* be true if a biased estimator, such as r_i, were substituted for the unbiased estimator $\hat{\rho}_i$. The reason for this is that the parameter value and the sampling error of a biased estimator can be correlated. This correlation implies that there is a nonzero covariance between the sampling error and the parameter value. Therefore the right-hand side of (26) would be in error by a factor proportional to that covariance. Furthermore, this covariance can be large compared with the sampling error of the estimator. This subtle difficulty is a serious problem in earlier attempts to study validity using variance components.

F.2 TESTING THAT THE VARIANCE OF POPULATION CORRELATIONS IS ZERO

If the variance $\sigma^2(P)$ of the distribution of the population correlations is zero, then $\rho_1 = \cdots = \rho_k$ for any sample of k population correlations. Thus the null hypothesis

$$\sigma^2(P) = 0$$

in the random effects model for correlations is the same as the hypothesis of homogeneity of correlations in the fixed effects model, so that tests for homogeneity of correlations given in Section D can be used as tests that $\sigma^2(P) = 0$ in the random effects model. As a test we use the weighted sum of squares of $z_1, \ldots, z_k$ [see (16)]. Note that this test statistic has the same distribution for fixed and random models only if the null hypothesis $\rho_1 = \cdots = \rho_k$, or, equivalently, $\sigma^2(P) = 0$, is true. The test statistic has a different distribution under the two models if the null hypothesis is false. This corresponds to the fact that the mean square ratio has the same null distribution (a central F) for fixed and random models in one-way analysis of variance. The nonnull distributions of the mean square ratios are quite different under the two models.

F.3 AN UNBIASED ESTIMATE OF THE CORRELATION VARIANCE COMPONENT

The expression (26) for the unconditional sampling variance $\sigma^2(\hat{\rho}_i)$ of $\hat{\rho}_i$ shows that this variance depends only on the parameter variance component $\sigma^2(P)$ and on the conditional sampling variance $\sigma^2(\hat{\rho}_i | P)$ of $\hat{\rho}_i$. In principle we can estimate $\sigma^2(\hat{\rho}_i)$ from a sample of $\hat{\rho}_i$ values. If an unbiased estimator of $\sigma^2(\hat{\rho}_i | P)$ were available then an estimate of $\sigma^2(P)$ could be obtained by subtraction.

The rationale for estimating $\sigma^2(P)$ is the same as that for the estimation of variance components in random effects analysis of variance. In random effects analysis of variance, the expected values of the mean squares are expressed in terms of variance components. The expected values of the mean squares are then replaced with their sample values and the equations are solved for the variance components. This process gives unbiased estimates of the variance components. The expected value of the sample variance of the $\hat{\rho}_1, \ldots, \hat{\rho}_k$ is expressed as a function of variance components including the conditional sampling variances of $\hat{\rho}_1, \ldots, \hat{\rho}_k$. Unbiased estimators of conditional sampling variances are obtained. Then the expected value of the unconditional variance of the $\hat{\rho}_i$ is replaced by the observed variance of the $\hat{\rho}_i$, and the equations are solved for $\sigma^2(P)$. This process results in an unbiased estimator of $\sigma^2(P)$.

Given the unbiased estimators $\hat{\rho}_1, \ldots, \hat{\rho}_k$ of the correlations $\rho_1, \ldots, \rho_k$, define the variance

$$s^2(\hat{\rho}) = \sum_{i=1}^{k} \frac{(\hat{\rho}_i - \text{avg } \hat{\rho})^2}{k - 1}, \tag{27}$$

TABLE 4

Results of Seven Studies of the Relationship between Teacher Indirectness and Student Achievement

Study	n	r	$\hat{\rho}$	$\hat{\rho}^2 - r^2$
1	15	−0.073	−0.076	0.0005
2	16	0.308	0.319	0.0067
3	15	0.481	0.496	0.0151
4	16	0.428	0.441	0.0117
5	15	0.180	0.187	0.0027
6	17	0.290	0.299	0.0056
7	15	0.400	0.414	0.0114
Mean		0.288	0.297	0.0077

where avg $\hat{\rho}$ is the unweighted mean of $\hat{\rho}_1, \ldots, \hat{\rho}_k$. A direct argument gives the expected value of $s^2(\hat{\rho})$ as

$$E(s^2(\hat{\rho})) = \sigma^2(P) + \frac{1}{k}\sum_{i=1}^{k} \sigma^2(\hat{\rho}_i | P). \tag{28}$$

An elaborate argument yields an approximate unbiased estimator of $\sigma^2(P)$ (to order $1/n$):

$$\hat{\sigma}^2(P) \doteq s^2(\hat{\rho}) - \frac{1}{k}\sum_{i=1}^{k} (\hat{\rho}_i^2 - r_i^2). \tag{29}$$

Variance components are the principal parameters in random effects models which are to be estimated. In the case of effect sizes we obtain large sample estimates of the variance of their estimates. Parallel estimators could be obtained in the random effects models for correlations. However, in this case we do not obtain such estimators for two reasons. First, the estimates of variances of variance components will be quite complicated. Second, the poor quality of the normal approximation to the distribution of the untransformed sample correlation suggests that the variance component estimates will be far from normally distributed unless sample sizes are unrealistically large.

Example

The results of seven studies of the relationship of teacher indirectness to student achievement were discussed in Section D of Chapter 2. Because the studies differ in many ways (type of subject matter taught, amount of teacher experience, etc.) we use a random effects model for the analysis of the correlation coefficients produced by the seven studies. The sample size, correlation coefficient r, approximate unbiased estimator $\hat{\rho}$, and $\hat{\rho}^2 - r^2$ for each study are given in Table 4. The variance of the $\hat{\rho}$ values is 0.0377. Using (28) for the estimator of the parameter variance component, we obtain

$$\hat{\sigma}^2(P) = 0.0377 - 0.0077 = 0.0300,$$

which corresponds to $\hat{\sigma}(P) = 0.173$. Thus the parameter variance component is substantial even though the correlations do not differ greatly.

CHAPTER 12

Diagnostic Procedures for Research Synthesis Models

Thus far we have developed the requisite statistical theory for quantitative research synthesis for several statistical models. These models vary in complexity, but most of them depend on the assumption that indices of effect magnitude can be determined by known explanatory variables. These models are together classified as "fixed effects" models. For example, the effect sizes from studies were assumed to fit a categorical model in Chapter 7 and a general linear model in Chapter 8. Analogous models for correlation coefficients were discussed in Chapter 11. For each fixed effects model, a statistical test was provided to determine whether the model is correctly specified; that is, the statistical test measures whether the model accounts for the variability among the (population) effect sizes or correlations.

The statistical procedures for estimating parameters in fixed effects models are, strictly speaking, correct only if the effect sizes actually fit the hypothesized model. The fact that procedures for estimating parameters in linear models are reliable only when data conform reasonably well to the supposed model is a well-known dictum in statistics. For example, if one of the true independent variables is omitted when estimating the coefficients in a multiple linear regression, the estimates of the remaining coefficients are biased and may not even be consistent. This is the well-known problem of specification error. In standard normal theory regression analysis, there is no standard test for specification error, and it is somewhat remarkable that tests for specification error can be derived for fixed effects models in research synthesis. These tests for model specification play an important role in strengthening interpretations of statistical analyses in research synthesis.

Alternatives to fixed effects models in quantitative research synthesis are the random effects models proposed in Chapters 9 and 11. The random effects models put fewer restrictions on the variation that may be observed in effect sizes or correlations. The price for the reduction in the number of restrictions, of course, is that the inferences made are weaker. In random effects models, studies do not replicate findings in any strong way. The actual values of population effect magnitudes are left to vary at random. The substantive interpretation of such analyses necessarily involves statements such as "Sometimes the treatment works better than others." For some applications of quantitative research synthesis the phenomenon under investigation may behave in such a way as to require random effects models. That is, random variation from study to study in the true effect of the treatment might be a substantively reasonable explanation of the phenomenon. In other cases, we might believe that some additional and unmeasured variable or variables affect the size of the effect magnitude. For example, in studies where there are preexisting differences between subjects in the experimental and control groups, effect sizes based on posttest scores are likely to reflect those initial differences. If no measure of the extent of preexisting differences is available, then the effect sizes may have to be treated with random effects models even though they are determined by a variable that can, in principle, be measured. Qualitatively stronger conclusions usually can be obtained by analysis under fixed effects models. It therefore behooves the investigator to try rather hard to obtain reasonable fixed models that provide adequate representations of the obtained effect sizes and correlations.

In this chapter we describe some methods for discovering potential sources of poor fit to fixed effects models. These diagnostic procedures provide methods for recognizing one or more estimates that deviate greatly from their expected values if the model were correct. These procedures often point to studies that differ from others in ways that are remediable (e.g., they may represent mistakes in coding or calculation). Sometimes the diagnostic procedures point to sets of studies that differ in a collective way that suggests a new explanatory variable. Sometimes a study is an "outlier" that cannot be explained by an obvious characteristic of the study.

The analysis of data containing some observations that are outliers (in the sense of not fitting the model of the other studies) is a complicated task.

It invariably requires the use of good judgment and the making of decisions that are, in some sense, compromises. There are two extreme positions on dealing with outliers:

(1) Data are "sacred," and no datum point (study) should ever be set aside for any reason.
(2) Data should be tested for outliers, and data points (studies) that fail to conform to the hypothesized model should be removed.

Neither extreme seems appropriate. There are cases where setting aside a small proportion of the data (less than 15–20 percent) has certain advantages. If nearly all the data can be modeled in a simple, straightforward way, it is certainly preferable to do so, even at the risk of requiring elaborate descriptions of the studies that are set aside.

We emphasize that studies that are set aside should not be ignored; often these studies reveal patterns that are interesting in and of themselves. Occasionally these deviant studies share a common characteristic that should be added to the model as a predictor. One of the reasons we prefer to model most of the data is that the results of the studies that are identified statistically as outliers often do not deviate enough to disagree with the substantive result of the model. That is, an effect size estimate may exhibit a statistically significant difference from those of other studies, yet fail to differ from the rest to an extent that would make a practical or substantive difference. However, it is crucial that all data be reported and that the deleted data be clearly noted.

A. HOW MANY OBSERVATIONS SHOULD BE SET ASIDE

The issue of how many observations can be set aside without compromising the statistical analysis is a difficult one with no obvious solution. One approach is to examine the methods used in the analysis of relatively good data. If the relatively well defined data in the physical sciences require deletion of outliers, we certainly expect that data in the social sciences, which are more amorphous, will contain at least as many outliers. Alternatively, we can examine the methods of insightful data analysts, and use these as a guide to both what is possible and what is necessary. However, we should recall that data sets differ and data analysts vary in their views.

There is considerable evidence that real data contain occasional observations that do not fit simple models well. The early developers of statistical methodology certainly believed that the exclusion of a certain amount of data from statistical analyses, solely on the basis of deviant values, was a good practice. Legendre, who is credited with the invention of the important statistical idea of least squares, recommended (in 1805) the use of his method after rejecting all observations whose errors "are found to be such that one judges them to be too large to be admissible" (Stigler, 1973). Edgeworth (1887), another important contributor to the foundations of data analysis, reached the same conclusion:

> The Method of Least Squares is seen to be our best course when we have thrown overboard a certain portion of our data—a sort of sacrifice which has been often made by those who sail upon the stormy seas of Probability (p. 269).

Modern exploratory data analysts generally agree that it is usually necessary to eliminate a few data points from any set of data in order to describe the data by a reasonably simple model (see, e.g., Tukey, 1977). How much data needs to be sacrificed to obtain a reasonable fit to a model is regarded by all as a difficult question. Stigler (1977) examined several historical data sets in the physical sciences, and concluded that "outliers are present in small quantities, but a small amount of trimming (no more than 10 percent) may be the best way of dealing with them" (p. 1070). Stigler's 10 percent trimming actually means discarding 10 percent of the highest observations and 10 percent of the lowest observations, so his recommendation amounts to discarding not more than 20 percent of the total observations.

Other statisticians have suggested that the best data analyses require deletion of a greater proportion of the data. Rocke, Downs and Rocke (1982) examine both historical and modern data sets in the physical sciences. They argue that "for greatest efficiency, one must either trim 40–50 percent of the data or employ one of the modern robust estimators" (p. 99). If this amount of outlier rejection is required in physical science data, it is all the more reasonable to expect that deletion of 10–20 percent of the data will often be necessary to fit data to models in the social sciences. This has often proved to be the case in modern psychometrics (Wright and Stone, 1979).

Whenever data are deleted solely to improve the fit of a model, we should determine whether the deletion of the data changes other conclusions besides the assessment of goodness of fit of the model. If the deletion of a few observations improves the fit of a simple model, but does not greatly affect the overall mean, for example, then such a deletion is probably well justified.

The diagnostic procedures given in this chapter are an aid in fitting observed estimates of effect magnitudes to reasonably simple models.

B. DIAGNOSTIC PROCEDURES FOR HOMOGENEOUS EFFECT SIZE MODELS

In this section we discuss diagnostic procedures for the model that all studies share the same population effect size, which is the simplest of the models used in quantitative research synthesis. If this model is consistent with the data, the substantive implications are easy to identify. The homogeneous effect size model implies that each study in essence confirms or replicates the findings of other studies. The effect size associated with a study deviates from this model when it differs from those of other studies. There are several methods to locate one or more deviant effect sizes. These methods usually, but not always, identify the same effect sizes as aberrant. In Section B.1 we examine graphic techniques for displaying effect size estimates, and in Section B.2 discuss the use of residuals to identify outliers. In Section B.3 the sensitivity of the homogeneity test statistic to deviant observations is used to identify estimates that influence the homogeneity statistic.

B.1 GRAPHIC METHODS FOR IDENTIFYING OUTLIERS

One of the simplest methods for finding deviant estimates in effect size data sets is to plot the effect sizes on one set of axes. A convenient technique is to sort the studies into groups on the basis of substantive characteristics and then plot the estimates on separate horizontal lines. In general, each estimate has a different variance, so that it is important to plot a confidence interval along with the estimate for each study. Plotting the overall weighted mean of the estimates provides a guide to the "center" of the distribution of effect size estimates.

Figure 1 presents a plot of the results of 11 well-controlled studies of the effects of open education on student attitude toward school (data of Table 6 of Chapter 2). Each study is labeled with an identification number, and the studies are grouped according to the method used to determine openness of classrooms and the grade level of the students involved in the study. The value of the test statistic Q given in (25) of Chapter 6 for homogeneity of effect size is $Q = 23.16$, which exceeds 18.31, and 95 percentage point of the chi-square distribution with 10 degrees of freedom. Examination of Fig. 1 suggests a potential cause of the heterogeneity. The estimate from study 1 appears to be somewhat deviant from those of the remaining studies. Only two of the other estimates have 95-percent confidence intervals that include the effect size estimate from study 1, and the entire confidence interval for δ based on study 1 seems shifted to the left. Studies 3 and 11 also seem to be somewhat deviant from the other studies.

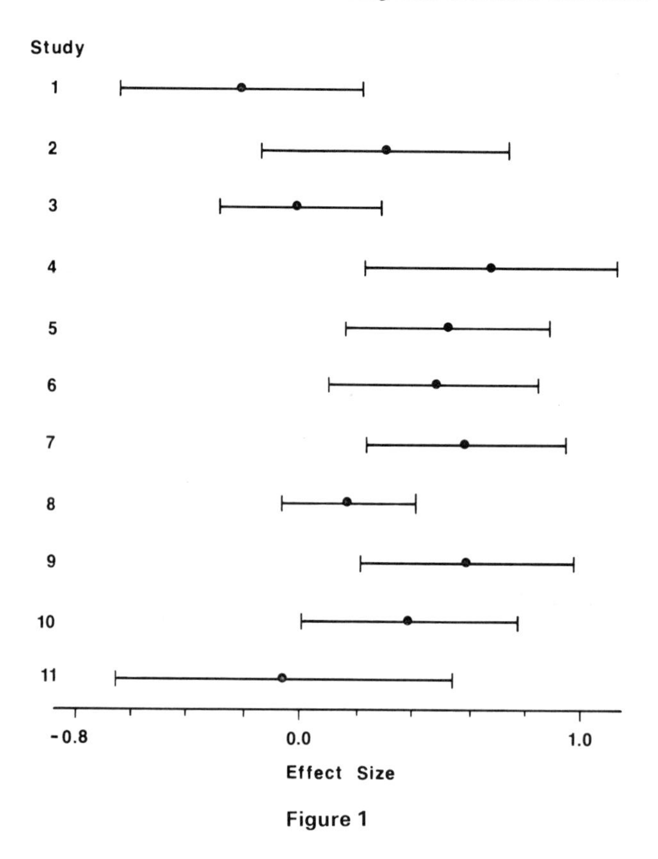

Figure 1

Thus one might imagine that the effect sizes of the studies would be considerably more homogeneous with study 1 (or study 3 or 11) deleted. In fact, this is true. When study 1 is deleted, the homogeneity statistic drops to $Q = 16.99$, which is to be compared with 16.92, the 95 percentage point of the chi-square distribution with nine degrees of freedom. Similarly, the homogeneity statistic drops to $Q = 17.50$ with study 3 deleted, and to $Q = 21.87$ with study 11 deleted. Thus, the 10 effect sizes with either study 1 or study 3 deleted are borderline homogeneous. We might be tempted to stop and declare that the remaining effect sizes are sufficiently homogeneous for the purposes of calculating an average effect size. Alternatively, we might delete another observation. If we first delete study 1, then a likely candidate for the second deletion is study 3. If studies 1 and 3 are both deleted and the homogeneity statistic is recalculated, the value of Q would be 9.98, which is very consistent with the hypothesis of homogeneity among eight effect sizes. Note that the deletion of one or both of these "deviant" studies has the effect of slightly increasing the estimate of effect size calculated from the remaining studies. The change due to deletion of one study is not large enough to affect substantive interpretation, however.

B.2 THE USE OF RESIDUALS TO LOCATE OUTLIERS

Residuals are useful for identifying outliers from statistical models. Using the notation of Chapter 6, suppose $d_1, \ldots, d_k$ are effect size estimates (unbiased) from k studies, and d_+ is the weighted mean effect size defined by Eq. (6) of Chapter 6. The residual of the ith study from the weighted mean is

$$d_i - d_+.$$

Because d_+ incorporates information from each d_i, the residuals are dependent. This dependency creates technical problems, so instead, we use a different residual, namely, the difference between d_i and a pooled estimate of δ that does not include the ith study.

The computation of this residual is simplified by defining several composites. Let TW denote the total of the weights (reciprocals of the variances), TWD the total of the weighted d's, and TWDS the total of the weighted d-squares:

$$\mathrm{TW} = \sum_{i=1}^{k} \frac{1}{\hat{\sigma}^2(d_i)}, \tag{1}$$

$$\mathrm{TWD} = \sum_{i=1}^{k} \frac{d_i}{\hat{\sigma}^2(d_i)}, \tag{2}$$

$$\mathrm{TWDS} = \sum_{i=1}^{k} \frac{d_i^2}{\hat{\sigma}^2(d_i)}, \tag{3}$$

where $\hat{\sigma}^2(d_i)$ is the estimated variance of d_i given in (15) of Chapter 5. It then follows that

$$d_+ = \mathrm{TWD}/\mathrm{TW},$$

and the estimated sampling variance $\hat{\sigma}^2(d_+)$ is

$$\hat{\sigma}^2(d_+) = 1/\mathrm{TW}.$$

We also define the values $\mathrm{TW}_{(i)}$, $\mathrm{TWD}_{(i)}$, and $\mathrm{TWDS}_{(i)}$ with the ith study omitted:

$$\mathrm{TW}_{(i)} = \mathrm{TW} - 1/\hat{\sigma}^2(d_i), \tag{4}$$

$$\mathrm{TWD}_{(i)} = \mathrm{TWD} - d_i/\hat{\sigma}^2(d_i), \tag{5}$$

$$\mathrm{TWDS}_{(i)} = \mathrm{TWDS} - d_i^2/\hat{\sigma}^2(d_i). \tag{6}$$

Then the weighted mean $d_{+(i)}$ with the ith study omitted is

$$d_{+(i)} = \mathrm{TWD}_{(i)}/\mathrm{TW}_{(i)}. \tag{7}$$

It is now relatively easy to calculate the weighted mean with each study omitted by simply using the sums TW and TWD.

The residual of concern is

$$e_i = d_i - d_{+(i)}, \tag{8}$$

which reflects the discrepancy between the ith estimate of effect size and a composite of the other observations.

Example

The results of 11 studies of the effects of open versus traditional education on student attitude toward school are summarized in Table 1. The effect size estimate, the sample sizes, the statistics $\mathrm{TW}_{(i)}$ and $\mathrm{TWD}_{(i)}$, the weighted means with each study removed, and the raw residuals for each study are also given in Table 1. Note that study 1 has the largest residual and that study 3 has the third largest residual.

Although this definition of residual is appealing, it suffers from the defect that the variance of e_i is not constant across studies. Estimates from studies with large sample sizes will generally have residuals with small variances, and residuals from studies with small sample sizes will generally tend to have large variances. Thus it is difficult to tell when a residual e_i is "large."

This difficulty is resolved by using a standardized residual. That is, we obtain the variance of e_i and then use this variance to standardize the residual.

TABLE 1

Effect Size Estimates and Computations for Residuals for 11 Well-Controlled Studies of the Effects of Open Education on Attitude toward School

Study	n^E	n^C	d	$\mathrm{TW}_{(i)}$	$\mathrm{TWD}_{(i)}$	$d_{+(i)}$	e_i
1	40	40	−0.254	304.396	−97.824	0.321	−0.575
2	40	40	0.261	304.405	87.609	0.288	−0.027
3	90	90	−0.043	279.247	−94.719	0.339	−0.382
4	40	40	0.649	305.237	80.454	0.264	0.385
5	79	49	0.503	294.872	78.041	0.265	0.238
6	84	45	0.458	295.616	79.676	0.265	0.188
7	78	55	0.577	293.232	74.895	0.255	0.322
8	131	138	0.158	257.241	82.199	0.320	−0.162
9	38	110	0.588	296.895	76.708	0.258	0.330
10	38	93	0.392	297.679	82.374	0.277	0.115
11	20	23	−0.055	313.542	−93.373	0.298	−0.353

Since d_i and $d_{+(i)}$ are independent, the approximate estimated variance $\hat{\sigma}^2(e_i)$ is given by

$$\hat{\sigma}^2(e_i) = 1/\mathrm{TW}_{(i)} + \hat{\sigma}^2(d_i), \tag{9}$$

where $\mathrm{TW}_{(i)}$ is given in (4) and $\hat{\sigma}^2(d_i)$ is given in Eq. (15) of Chapter 5. The ith standardized residual $\tilde{e}_i$ is then

$$\tilde{e}_i = e_i/\hat{\sigma}(e_i). \tag{10}$$

When each study has the same population effect size and a moderately large sample size, the standardized residuals $\tilde{e}_1, \ldots, \tilde{e}_k$ have approximately standard normal distributions. Therefore, residuals larger than about 2.00 in absolute magnitude occur only about 5 percent of the time when the effect sizes are homogeneous. Standardized values as large as 3.00 in absolute magnitude are quite rare if the effect sizes are homogeneous. Therefore, standardized residuals that have large absolute values pinpoint a set of studies with different effect sizes.

Example

Table 2 gives a residual analysis for the 11 studies of the effects of open education on attitude toward school discussed in Section B.1. The residuals, weighted means with each study removed, and standardized residuals are given. Note that the largest residual and the largest standardized residual are those of study 1, and that larger residuals do not always correspond to larger standardized residuals. For example, study 4 has the second largest residual but the fifth largest standardized residual.

TABLE 2

Effect Size Estimates and Computations for Residuals from 11 Well-Controlled Studies of the Effects of Open Education on Attitude toward School

Study	n_i^E	n_i^C	d_i	$\hat{\sigma}^2(d_i)$	$\mathrm{TW}_{(i)}$	e_i	$\hat{\sigma}^2(e_i)$	$\tilde{e}_i$
1	40	40	−0.254	0.0504	304.396	−0.575	0.0536	−2.483
2	40	40	0.261	0.0504	304.405	−0.027	0.0542	−0.116
3	90	90	−0.043	0.0222	279.247	−0.382	0.0258	−2.379
4	40	40	0.649	0.0526	305.237	0.385	0.0633	1.530
5	79	49	0.503	0.0340	294.872	0.238	0.0373	1.232
6	84	45	0.458	0.0349	295.616	0.188	0.0381	0.963
7	78	55	0.577	0.0322	293.232	0.322	0.0358	1.703
8	131	138	0.158	0.0149	257.241	−0.162	0.0189	−1.178
9	38	110	0.588	0.0366	296.895	0.330	0.0400	1.649
10	38	93	0.392	0.0377	297.679	0.115	0.0408	0.569
11	20	23	−0.055	0.0935	313.542	−0.353	0.0967	−1.135

It appears that study 1 with $\tilde{e}_i = -2.48$ and study 3 with $\tilde{e}_i = -2.38$ are potential outliers. These are the same two studies that we identified in Section B.1 by using graphic methods.

B.3 THE USE OF HOMOGENEITY STATISTICS TO LOCATE OUTLIERS

Another procedure for locating outliers is to calculate the homogeneity test statistic Q when each observation in turn is deleted. The value of the homogeneity statistic will usually show a dramatic change when an outlier is deleted. The homogeneity statistic with the ith estimate deleted is easily calculated using the quantities $\mathrm{TW}_{(i)}$, $\mathrm{TWD}_{(i)}$, and $\mathrm{TWDS}_{(i)}$ defined by (4), (5), and (6). We write $Q_{(i)}$ to denote the homogeneity statistic with the ith study deleted:

$$Q_{(i)} = \mathrm{TWDS}_{(i)} - (\mathrm{TWD}_{(i)})^2/\mathrm{TW}_{(i)}. \tag{11}$$

Example

The studies of the relationship between open education and attitude toward school were also analyzed via the homogeneity statistic method. Table 3 presents the values of the homogeneity statistics $Q_{(i)}$ with each study removed from the collection. This method also identifies study 1 as a potential outlier, since the value of the homogeneity statistic with study 1 removed is the smallest of the $Q_{(i)}$ values. Note that study 3 has the next smallest $Q_{(i)}$

TABLE 3

Effect Size Estimates and Computations for Homogeneity Statistics with Each Study Deleted for 11 Well-Controlled Studies of the Effects of Open Education on Attitude toward School

Study	n^E	n^C	d	$\mathrm{TW}_{(i)}$	$\mathrm{TWD}_{(i)}$	$\mathrm{TWDS}_{(i)}$	$Q_{(i)}$
1	40	40	−0.254	304.396	97.824	48.431	16.993
2	40	40	0.261	304.405	87.609	48.360	23.146
3	90	90	−0.043	279.247	94.719	49.627	17.499
4	40	40	0.649	305.237	80.454	41.708	20.502
5	79	49	0.503	294.872	78.014	42.281	21.641
6	84	45	0.458	295.616	79.676	43.707	22.232
7	78	55	0.577	293.232	74.895	39.388	20.259
8	131	138	0.158	257.241	82.199	48.038	21.772
9	38	110	0.588	296.895	76.708	40.258	20.439
10	38	93	0.392	297.679	82.374	45.630	22.835
11	20	23	−0.055	313.542	93.373	49.678	21.872

value. Therefore this method suggests the same two potential outliers as did the graphic method presented in Section B.1 and the method based on standardized residuals presented in Section B.2. These methods will usually identify the same observations as outliers, and indeed the complete ordering of the $Q_{(i)}$ statistic is the same as that of the standardized residuals.

C. DIAGNOSTIC PROCEDURES FOR CATEGORICAL MODELS

Techniques for fitting models with categorical predictor variables to effect sizes were discussed in Chapter 7. The test of model specification involves the within-class fit statistic Q_W described in Section C.3 of Chapter 7. This within-class fit statistic is just the sum

$$Q_W = Q_{W_1} + \cdots + Q_{W_p}$$

of the within-class fit statistics for each of the classes. The overall fit of the model is determined by the fit of the model in each class. Because each subclass is supposed to have homogeneous effect sizes, the methods described in Section B can be applied to each subclass separately.

Occasionally it may happen that the within-class fit statistics $Q_{W_1}, \ldots, Q_{W_p}$ are all nonsignificant at some prescribed significance level α, but the overall within-class fit statistic Q_W is significant at level α. This situation typically occurs when most of the within-class fit statistics are large, but individually not quite large enough to be significant. The increased sensitivity of the pooled test detects the departure from an adequate fit of the model. Such a situation suggests that the problem may not be one of outliers, but may be the effect of another variable influencing the effect sizes. In this case, another classification variable may be needed to obtain an adequate model for effect sizes.

D. DIAGNOSTIC PROCEDURES FOR GENERAL LINEAR MODELS

The issue of diagnosing problems in regression analysis is an extremely complicated one and is associated with a voluminous literature. We do not attempt even to survey this literature, but present approaches that have proved particularly useful for detecting outliers in quantitative research synthesis. Surveys of classical methods for identifying outliers in linear models are given, for example, in books by Belsley, Kuh, and Welsch (1980), Barnett and Lewis (1978), Cook and Weisberg (1982), and Miller (1981). Some of this literature also includes methods for identifying sources of collinearity. Many of the methods suggested can be applied directly to problems in fitting linear models in research synthesis. Others can be used after very minor adaptations. For example, analyses of leverage by the so-called hat matrix (see Hoaglin & Welsch, 1978) can be applied with the minor modification of incorporating the variance matrix $\mathbf{\Sigma}_d$.

D.1 THE USE OF RESIDUALS TO LOCATE OUTLIERS IN GENERAL LINEAR MODELS

Residuals are also important for locating outliers in general linear models. Using the notation of Chapter 8, let $\boldsymbol{\delta} = (\delta_1, \ldots, \delta_k)'$ and $\mathbf{d} = (d_1, \ldots, d_k)'$ be vectors of population effect sizes and their sample estimates for the k studies. Further, suppose that each effect size depends on parameters $\beta_1, \ldots, \beta_p$ in a regression format. That is, $\boldsymbol{\delta} = \mathbf{X}\boldsymbol{\beta}$, where

$$\mathbf{X} = \begin{bmatrix} x_{11} & \cdots & x_{1p} \\ x_{21} & \cdots & x_{2p} \\ \vdots & & \vdots \\ x_{k1} & \cdots & x_{kp} \end{bmatrix}$$

denotes the design matrix, and $\boldsymbol{\beta} = (\beta_1, \ldots, \beta_p)'$ is a vector of regression coefficients.

The residual examined here is the difference between the sample estimate d_i of δ_i and an estimate of δ_i that does not incorporate information from d_i into the estimate of $\boldsymbol{\beta}$. More specifically, we estimate δ_i from $x_{i1}, \ldots, x_{ip}$ and and estimate $\hat{\boldsymbol{\beta}}_{(i)}$ of $\boldsymbol{\beta}$ calculated with the ith study removed. This amounts to a standard regression analysis with $k - 1$ studies and using the regression coefficient so estimated to predict δ_i from $x_{i1}, \ldots, x_{ip}$.

Because we omit one study with each regression, we need to devise an adequate notation to indicate the computations when a particular study is omitted. The notation we adopt is to denote the omitted study by adding a parenthesis. Thus, for example, d_j is the estimate of the jth effect size, and $d_{j(i)}$ is the estimate (the jth fitted value) with study i omitted. Similarly, $\mathbf{X}_{(i)}$ is the design matrix without the ith study.

For each regression we obtain estimates of $\beta_1, \ldots, \beta_p$. We denote the estimates by $\hat{\boldsymbol{\beta}}_{(i)} = (\hat{\beta}_{1(i)}, \ldots, \hat{\beta}_{p(i)})'$ to indicate that study i is omitted. Each analysis also yields an estimate $\mathbf{d}_{(i)} = (d_{1(i)}, \ldots, d_{k(i)})' = \mathbf{X}\hat{\boldsymbol{\beta}}_{(i)}$ of $\boldsymbol{\delta} = (\delta_1, \ldots, \delta_k)'$, so that we actually have k different vectors of estimates $\mathbf{d}_{(1)}, \ldots, \mathbf{d}_{(k)}$. We also have the observed values $\mathbf{d} = (d_1, \ldots, d_k)'$. The difference

$$e_{j(i)} = d_{j(i)} - d_j$$

is the residual of the estimate of the jth fitted value when the ith study is omitted versus the estimate of effect size actually observed for the jth study. This leads to the following set of residuals:

Study omitted	Study 1		Study k
1	$e_{1(1)} = d_{1(1)} - d_1, \ldots,$		$e_{k(1)} = d_{k(1)} - d_k$
$\vdots$	$\vdots$		$\vdots$
k	$e_{1(k)} = d_{1(k)} - d_1, \ldots,$		$e_{k(k)} = d_{k(k)} - d_k$

Generally, a study that is an outlier will exhibit a large difference between the obtained and fitted values; hence we would usually expect outliers to yield large values of $e_{i(i)}$. However, studies with large sample sizes tend to have residuals $e_{i(i)}$ with small variances, whereas studies with small sample sizes have residuals with large variances. Thus, as before, the absolute magnitude of residuals alone is not necessarily a useful guide as to when residuals are large enough to be unusual.

To standardize these residuals we need to determine the approximate estimated variance $\hat{\sigma}^2(e_{i(i)})$:

$$\hat{\sigma}^2(e_{i(i)}) = \mathbf{x}_i'(\mathbf{X}_{(i)}'\hat{\mathbf{\Sigma}}_{d(i)}^{-1}\mathbf{X}_{(i)})^{-1}\mathbf{x}_i + \hat{\sigma}^2(d_i),$$

where

$$\mathbf{x}_i = (x_{i1}, \ldots, x_{ip})', \qquad \mathbf{X}_{(i)} = (\mathbf{x}_1, \ldots, \mathbf{x}_{i-1}, \mathbf{x}_{i+1}, \ldots, \mathbf{x}_k)',$$
$$\hat{\mathbf{\Sigma}}_{d(i)} = \text{Diag}[\hat{\sigma}^2(d_1), \ldots, \hat{\sigma}^2(d_{i-1}), \hat{\sigma}^2(d_{i+1}), \ldots, \hat{\sigma}^2(d_k)],$$

and $\hat{\sigma}^2(d_i)$ is given by Eq. (15) of Chapter 5. The standardized residual $\tilde{e}_{i(i)}$ corresponding to $e_{i(i)}$ is given by

$$\tilde{e}_{i(i)} = e_i/\hat{\sigma}(e_{i(i)}). \tag{12}$$

When the effect sizes for each study are determined by the model $\boldsymbol{\delta} = \mathbf{X}\boldsymbol{\beta}$, and each study has a moderately large sample size, the standardized residuals $\tilde{e}_{1(1)}, \ldots, \tilde{e}_{k(k)}$ will have approximately standard normal distributions. Therefore, standardized residuals larger than about 2.00 in absolute magnitude will happen only about 5 percent of the time when the model holds. Standardized residuals as large as 3.00 in absolute magnitude would be quite unusual when the population effect sizes of all the studies are determined by the linear model.

Example

Some data from seven studies of sex differences in visual–spatial ability were given in Table 3 of Chapter 2. Previous analyses suggested that the magnitude of the sex difference is linearly related to the year in which the study was published, and that sex differences have been decreasing over time (Rosenthal & Rubin, 1982a). We began by fitting a linear model that uses an intercept and year of publication to predict effect size. The data vector is

$$\mathbf{d} = (0.60, 0.41, 0.52, 0.83, 0.04, 0.48, 0.31)',$$

and the design matrix is

$$\mathbf{X} = \begin{bmatrix} 1 & 1 & 1 & 1 & 1 & 1 & 1 \\ 61 & 67 & 67 & 72 & 75 & 75 & 78 \end{bmatrix}'.$$

Standardized residuals were calculated using SAS Proc Matrix. The calculation of the standardized residuals $\tilde{e}_{1(1)}$, $\tilde{e}_{2(2)}$, $\tilde{e}_{3(3)}$, $\tilde{e}_{4(4)}$, $\tilde{e}_{5(5)}$, $\tilde{e}_{6(6)}$, and $\tilde{e}_{7(7)}$ is illustrated in Table 4. Most of the standardized residuals are relatively small, but $\tilde{e}_{4(4)} = -15.06$, which is much larger than the reference values of 2 to 3. Thus study 4 probably does not conform to the same model as do the other studies. As we shall see in Section D.3, this study was also identified as an outlier using other techniques, and closer examination of the study suggests that there are good reasons to expect the estimate from this study to behave as an outlier.

TABLE 4

Data from Seven Studies of Sex Differences in Visual–Spatial Ability and Calculations for the Analysis of Residuals

Study	N_i	d_i	$d_{i(i)}$	$e_{i(i)}$	$\mathbf{x}_i'(\mathbf{X}_{(i)}'\hat{\Sigma}_{d(i)}^{-1}\mathbf{X}_{(i)})^{-1}\mathbf{x}_i$	$\hat{\sigma}^2(d_i)$	$\hat{\sigma}^2(e_{i(i)})$	$\tilde{e}_{i(i)}$
1	128	0.60	0.43	−0.17	0.0021	0.0327	0.0348	−0.91
2	6167	0.41	0.34	−0.07	0.0030	0.0067	0.0097	−0.71
3	355	0.52	0.13	−0.39	0.0004	0.0117	0.0121	−3.54
4	2925	0.83	0.14	−0.69	0.0005	0.0016	0.0021	−15.06
5	105	0.04	0.13	0.09	0.0009	0.0381	0.0390	0.46
6	102	0.48	0.14	−0.34	0.0011	0.0403	0.0414	−1.67
7	1233	0.31	0.30	−0.01	0.0060	0.0033	0.0093	−0.10

D.2 CHANGES IN REGRESSION COEFFICIENTS

Another method for finding outliers is based on the fact that outliers often change the value of the estimated regression coefficients. Therefore, examination of the difference $\hat{\boldsymbol{\beta}} - \hat{\boldsymbol{\beta}}_{(i)}$ between estimates of $\hat{\boldsymbol{\beta}}$ including and not including data from the ith study is often useful for finding influential observations. This difference is analogous to the difference $d_+ - d_{+(i)}$ discussed in Section B.2, and it is used in much the same way.

D.3 THE USE OF MODEL SPECIFICATION STATISTICS TO LOCATE OUTLIERS

An alternative to locating outliers is to examine values of the model specification test statistic. The basic idea is to compute the value of this statistic as each observation, in turn, is deleted. The model misspecification statistic Q_E will usually change dramatically when an outlier is deleted, but is altered less dramatically when other studies are deleted. The model specification statistic $Q_{E(i)}$, with the ith study removed, is given by

$$Q_{E(i)} = \mathbf{d}'_{(i)}\hat{\mathbf{\Sigma}}^{-1}_{d(i)}\mathbf{d}_{(i)} - \hat{\boldsymbol{\beta}}'_{(i)}\hat{\mathbf{X}}'_{(i)}\hat{\mathbf{\Sigma}}^{-1}_{d(i)}\mathbf{X}_{(i)}\hat{\boldsymbol{\beta}}_{(i)}, \tag{13}$$

where $\hat{\boldsymbol{\beta}}_{(i)}$ are the estimates of $\beta_1, \ldots, \beta_p$ when the ith study is omitted.

Example

Some data from seven studies of sex differences in visual–spatial ability were given in Table 3 of Chapter 2. Previous analyses suggested that the magnitude of the sex difference effect size is related to the year in which the study was done; that is, sex differences have been decreasing over time (Rosenthal & Rubin, 1982a). We begin by fitting a linear model that uses an intercept and year of publication to predict effect size.

The value of the test statistic for model specification is given by Eq. (10) of Chapter 8, and was computed from a statistical package (SAS Proc GLM) to be $Q_E = 99.951$, which greatly exceeds the 95-percent critical value of the chi-square distribution with $7 - 2 = 5$ degrees of freedom. To explore the question of whether outliers are present, we compute $Q_{E(1)}, \ldots, Q_{E(7)}$:

Study	1	2	3	4	5	6	7
$Q_{E(i)}$	92.96	99.82	99.17	4.15	46.99	99.86	40.39

These fit statistics show that the fit of the model improves dramatically when study 4 (Backman, 1972) is deleted. Note that the effect size associated with this study is larger than that of the other six studies. Subsequent investigation showed that the statistics used to calculate the effect size estimate in the Backman study were the means and standard deviations of group means (Becker & Hedges, 1984). The means of randomly selected groups (as in the Backman study) are less variable than the observations on which they are based. Consequently the effect size calculated from group means in the

Backman study is expected to be larger than the effect sizes of the other studies based on ungrouped data. That is, the effect size associated with the Backman study is not comparable to those of other studies. Modeling the data in the remaining studies, we find that the value of $Q_{E(4)} = 4.15$ is not significant. In fact, it is quite near the mean of the chi-square distribution with four degrees of freedom. Thus the model that predicts effect size from year of publication is quite consistent with the data from the six studies that calculated effect size from ungrouped data.

It is interesting to note that two published analyses of these data (Hyde, 1981; Rosenthal & Rubin, 1982a), which did not analyze the goodness of fit of their data analysis models, failed to identify the Backman data point as basically incomparable with the others. The present analysis shows that tests of goodness of fit often help to achieve a better understanding of the data.

E. DIAGNOSTIC PROCEDURES FOR COMBINING ESTIMATES OF CORRELATION COEFFICIENTS

Diagnostic procedures for models involving correlation coefficients are completely analogous to those for fixed effects models for effect sizes. The statistical procedures described in Sections C.1, D.1, and E of Chapter 11 use $z_1, \ldots, z_k$, the z-transforms of k independent correlations $r_1, \ldots, r_k$. The diagnostic procedures described in this chapter are applied to models for $z_1, \ldots, z_k$ by substituting z_i for d_i, z_+ for d_+, and $n_i - 3$ for $1/\hat{\sigma}^2(d_i)$ wherever they occur in formulas (1)–(13). The interpretation of the statistic so obtained is exactly the same as in the case of effect sizes that is described in this chapter.

F. TECHNICAL COMMENTARY*

Computational Formulas for $\hat{\boldsymbol{\beta}}_{(i)}$, $\hat{\boldsymbol{\Sigma}}_{(i)}$, and $Q_{E(i)}$: In Section D we presented diagnostic procedures using standardized residuals $\tilde{e}_{i(i)}$ and the homogeneity statistic $Q_{E(i)}$ with the ith study removed. These statistics, computed for each study, involved the estimated regression coefficient $\hat{\boldsymbol{\beta}}_{(i)}$ with the ith study removed, and the covariance matrix $\hat{\boldsymbol{\Sigma}}_{(i)}$ of $\hat{\boldsymbol{\beta}}_{(i)}$. For large models (large p) or large numbers of studies (large k), computing these diagnostic statistics from the defining formulas would consume a great deal of computer time, since the inverse of a new $p \times p$ matrix $\mathbf{X}'_{(i)}\hat{\boldsymbol{\Sigma}}_{d(i)}^{-1}\mathbf{X}_{(i)}$ must be computed for each study.

A simple computational formula permits calculation of all the necessary inverses from $(\mathbf{X}'\hat{\boldsymbol{\Sigma}}_d^{-1}\mathbf{X})^{-1}$.

Define $\mathbf{A} = \mathbf{X}'\hat{\boldsymbol{\Sigma}}_{d(i)}^{-1}\mathbf{X}$ and $\mathbf{A}_{(i)} = \mathbf{X}'_{(i)}\hat{\boldsymbol{\Sigma}}_{d(i)}^{-1}\mathbf{X}_{(i)}$. Then $\mathbf{A}_{(i)} = \mathbf{A} - \mathbf{x}_i\mathbf{x}'_i/\hat{\sigma}^2(d_i)$. Further, by a well-known computational formula,

$$\mathbf{A}_{(i)}^{-1} = \mathbf{A}^{-1} + \frac{\mathbf{A}^{-1}\mathbf{x}_i\mathbf{x}'_i\mathbf{A}^{-1}}{\hat{\sigma}^2(d_i) - \mathbf{x}'_i\mathbf{A}^{-1}\mathbf{x}_i}.$$

Therefore,

$$\hat{\boldsymbol{\Sigma}}_{(i)} = \mathbf{A}_{(i)}^{-1},$$

$$\hat{\boldsymbol{\beta}}_{(i)} = \mathbf{A}_{(i)}^{-1}\mathbf{X}'_{(i)}\hat{\boldsymbol{\Sigma}}_{d(i)}^{-1}\mathbf{d}_{(i)},$$

$$Q_{\mathrm{E}(i)} = \mathbf{d}'_{(i)}\hat{\boldsymbol{\Sigma}}_{d(i)}^{-1}\mathbf{d}_{(i)} - \hat{\boldsymbol{\beta}}'_{(i)}\mathbf{X}'_{(i)}\hat{\boldsymbol{\Sigma}}_{d(i)}^{-1}\mathbf{X}_{(i)}\hat{\boldsymbol{\beta}}_{(i)},$$

which gives all of the diagnostic statistics used in Section D.

CHAPTER 13

Clustering Estimates of Effect Magnitude

Methods for detecting variation in effect sizes have already been provided. In Chapter 5 a homogeneity statistic was used to determine whether the effect sizes from a series of studies exhibit any variability beyond that which could be expected due to sampling error. Statistical methods for modeling the variability in effect sizes using known explanatory variables were given in Chapters 7 and 8. Similar methods for modeling the variability in a series of correlation coefficients were given in Chapter 11. These methods permit the investigator to "account for" variability in effect sizes or correlations using various coded characteristics of studies as explanatory variables in a general linear model. However, these methods require that the explanatory variables be known a priori.

Sometimes the factors that influence effect sizes or correlations are not known a priori. That is, there is believed to be variation in effect sizes, but it is not obvious which characteristics of studies would be useful in explaining this variation. In this situation, the methods previously mentioned for fitting models to effect sizes or correlations are not suitable. Because of this, we examine the estimates of effect size directly to gain some insight about those characteristics that might explain the variations in effect sizes.

Specifically, suppose that $\theta_1, \ldots, \theta_k$ represent indices or parameters of a true (population) effect magnitude. If the true effect magnitude is not the same for all studies in a collection, i.e., if $\theta_1 = \theta_2 = \cdots = \theta_k$ is false, then we may wish to partition these heterogeneous populations into more homogeneous groups in which the θ's within clusters are close, but the θ's between clusters are separated.

In this chapter we discuss two methods for grouping estimates of effect magnitude into homogeneous classes proposed by Hedges and Olkin (1983a). One procedure decomposes a set of effect magnitude indices into disjoint or nonoverlapping classes, whereas in another procedure the decomposition is into overlapping groups. In either case methods are provided for determining statistical significance levels of the clusters. Note that methods given in Chapter 7 for determining the significance of groupings specified a priori do not apply to groups that are generated on the basis of the clustering procedures. Significance levels of a posteriori clusterings must be obtained by special techniques. This is similar to obtaining multiple comparisons in the analysis of variance, where special methods are needed to assess the significance levels of post hoc contrasts.

When homogeneous clusters have been obtained, the investigator can then estimate a pooled effect magnitude within each cluster. Procedures for estimating a common correlation from several independent sample correlations were given in Chapter 11, and procedures for estimating a common effect size from several independent effect size estimates were given in Chapter 6. Examination of the pooled effect magnitude can lead to hypotheses about the relationship between study characteristics and effect magnitude—hypotheses that can be tested in future studies.

Both procedures are based on clustering theory for standard normal random variables. However, correlations and standardized mean differences are generally not normally distributed except in large samples. We use large sample theory to transform the estimators to approximately standard normal variates. We first explain the clustering procedures for arbitrary unit normal variates, and then in subsequent sections show how to apply the methods to correlations and effect sizes.

A. THEORY FOR CLUSTERING UNIT NORMAL RANDOM VARIABLES

Suppose that $U_1, \ldots, U_k$ is a sample from a normal distribution with unit variance and $U_{(1)} \le U_{(2)} \le \cdots \le U_{(k)}$ are the ordered values arranged in ascending order of magnitude. These order statistics define $k - 1$ gaps or differences between adjacent values

$$\begin{aligned} g_{12} &= U_{(2)} - U_{(1)}, \\ g_{23} &= U_{(3)} - U_{(2)}, \\ &\ \vdots \qquad \vdots \\ g_{k-1,k} &= U_{(k)} - U_{(k-1)}. \end{aligned}$$

Because these gaps are random variables, some will be larger than a prespecified length L and some will be smaller. In particular, we could (in principle) determine a number L_α such that the expected proportion of the $k - 1$ gaps that exceed L_α is some preset significance level α.

Suppose the k variates have the same mean. Then if we declare that a gap greater than L_α is significantly large, we will be incorrect 100α percent of the time. On the other hand, when all of the variates do not have the same mean, then at least one gap is larger. This provides the basis for a significance test for clusterings based on gaps. That is, if the k variates are divided into clusters at a gap of length L_α, the clustering is significant at the 100α-percent level of significance in the sense that the probability of at least one gap larger than L_α is equal to α when all variates have the same distribution. Table 1 provides a tabulation of the minimum significant gap lengths for various k and significance levels α. Thus, for example, in a set of 50 studies, at a 5-percent level of significance, gaps of length greater than or equal to 1.34 are significant.

A.1 DISJOINT CLUSTERING

The procedure for forming disjoint clusters from k variates consists in successively splitting the variates into clusters at one of the gaps. The first step is to partition the k variates into two groups at the largest gap. The next division occurs at the second largest gap and divides one of the two groups obtained in the first step. The process continues until there are k "groups"; i.e., all the variates are in a single group. Alternatively, we stop the clustering process when the gaps no longer exceed the critical value L_α for some significance level α.

Example

The process is illustrated by an example with $k = 5$ and the observations arranged in ascending order:

$$U_{(1)} = 1.7, \qquad U_{(2)} = 3.4, \qquad U_{(3)} = 6.9, \qquad U_{(4)} = 8.5, \qquad U_{(5)} = 10.3.$$

TABLE 1

Significance Gap Length between Unit Normal Order Statistics for Samples of Size k and Statistical Significance Level α

	α							
k	0.40	0.30	0.20	0.10	0.05	0.01	0.005	0.001
3	1.37	1.56	1.80	2.17	2.51	3.19	3.46	4.01
4	1.34	1.49	1.69	2.00	2.29	2.90	3.14	3.65
5	1.28	1.41	1.58	1.87	2.14	2.70	2.93	3.42
6	1.22	1.34	1.50	1.77	2.02	2.57	2.79	3.26
7	1.17	1.28	1.43	1.69	1.93	2.47	2.68	3.15
8	1.13	1.23	1.38	1.62	1.86	2.39	2.60	3.07
9	1.09	1.19	1.33	1.57	1.81	2.33	2.54	3.00
10	1.05	1.15	1.29	1.53	1.76	2.28	2.49	2.94
11	1.03	1.12	1.25	1.49	1.72	2.23	2.44	2.89
12	1.00	1.09	1.23	1.46	1.69	2.20	2.40	2.85
13	0.98	1.07	1.20	1.43	1.66	2.17	2.37	2.81
14	0.96	1.05	1.18	1.41	1.64	2.14	2.34	2.78
15	0.94	1.03	1.16	1.39	1.62	2.11	2.31	2.75
20	0.87	0.96	1.09	1.31	1.53	2.02	2.21	2.64
30	0.80	0.89	1.01	1.23	1.44	1.90	2.09	2.50
40	0.76	0.84	0.97	1.17	1.38	1.83	2.02	2.42
50	0.73	0.82	0.93	1.14	1.34	1.78	1.97	2.36
60	0.71	0.79	0.91	1.11	1.31	1.75	1.93	2.32
70	0.70	0.78	0.89	1.09	1.28	1.72	1.90	2.28
80	0.68	0.76	0.88	1.07	1.27	1.70	1.87	2.25
90	0.67	0.75	0.86	1.06	1.25	1.67	1.85	2.23
100	0.66	0.74	0.85	1.04	1.23	1.66	1.83	2.21
120	0.65	0.72	0.83	1.02	1.21	1.63	1.80	2.17
140	0.64	0.71	0.82	1.01	1.19	1.60	1.77	2.15
160	0.63	0.70	0.81	0.99	1.18	1.59	1.75	2.12
180	0.62	0.69	0.80	0.98	1.16	1.57	1.74	2.10
200	0.61	0.78	0.79	0.97	1.15	1.55	1.72	2.08
250	0.60	0.67	0.77	0.95	1.13	1.53	1.69	2.05
300	0.59	0.66	0.76	0.93	1.11	1.50	1.66	2.02
350	0.58	0.65	0.75	0.92	1.10	1.48	1.65	2.00
400	0.57	0.64	0.74	0.91	1.08	1.47	1.63	1.98
450	0.56	0.63	0.73	0.90	1.07	1.46	1.61	1.96
500	0.56	0.63	0.72	0.89	1.06	1.45	1.60	1.95
550	0.55	0.62	0.72	0.89	1.06	1.44	1.59	1.94
600	0.55	0.62	0.71	0.88	1.05	1.43	1.58	1.93
650	0.54	0.61	0.71	0.87	1.04	1.42	1.57	1.92
700	0.54	0.61	0.70	0.87	1.04	1.41	1.57	1.91
750	0.54	0.60	0.70	0.86	1.03	1.40	1.56	1.90
800	0.53	0.60	0.70	0.86	1.03	1.40	1.55	1.89
850	0.53	0.60	0.69	0.86	1.02	1.39	1.54	1.88
900	0.53	0.59	0.69	0.85	1.02	1.39	1.54	1.88
950	0.53	0.59	0.69	0.85	1.01	1.38	1.53	1.87
1000	0.53	0.59	0.68	0.85	1.01	1.38	1.53	1.87

The gap lengths are

$$g_{12} = 1.7, \quad g_{23} = 3.5, \quad g_{34} = 1.6, \quad g_{45} = 1.8.$$

The first step is to separate the five variates into two groups at the largest gap between $U_{(2)}$ and $U_{(3)}$ to form the groups $\{U_{(1)}, U_{(2)}\}$ and $\{U_{(3)}, U_{(4)}, U_{(5)}\}$. The next step is to separate the variates into groups at the next largest gap between $U_{(4)}$ and $U_{(5)}$ to yield the groups $\{U_{(1)}, U_{(2)}\}$, $\{U_{(3)}, U_{(4)}\}$, and $\{U_{(5)}\}$. The third step separates $U_{(1)}$ from $U_{(2)}$ to yield the groups $\{U_{(1)}\}$, $\{U_{(2)}\}$, $\{U_{(3)}, U_{(4)}\}$, and $\{U_{(5)}\}$. This resulting hierarchical cluster map is illustrated in Fig. 1.

Examination of Table 1 shows that with $k = 5$ and $\alpha = 0.05$ the minimum significant gap length is $L_{0.05} = 2.14$. Consequently, only the first step clustering is significant at the $\alpha = 0.05$ level. In this example, we see that the overall mean is 6.2, whereas the means of the two clusters are 2.6 and 8.6. Thus the overall mean is a poor representation of the typical value in either of the clusters.

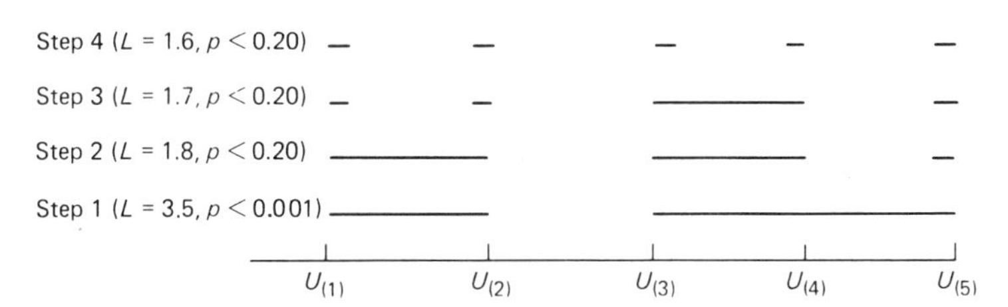

Fig. 1 Cluster map of hierarchical decomposition of five variates into disjoint clusters.

A.2 OVERLAPPING CLUSTERING

A different clustering scheme can be used to divide the studies into overlapping clusters. The procedure is analogous to the Bonferroni multiple comparison test described by Dunn (1961) for testing the difference between means in the analysis of variance. The procedure used to form these clusters at a desired (overall) significance level α depends on the computed $100(1 - \alpha^*)$-percent critical value $c(k)$ of the range of k observations from a standard normal distribution, where $\alpha^* = 1 - (1 - \alpha^{1/(k-1)})$. Table 2 provides critical values $c(k)$ of the range of k observations from a standard normal distribution for various α levels.

The procedure starts with a comparison between the largest and the smallest of the k varieties, namely, $U_{(k)} - U_{(1)}$. If this difference is larger than the critical value $c(k)$, test the difference $U_{(k)} - U_{(2)}$ using the critical value $c(k - 1)$. If this is significant, test $U_{(k)} - U_{(3)}$ using the critical value $c(k - 2)$.

TABLE 2

Critical Values $c(k)$ for Bonferroni Multiple Comparisons of the Range of Standard Normal Variates for Overall Level of Significance α

	α							
k	0.40	0.30	0.20	0.10	0.05	0.01	0.005	0.001
2	1.19	1.47	1.81	2.33	2.77	3.64	3.97	4.65
3	2.33	2.57	2.87	3.30	3.68	4.42	4.71	5.32
4	2.95	3.17	3.43	3.83	4.17	4.85	5.11	5.73
5	3.37	3.57	3.82	4.18	4.50	5.15	5.39	5.95
6	3.68	3.86	4.10	4.44	4.75	5.38	5.62	6.15
7	3.92	4.10	4.32	4.65	4.95	5.55	5.80	6.30
8	4.12	4.29	4.50	4.83	5.11	5.71	5.93	6.45
9	4.28	4.45	4.66	4.97	5.25	5.82	6.05	6.54
10	4.33	4.59	4.79	5.10	5.37	5.94	6.16	6.65
11	4.55	4.71	4.91	5.21	5.48	6.04	6.26	6.74
12	4.67	4.82	5.02	5.31	5.57	6.16	6.34	6.82
13	4.77	4.92	5.11	5.40	5.66	6.20	6.41	6.89
14	4.86	5.01	5.20	5.48	5.74	6.27	6.48	6.95
15	4.94	5.09	5.27	5.55	5.80	6.34	6.55	7.01
20	5.28	5.42	5.59	5.86	6.10	6.59	6.81	7.26
50	6.22	6.33	6.48	6.72	6.93	7.38	7.55	7.97

Continue testing $U_{(k)}$ against successively larger variates until one of the differences is less than the appropriate critical value. The procedure is then repeated by omitting $U_{(k)}$ and letting $U_{(k-1)}$ take the role of $U_{(k)}$. This procedure is continued until we are left with $U_{(2)}$ and $U_{(1)}$.

Remark: The above procedure is forward clustering in the sense that we start with the small ordered values $U_{(1)}, U_{(2)}, \ldots$ and compare these with the largest $U_{(k)}$. However, we could also define a backward clustering in which we compare the large ordered values $U_{(k)}, U_{(k-1)}, \ldots$ with the smallest $U_{(1)}$. These two procedures need not yield the same separations. Disagreements between forward and backward stepwise solutions occur frequently. One common example is that of forward and backward stepwise regression.

Example

Using the data of the previous example,

$$U_{(1)} = 1.7, \qquad U_{(2)} = 3.4, \qquad U_{(3)} = 6.9, \qquad U_{(4)} = 8.5, \qquad U_{(5)} = 10.3,$$

we calculate

$$U_{(5)} - U_{(1)} = 8.6,\quad U_{(4)} - U_{(1)} = 6.8,\quad U_{(3)} - U_{(1)} = 5.2,\quad U_{(2)} - U_{(1)} = 1.7,$$

$$U_{(5)} - U_{(2)} = 6.9,\quad U_{(4)} - U_{(2)} = 5.1,\quad U_{(3)} - U_{(2)} = 3.5,$$

$$U_{(5)} - U_{(3)} = 3.4,\quad U_{(4)} - U_{(3)} = 1.6,$$

$$U_{(5)} - U_{(4)} = 1.8.$$

Since $U_{(5)} - U_{(1)} = 8.6$ is larger than $c(5) = 5.15$ (at the 1-percent level of significance), compare $U_{(5)} - U_{(2)} = 6.9$ with $c(4) = 4.85$. Since the gap exceeds the critical point, compute $U_{(5)} - U_{(3)} = 3.4$, which is smaller than $c(3) = 4.42$. Therefore, $U_{(5)}$ is significantly different from $U_{(1)}$ and $U_{(2)}$, but not $U_{(3)}$ and $U_{(4)}$.

Next note that $U_{(4)} - U_{(1)} = 6.8$ is larger than $c(4) = 4.85$. Since $U_{(4)} - U_{(2)} = 5.1$ is larger than $c(3) = 4.42$, we compute $U_{(4)} - U_{(3)} = 1.6 < c(2) = 3.64$. This implies that $U_{(4)}$ is significantly different from $U_{(1)}$ and $U_{(2)}$, but not from $U_{(3)}$.

The next step is to compare $U_{(3)} - U_{(1)} = 5.2$ with $c(3) = 4.42$ and $U_{(3)} - U_{(2)} = 3.5$ with $c(2) = 3.64$. Therefore $U_{(3)}$ differs significantly from $U_{(1)}$, but not from $U_{(2)}$. Finally, $U_{(2)} - U_{(1)} = 1.7 < c(2) = 3.64$, so $U_{(2)}$ does not differ significantly from $U_{(1)}$. In this case, there are three overlapping clusters, $\{U_{(1)}, U_{(2)}\}$, $\{U_{(2)}, U_{(3)}\}$, and $\{U_{(3)}, U_{(4)}, U_{(5)}\}$. These results are summarized in the following diagram, which lists the data and underscores the variates that do not differ significantly:

$U_{(1)}$	$U_{(2)}$	$U_{(3)}$	$U_{(4)}$	$U_{(5)}$
1.7	3.4	6.9	8.5	10.3

Two of the clusters obtained with the present method are the same as those obtained by the disjoint clustering procedure. The third cluster overlaps the other two. The means of the three clusters are 2.6, 5.2, and 8.6. The overall mean of 6.2 seems to be a poor reflection of typical values of all but the third cluster containing $U_{(2)}$ and $U_{(3)}$.

B. CLUSTERING CORRELATION COEFFICIENTS

Suppose that we have k populations with samples of size $n_1, \ldots, n_k$. Denote by $r_1, \ldots, r_k$ and $\rho_1, \ldots, \rho_k$ the sample and population correlation coefficients. The clustering procedures described previously cannot be directly applied to the correlations $r_1, \ldots, r_k$ because the variance of a correlation coefficient r_i depends on the population correlation coefficient ρ_i. However, if we apply Fisher's z-transform as discussed in (19) of Chapter 6, then the transformed correlations are approximately normal variates and the theory for clustering normal variates applies.

Before any clustering takes place, we need to test for the homogeneity of the correlations. A simple test for the homogeneity of the ρ_i was given in Section D of Chapter 11. If the homogeneity hypothesis is rejected, then the correlations can be considered as samples from different populations, in which case the clustering procedure can be used to create groups of studies for which the correlations are homogeneous.

For simplicity of exposition, suppose that the sample sizes are equal, i.e., $n_i = \cdots = n_k = n$. We later discuss how to modify this procedure when the sample sizes differ. Transform the data by letting

$$U_i = \sqrt{n - 3}\, z(r_i), \qquad i = 1, \ldots, k, \tag{1}$$

where $z(r)$ is Fisher's z-transformation [see (19) of Chapter 6 or (4) of Chapter 11]. (A table that permits ready evaluation of z given r or vice versa is provided in Appendix C. However, this computation is quite simple with a hand calculator.) The U_i are independently distributed with a common mean and unit variance, and we can directly apply the previous theory. When the sample sizes are unequal, the procedure becomes more complicated, and we later discuss a modification that permits us to handle this case.

Example

The techniques described above are applied to data of Erlenmeyer-Kimling and Jarvik (1963), who report correlations between the intelligence of monozygotic twins (MZ), dizygotic twins (DZ), siblings, and unrelated individuals. The sample sizes vary, but are always in excess of 103, so we use $n = 103$ as a conservative estimate of sample size for each study. Table 3 reports the correlation, z-transform of the correlation, and normal deviate U_i for each group. The mean $\bar{z}$ of the z-transformed correlations is $\bar{z} = 0.54$. The obtained value of the homogeneity test statistic using computational formula (17) of Chapter 11 is $Q = 104.7$, which is compared to 12.6, the 95-percent critical value of the chi-square distribution with six degrees of freedom. Therefore the hypothesis of equal correlations is rejected.

The adjacent gaps $g_{i,i+1} = U_{(i+1)} - U_{(i)}$ are

$$g_{12} = 2.50, \qquad g_{23} = 1.80, \qquad g_{34} = 1.20,$$

$$g_{45} = 0.80, \qquad g_{56} = 0.10, \qquad g_{67} = 7.00.$$

Comparing these gaps with the critical value of 1.93 obtained from Table 1 with $k = 7$ and the 5-percent level of significance, we see that the gaps between

TABLE 3

Data for the Example of Clustering Correlation Coefficients

	n	r	$z(r)$	$U = \sqrt{n-3}\, z(r)$
MZ, reared apart	103	0.87	1.33	13.3
DZ, same sex	103	0.56	0.63	6.3
Siblings, reared together	103	0.55	0.62	6.2
DZ, opposite sex	103	0.49	0.54	5.4
Siblings, reared apart	103	0.40	0.42	4.2
Unrelated, reared together	103	0.24	0.24	2.4
Unrelated, reared apart	103	−0.01	−0.01	−0.1

$U_{(7)}$ and $U_{(6)}$ and between $U_{(2)}$ and $U_{(1)}$ are significant. Thus the gap procedure yields the clustering

$$\{U_{(1)}\}, \qquad \{U_{(2)}, U_{(3)}, U_{(4)}, U_{(5)}, U_{(6)}\}, \qquad \{U_{(7)}\}.$$

This hierarchical clustering separates monozygotic twins and unrelated individuals as having different correlations from the other groups studied.

The overlapping clustering procedure yields different clusters. Starting with $U_{(7)} - U_{(6)} = 7.00 > 2.77 = c(2)$, we see that $U_{(7)}$ is significantly different from $U_{(6)}$ at the 5 percent level at significance. We then calculate $U_{(6)} - U_{(1)} = 6.40 > c(6)$, but $U_{(6)} - U_{(2)} = 3.90 < c(5)$, so $U_{(6)}$ is significantly different from $U_{(1)}$, but not from $U_{(2)}$. We also calculate $U_{(3)} - U_{(1)} = 4.30 > 3.68 = c(3)$, but $U_{(3)} - U_{(2)} = 1.80 < c(2)$, so $U_{(3)}$ is significantly different from $U_{(1)}$, but not from $U_{(2)}$. Finally, $U_{(2)} - U_{(1)} = 2.50 < 2.77 = c(2)$, so $U_{(2)}$ is not significantly different from $U_{(1)}$. The data therefore yield the clusters represented in the following diagram:

$$U_{(1)} \quad U_{(2)} \quad U_{(3)} \quad U_{(4)} \quad U_{(5)} \quad U_{(6)} \quad U_{(7)}$$

The overlapping clustering procedure, like the hierarchical procedure, separates monozygotic twins as one cluster that does not overlap other clusters. The overlapping clustering procedure, however, separates unrelated individuals into a cluster that overlaps a third cluster containing results of dizygotic twins and siblings. Obviously, the clusters produced by the two procedures in this example are not the same.

C. CLUSTERING EFFECT SIZES

Our starting point is a set of k control and experimental populations with population effect sizes $\delta_1, \ldots, \delta_k$. We assume (for simplicity) that balanced samples with $n_1^E = n_1^C, \ldots, n_k^E = n_k^C$ are taken from the k populations. Call these (common) group sample sizes $n_1, \ldots, n_k$, and denote the sample effect sizes by $d_1, \ldots, d_k$.

The clustering methods described in Section A cannot be directly applied to the sample effect sizes $d_1, \ldots, d_k$ because the variance of d_i depends on the population effect size δ_i. However, because $n_i^E = n_i^C$, the transformed versions of sample effect sizes discussed in Section C.2 of Chapter 6 have variances that do not depend on the δ_i. The transformation is

$$h(x) = \sqrt{2} \sinh^{-1}\left(\frac{x}{2\sqrt{2}}\right) = \sqrt{2} \log\left(\frac{x + \sqrt{x^2 + 8}}{\sqrt{8}}\right) \tag{2}$$

and is tabulated in Appendix D. The transformation removes the effect of δ on the variance, so that the variables

$$\sqrt{2n_i}\, h(d_i), \qquad i = 1, \ldots, k, \tag{3}$$

are approximately unit normal variates.

A simple test of the hypothesis of the homogeneity of the δ_i was given in Section D.2 of Chapter 2, and can be applied before any clustering of effect sizes is attempted. If the homogeneity hypothesis that $\delta_1 = \delta_2 = \cdots = \delta_k$ is rejected, then the effect sizes can be considered as samples from different populations, in which case the clustering procedures can be used to determine groups of studies (populations) for which the effect sizes are homogeneous.

Suppose that the sample sizes in the k studies are equal, i.e., $n_1 = \cdots = n_k = n$. In this case, transform the independent effect size estimates by letting $U_i = \sqrt{2nh(d_i)}$, $i = 1, \ldots, k$. When the population effect sizes are the same, $U_1, \ldots, U_k$ are independently distributed with a common mean and unit variance. When the sample sizes differ, a modification of this procedure is required, and we discuss this modification later in this chapter.

Example

The techniques described in this chapter are applied to data from nine studies of the effects of open education on the independence and self-reliance of students (Hedges, Giaconia, & Gage, 1981). Table 4 is a tabulation of sample standardized mean differences d, sample sizes n, and transformed values $h(d)$ for each of the studies. The average $\bar{h}$ of the transformed effect size estimates is $\bar{h} = -0.047$. The obtained value of the homogeneity test statistic using the computational formula (27) of Chapter 6 is $Q_1 = 16.057$. This value is compared to 12.6, the 95 percentage point of the chi-square

TABLE 4
Data for the Example of Clustering Standardized Mean Differences

Study	n	d	$h(d)$	$U = \sqrt{2n}\, h(d)$
1	30	0.699	0.346	2.680
2	30	0.091	0.045	0.352
3	30	−0.058	−0.029	−0.225
4	30	−0.079	−0.039	−0.306
5	30	−0.235	−0.117	−0.909
6	30	−0.494	−0.246	−1.904
7	30	−0.587	−0.291	−2.257

distribution with six degrees of freedom, to yield a significant result. Therefore, the hypothesis of homogeneity of effect sizes is rejected.

The adjacent gaps $g_{i,i+1} = U_{(i+1)} - U_{(i)}$ are

$$g_{12} = 0.353, \qquad g_{23} = 0.995, \qquad g_{34} = 0.603,$$

$$g_{45} = 0.081, \qquad g_{56} = 0.577, \qquad g_{67} = 2.328.$$

Comparing these gaps with 1.69, the $\alpha = 0.10$ critical value for $k = 7$, we see that only the gap between $U_{(7)}$ and $U_{(6)}$ is significant. Thus the gap procedure yields the clusters

$$\{U_{(1)}, U_{(2)}, U_{(3)}, U_{(4)}, U_{(5)}, U_{(6)}\} \qquad \{U_{(7)}\}.$$

The overlapping clustering procedure yields different clusters at the $\alpha = 0.10$ significance level. From

$$U_{(7)} - U_{(2)} = 4.58 > 4.44 = c(6),$$

$$U_{(7)} - U_{(3)} = 3.59 < 4.18 = c(5)$$

we see that $U_{(7)}$ is significantly different from $U_{(2)}$, but not from $U_{(3)}$. None of the other differences is significant. Thus the data yield the two clusters represented in the following diagram:

$$U_{(1)}\quad U_{(2)}\quad \underline{U_{(3)}\quad U_{(4)}\quad U_{(5)}\quad U_{(6)}\quad U_{(7)}}.$$

Note that these clusters differ from the clusters obtained using the gap method. The clusters obtained in this example overlap, although non-overlapping clusters are also possible with this method.

D. THE EFFECT OF UNEQUAL SAMPLE SIZES

Both clustering procedures depend on a transformation of the correlation coefficients or effect size estimates from each study into standard normal variates U_i, $i = 1, \ldots, k$. In each case, the expressions for the U_i involve n, the common sample size of the studies. For many collections of research studies, the sample sizes will not be exactly equal. When the sample sizes are not very different, a useful working procedure is to use some derived value in place of the common sample size n.

In the case of correlations, we recommend replacing the common sample size $n - 3$ by the square mean root

$$n' = \left(\frac{\sqrt{n_1 - 3} + \cdots + \sqrt{n_k - 3}}{k}\right)^2,$$

where $n_1, \ldots, n_k$ are the sample sizes of the k studies. (For a discussion of the rationale for using n' see Chapter 4.) Thus, the clustering procedure is actually applied to the variates $U_i = \sqrt{n'}\, z(r_i)$, $i = 1, \ldots, k$, which have approximately normal distributions with the same mean and unit variance when the correlations are equal.

In the case of standardized mean differences, we recommend replacing the common within-group sample size n by the square mean root

$$n_{\text{SMR}} = \left(\frac{\sqrt{n_1} + \cdots + \sqrt{n_k}}{k}\right)^2,$$

where $n_1, \ldots, n_k$ are the sample sizes of the k studies. Thus, the clustering procedure is applied to the variates

$$U_i = \sqrt{2n_{\text{SMR}}}\, h(d_i), \qquad i = 1, \ldots, k,$$

which have approximately normal distributions with the same mean and unit variances when the effect sizes are equal.

The use of the synthetic sample size n' and n_{SMR} when sample sizes are unequal causes the variances of some U_i to be slightly larger than unity, whereas other U_i will have variances smaller than unity. If the sample sizes are not very different, the variances will all be near unity and the actual significance levels of the clustering procedures will be close to the nominal significance levels based on Tables 1 and 2. If some of the studies have very different sample sizes, then some of the U_i will have variances much smaller or larger than unity.

An obvious practical question is whether differences among the variances of the U_i have a large effect on the statistical significance levels associated with the clustering procedures. This is a difficult question in general, but some insight can be obtained by investigating the situation when there is a single discrepant observation. This is also a practically important situation, since it is often the case that most studies of a phenomenon have similar sample sizes, and only a few studies are very much larger or much smaller than the rest.

D.1 THE EFFECT ON THE DISJOINT CLUSTERING PROCEDURE*

Suppose that $U_1, \ldots, U_k$ are independent normal variates, U_1 has variance σ^2, and $U_2, \ldots, U_k$ have variance 1, where σ^2 is not necessarily equal to unity. The ordered values are denoted $U_{(1)}, U_{(2)}, \ldots, U_{(k)}$ as before. Let

TABLE 5

Exact Probabilities That at Least One Gap among k Normal Order Statistics Exceeds the Critical Value for the Gap at Significance Level α When One Variate Has Variance σ^2 and $k - 1$ Variates Have Variance 1

		α			
k	σ	0.10	0.05	0.01	0.001
3	0.50	0.053	0.024	0.003	0.0002
3	0.75	0.070	0.035	0.005	0.0004
3	1.25	0.141	0.082	0.020	0.0030
3	1.50	0.190	0.120	0.038	0.0085
3	1.75	0.242	0.163	0.063	0.0193
5	0.50	0.076	0.038	0.007	0.0007
5	0.75	0.083	0.041	0.008	0.0007
5	1.25	0.130	0.071	0.017	0.0025
5	1.50	0.171	0.102	0.032	0.0074
5	1.75	0.218	0.141	0.055	0.0169
10	0.50	0.098	0.050	0.010	0.0009
10	0.75	0.094	0.048	0.009	0.0008
10	1.25	0.117	0.062	0.014	0.0016
10	1.50	0.147	0.084	0.024	0.0045
10	1.75	0.182	0.112	0.040	0.0106
20	0.50	0.104	0.051	0.010	0.0010
20	0.75	0.101	0.049	0.010	0.0010
20	1.25	0.109	0.055	0.012	0.0016
20	1.50	0.126	0.067	0.019	0.0041
20	1.75	0.148	0.085	0.030	0.0099

$E[N(L, k, \sigma)]$ denote the expected number, of the $k - 1$ gaps, that exceed a prespecified length L when all the variates have the same mean.

By evaluating $E[N(L, k, \sigma)]$ for various values of σ at the critical values L_α given in Table 1, we can study the effect of σ on the expected number of gaps that exceed L_α. Table 5 is a presentation of values of $E[N(L_\alpha, k, \sigma)]/(k - 1)$ for $k = 3, 5, 10, 20$, $\sigma = 0.50, 0.75, 1.25, 1.50, 1.75$, and $L_{0.10}$, $L_{0.05}$, $L_{0.01}$, and $L_{0.001}$, where L_α is the α critical value of the gap given in Table 1. We see that $E[N(L_\alpha, k, \sigma)]$ is sensitive to σ, and is more sensitive to changes in σ in small samples than in large samples. Furthermore, for $k \geq 5$, the values of $E(N(L_\alpha, k, \sigma)]$ are reasonably close to their nominal levels when $0.75 \leq \sigma \leq 1.25$ or $0.56 \leq \sigma^2 \leq 1.56$. There is also a trend for $E[N(L_\alpha, k, \sigma)]$ to be less sensitive to values of σ less than unity than to values of σ greater than unity. Since the variances of the transformed estimators of ρ and δ are inversely proportional to n, these results suggest that the significance levels of the disjoint clustering procedure will be reasonably accurate when $k \geq 5$, and the most deviant sample sizes are between 80 percent ($=1/1.56$) and 133 percent ($=1/0.56$) of the others.

Therefore, we see that the significance levels of the disjoint clustering procedure are not very sensitive to unequal sample sizes provided that the sample sizes are all within 20 percent of the average. One can obtain an upper bound for the correct significance level by using the smallest sample size in place of n in the definition of the U_i, thereby obtaining a conservative significance level. Similarly, by using the largest sample size in place of n in the definition of U_i, we obtain a liberal significance level. These two significance levels provide bounds for the true significance level. Indeed, when sample sizes do not greatly differ, the use of n' or n_{SMR} will provide a reasonable approximation.

D.2 THE EFFECT ON THE OVERLAPPING CLUSTERING PROCEDURE

With the same assumptions as in Section D.1, denote the range by $W = U_{(k)} - U_{(1)}$. When all of the variates have the same mean, it is possible to obtain the cumulative distribution function for the range W as a function of k and σ. By evaluating the actual probability that the range of the k variates exceeds the critical values used to compute the values in Table 2, we can study the effects of σ on the accuracy of the significance levels for the overlapping clustering procedure. Table 6 is a tabulation of the actual significance levels of the range for several values of k and σ, and the various nominal significance levels α (i.e., significance levels assuming $\sigma = 1$). The significance levels of the overlapping clustering procedure, which depend on

TABLE 6

Exact Bonferroni Probabilities That the Range of k Normal Variates Exceeds the α Critical Value of the Range of k Standard Normal Variates When One Variate Has Variance σ^2 and $k - 1$ Variates Have Variance 1

		α			
k	σ	0.10	0.05	0.01	0.001
3	0.50	0.047	0.021	0.004	0.0004
3	0.75	0.064	0.029	0.005	0.0004
3	1.25	0.158	0.089	0.024	0.0037
3	1.50	0.230	0.146	0.052	0.0118
3	1.75	0.309	0.214	0.094	0.0288
5	0.50	0.066	0.032	0.006	0.0006
5	0.75	0.073	0.035	0.007	0.0006
5	1.25	0.161	0.090	0.024	0.0036
5	1.50	0.254	0.163	0.058	0.0135
5	1.75	0.363	0.259	0.117	0.0377
10	0.50	0.082	0.042	0.010	0.0042
10	0.75	0.084	0.043	0.010	0.0042
10	1.25	0.156	0.089	0.027	0.0109
10	1.50	0.272	0.181	0.080	0.0449
10	1.75	0.427	0.322	0.187	0.1231

the distribution of the range, are quite sensitive to σ. Even for fairly large k, it may be necessary to use a low nominal significance level, such as $\alpha = 0.01$, to assure a true significance level of 0.05.

D.3 AN ALTERNATIVE METHOD FOR HANDLING UNEQUAL SAMPLE SIZES

When the sample sizes of the studies are markedly discrepant we do not wish to treat all studies as if they had an average value of the sample size. An alternative procedure for assessing significance levels of clusterings makes use of the generalization of the Scheffé procedure for testing post hoc contrasts among effect sizes described in Chapter 7. Both the hierarchical and the overlapping clustering procedures involve testing the significance of differences among the estimates by using order statistics. Therefore, we can use procedures for testing post hoc comparisons to obtain significance tests for these differences. For this procedure, it is easier to work directly with the transformed estimates of effect magnitude.

The clustering procedures are applied in the usual way except that the significance of any difference between estimates is established as follows. To simultaneously test the significance of the difference between all pairs correlations ρ_i and ρ_j, compare the value of the statistic

$$\frac{[z(r_i) - z(r_j)]^2(n_i - 3)(n_j - 3)}{n_i + n_j - 6} \tag{4}$$

with the critical value of the chi-square distribution with $k - 1$ degrees of freedom. The hypothesis of equality of correlations is rejected and the gap declared significant if the statistic exceeds the critical value.

Example

For the example of clustering correlation coefficients discussed in Section B, the sample sizes, the correlations, and their z-transforms are given in Table 3. We cluster in the same way as before except that we test the statistical significance of the difference between correlations by comparing the test statistic (4) with 12.59, the 95-percent point of the chi-square distribution with six degrees of freedom. To test the significance of the hierarchical clustering, first calculate the statistic for the largest gap (between 1.33 and 0.63):

$$(1.33 - 0.63)^2(100)(100)/200 = 24.5.$$

Because 24.5 exceeds 12.59, the first division into clusters is statistically significant at the $\alpha = 0.05$ level. The significance of the next largest gap (between 0.24 and -0.01) is tested via

$$[0.24 - (-0.01)]^2(100)(100)/200 = 3.125.$$

Because 3.125 is smaller than 12.59, this gap, the corresponding clustering, and subsequent divisions into clusters are not significant at the $\alpha = 0.05$ level.

To simultaneously test the significance of the difference between all pairs of effect sizes δ_i and δ_j, compare the value of the statistic

$$\frac{2[h(d_i) - h(d_j)]^2 n_i n_j}{n_i + n_j} \tag{5}$$

with the critical value of the chi-square distribution with $k - 1$ degrees of freedom. The hypothesis of equality of effect sizes is rejected and the gap is declared significant if the statistic exceeds the critical value. In the case of either correlations or effect sizes, the simultaneous significance level of all of the tests will not exceed α, and hence each of the clusterings has a significance level α.

Example

For the example of clustering effect sizes discussed in Section C, the sample sizes, effect size estimates, and transformed estimates (h_i values) are given in Table 4. The clustering procedures are the same as before except that we test the statistical significance of differences between effect sizes by comparing the test statistic (5) with 10.64, the 90-percent point of the chi-square distribution with $7 - 1 = 6$ degrees of freedom. To test the significance of the largest gap, compute

$$2(0.346 - 0.045)^2(30)(30)/60 = 2.72$$

which is not significant. Smaller gaps will also be nonsignificant, and consequently none of the hierarchical clusterings is significant at the $\alpha = 0.10$ level of significance.

If the sample sizes are approximately equal, the testing procedure described is not as sensitive as the procedure described previously. If the sample sizes are very different, the Scheffé procedure described in this section can be considerably more powerful than the most conservative procedure using the smallest sample size. (We say that sample sizes are approximately equal if all the sample sizes are within 20 percent of each other.)

E. RELATIVE MERITS OF THE CLUSTERING PROCEDURES

The two clustering procedures can give different answers, owing in part to the fact that the procedures are designed to achieve different goals. The disjoint clustering procedure seeks sharp divisions into clusters, whereas the overlapping clustering procedure seeks less distinct divisions into clusters.

If the investigator seeks a sharp division into disjoint clusters, then the disjoint clustering procedure should be used. Sharp divisions are appropriate, for example, when a single underlying explanatory variable accounts for the clustering. In this case, there may be one cluster for each value of the underlying variable, and the studies in a cluster will share a common value of that underlying variable. Alternatively, the investigator may not believe that sharp divisions are realistic. Less distinct divisions are appropriate, for example, when there are several underlying explanatory variables. In this case studies can differ on each of these variables, leading to less distinct groupings.

Statistical considerations can also affect the choice of a clustering procedure. In the preceding section we demonstrated that the significance levels of the disjoint clustering procedure are more robust to unequal sample sizes. Therefore, if the sample sizes of the studies differ considerably, one might tend to choose the disjoint clustering procedure. A second statistical consideration is the power of the tests for clusterings. The overlapping clustering procedure is based on critical values for the range of standard normal variates adjusted using the Bonferroni inequality.

If the number k of studies is large, then the adjusted critical value of the range is very large. We speculate (based on similar results for multiple comparison tests based on the Bonferroni inequality) that the power of the overlapping clustering procedure to detect small differences among clusters is not high when k is large. In this case the disjoint clustering procedure may be more powerful for detecting certain types of clusterings.

F. COMPUTATION OF TABLES*

The statistical significance of clusters obtained by the methods presented is evaluated by using the tables of critical values. [Portions of the tables are contained in Hedges and Olkin (1983a).] Because these critical values are not generally available, we outline the methods used for computing the critical values in this section.

F.1 THE SIGNIFICANCE OF GAPS

If $U_1, \ldots, U_k$ are k independent standard normal variates that define $k - 1$ adjacent gaps, Tukey (1949) showed that the expected proportion of gaps that exceed a prespecified length L is

$$E[N(L, k, 1)] = k \int_{-\infty}^{\infty} \{[\phi(x) + \bar{\Phi}(x + L)]^{k-1} - [\Phi(x)]^{k-1}\}\, d\Phi(x), \quad (6)$$

where $\Phi(x)$ is the standard normal cumulative distribution function, and $\bar{\Phi}(x) = 1 - \Phi(x)$. The values in Table 1 are obtained by solving (6) numerically for each given k and $E[N(L, k, 1)] = \alpha$.

If U_1 has variance σ^2 and $U_2, \ldots, U_k$ have variance 1, then the k normal variates still define $k - 1$ gaps. A direct argument shows that the expected number of gaps that exceed a prespecified length L is

$$\begin{aligned} E[N(L, k, \sigma)] = {} & \frac{1}{\sigma} \int_{-\infty}^{\infty} \{[\Phi(y) + \bar{\Phi}(y + L)]^{k-1} - [\Phi(y)]^{k-1}\}\, d\Phi\left(\frac{y}{\sigma}\right) \\ & + (k-1) \int_{-\infty}^{\infty} \left\{[\Phi(y) + \bar{\Phi}(y + L)]^{k-2}\left[\Phi\left(\frac{y}{\sigma}\right) + \Phi\left(\frac{y+L}{\sigma}\right)\right]\right. \\ & \left. - \Phi\left(\frac{y}{\sigma}\right)[\Phi(y)]^{k-2}\right\} d\Phi(y). \end{aligned} \quad (7)$$

Note that expression (7) reduces to expression (6) when $\sigma = 1$. For given values of k, σ, and L_α corresponding to critical values from Table 1, the values of $E[N(L, k, \sigma)]/(k - 1)$ given in Table 5 were obtained by numerical integration of the right-hand side of (7).

F.2 SIGNIFICANCE VALUES FOR THE OVERLAPPING CLUSTERING PROCEDURE

The procedure used to form clusters at a desired significance level α depends on the computed $100(1 - \alpha^*)$-percent critical value $c(k)$ of the range of k observations from a standard normal distribution, where $\alpha^* = 1 - (1 - \alpha)^{1/(k-1)}$. The values in Table 2 were obtained by first determining α^* for each α level and then using tables of the distribution of the range (Harter, 1961) at the α^* level.

If U_1 has variance σ^2 and $U_2, \ldots, U_k$ have variance 1, where $U_1, \ldots, U_k$ are independent normal variates, then a direct argument yields the cumulative distribution function of the range w of the k variates as

$$F(w) = \frac{1}{\sigma}\int_{-\infty}^{\infty}\left[[\Phi(x + w) - \Phi(x)]^{k-1}\, d\Phi\left(\frac{x}{\sigma}\right)\right] + (k - 1)\int_{-\infty}^{\infty}\left[[\Phi(x + w) - \Phi(x)]^{k-2}\Phi\left(\frac{x + w}{\sigma}\right) - \Phi\left(\frac{x}{\sigma}\right)\right] d\Phi(x). \tag{8}$$

When $\sigma = 1$, (8) reduces to the usual expression for the cumulative distribution function of the range of standard normal variates (see, e.g., Harter, 1961). The values in Table 6 were obtained by evaluating (8) numerically at the nominal critical values of the range.

Note that the gaps between normal order statistics are not uniformly distributed, so that when the null hypothesis is true (e.g., all of the studies share a common effect size), the gaps that are (falsely) declared significant tend to occur between the more extreme observations. This phenomenon is an intrinsic property of some multiple comparison tests based on order statistics.

CHAPTER 14

Estimation of Effect Size When Not All Study Outcomes are Observed

The properties of statistical procedures described in previous chapters depend on the availability of unrestricted samples of effect size estimates. If the effect size estimates available to the investigator are systematically biased, then special statistical procedures are needed to take account of the biasing mechanism.

Unfortunately, there is often reason to believe that nonrepresentative sampling is prevalent in quantitative research synthesis. This sampling bias stems from a "prejudice against the null hypothesis" by some social and behavioral science researchers (see, e.g., Greenwald, 1975) in the sense that only research results that are statistically significant are reported, and effect size estimates that correspond to statistically nonsignificant mean differences may not be available for inclusion in a meta-analysis. The problem created by the censoring of effect size estimates corresponding to statistically nonsignificant results is discussed in this chapter. Some evidence on the existence of the sampling bias is presented in Section A. A statistical model for studying the effects of the sampling bias is presented in Section B, and the consequences of such bias on the estimation of effect size are examined. Maximum likelihood estimates of effect size under the model of sampling bias are obtained in Section C. The combination of estimates from several experiments under the model of sampling bias is discussed in Section D. Section E contains a discussion of ways in which the results of this chapter can be used to draw conclusions when sampling bias may exist.

A. THE EXISTENCE OF SAMPLING BIAS IN OBSERVED EFFECT SIZE ESTIMATES

One readily demonstrable source of bias in reported estimates of effect size stems from the failure to report the details of statistical analyses when mean differences are not statistically significant. Calculation of an effect size estimate requires the values of sample means and standard deviations or the value of a t- or F-test statistic for the difference between means. Authors sometimes fail to report either the test statistics or the descriptive statistics, and simply report "no significant difference."

One consequence of this practice is that the data needed to calculate effect size estimates are sometimes selectively omitted in research reports when statistically nonsignificant results are obtained. For example, as much as 40 percent of the potential effect size estimates in two recent meta-analyses (Eagly & Carli, 1981; Strube, 1981) could not be calculated because of incomplete reporting of nonsignificant results.

Another potential source of bias in estimates of effect size stems from the purported tendency of journal editors not to publish research studies that fail to produce statistically significant results. Sterling (1959) argued that nonsignificant results are rarely published, and therefore the published literature may be full of type I errors. A survey of three psychology journals by Sterling and a more extensive survey of psychology journals by Bozarth and Roberts (1972) showed that 97 and 94 percent, respectively, of the articles examined that used statistics rejected the statistical null hypothesis at the $\alpha = 0.05$ significance level. If the studies examined by Sterling and by Bozarth and Roberts are representative of all studies conducted in psychology, their results imply the existence of a large number of type I errors or average power in excess of 0.90 or some combination of these.

The overwhelming proportion of articles rejecting the null hypothesis in these surveys is indeed remarkable given the results of studies of the power of statistical tests in psychological research (Chase & Chase, 1976; Cohen, 1962). These surveys of statistical power suggest that the average power of statistical tests in psychological research is between 0.25 and 0.85, depending on the assumed magnitude of effects. The surveys of statistical power therefore suggest that only 25–85 percent of the studies in psychology journals would be expected to yield statistically significant results.

Thus, there is reason to believe that journal articles contain a greater than expected proportion of statistically significant results, and relatively fewer nonsignificant results than would be expected from a random sample of all studies actually conducted. It is important to recognize that the empirical evidence about the existence of editorial bias is not overwhelming, and that surveys of statistical power involve assumptions about effect size, for example, that may not be tenable. Moreover, the data from surveys of statistical power involve a somewhat different journal data base than do the studies of Sterling and of Bozarth and Roberts.

Other sources provide direct evidence of an intent of editorial policy to discourage publication of research that fails to yield statistically significant results (see Bakan, 1966; Sidman, 1960). For example, an editorial by Melton (1962) explicitly includes statistical significance as one of the most important criteria used to select manuscripts for the *Journal of Experimental Psychology*. A more recent survey of reviewers for major journals in psychology suggests that statistical significance remains an important criterion in the review of manuscripts for publication (Greenwald, 1975).

Note that the perception that an editorial policy that uses statistical significance as a criterion for publishing manuscripts may have nearly the same effect as an actual policy to that effect. If authors believe that manuscripts that do not contain statistically significant results will not be published, they are unlikely to submit such manuscripts for publication.

B. CONSEQUENCES OF OBSERVING ONLY SIGNIFICANT EFFECT SIZES

If effect size estimates corresponding to statistically nonsignificant results are less likely to be sampled, or be available for sampling, then the sample of effect size estimates will be biased. Because the test statistic for the t-test is monotonically related to the effect size, studies that produce statistically significant results tend to have effect sizes that are larger in absolute magnitude. Thus, for positive population effect sizes, overrepresentation of studies producing significant differences tends to bias the sample toward larger effect sizes, and estimates from such a sample will overestimate the absolute magnitude of the population effect size.

An extreme form of bias toward significant results is examined by Lane and Dunlap (1978) and Hedges (1984) in the case that only experiments yielding statistically significant results are observed. Lane and Dunlap simulated the results of a large number of two-group experiments. They selected the experiments that yielded statistically significant mean differences for further study. As expected, they found that the mean difference estimated from the experiments yielding significant results overestimated the population mean difference. Similarly, they found that estimates of Hays's magnitude index ω^2 were also upwardly biased. Hedges (1984) studied the same situation analytically, confirmed Lane and Dunlap's empirical findings, and showed that estimates of effect size were also upwardly biased when $\delta > 0$.

B.1 MODEL AND NOTATION

Suppose that effect size estimates arise in connection with two-group experiments having a common sample size $n^E = n^C = n$. Denote the population and sample effect sizes (standardized mean differences) by δ and g as in Chapter 5. When necessary, subscripts are used to distinguish parameters and estimates of different studies. Further, we assume that g is observed only if the F-test for the difference between means is statistically significant at some preset significance level α. That is, a sample effect size is observed only if

$$g^2 > 2F(\alpha, n)/n,$$

where $F(\alpha, n)$ is the $100(1 - \alpha)$ percentage point of the F-distribution with 1 and $2n - 2$ degrees of freedom.

Unless otherwise specified, the significance level α is chosen to be $\alpha = 0.05$. Denote the observed effect size by g^*. Note that in the present model there are no values of g^* smaller than $\sqrt{2F(\alpha, n)/n}$, since the corresponding mean differences are not statistically significant.

Limitation: The model proposed in this section involves strict censoring of all results that are not significant at the $\alpha = 0.05$ level. More complicated censoring schemes are possible. In many practical situations the censoring rule used is unknown, so that the model and results described in this section may not be applicable. However, it does provide a framework from which to make modifications. In other situations the censoring rule may be known. For example, if the censoring is a result of failure to report statistics for mean differences that are not statistically significant at the $\alpha = 0.05$ level, then the model described herein is appropriate. In any case it is important to note that the results depend on the particular model, and the application of the methods presented may be misleading if the model is not appropriate.

B.2 THE DISTRIBUTION OF THE OBSERVED EFFECT SIZE

The statistic g^* corresponds to the effect size estimate when only statistically significant results are observed. We need to obtain the distribution of

g^*, in order to study the properties of the usual estimator of effect size when restricted to significant results. Note that g^* is simply the effect size subject to censoring all values inside the interval $(-\sqrt{2F(\alpha, n)/n}, \sqrt{2F(\alpha, n)/n})$, so that its distribution is a truncated noncentral t-distribution (see Technical Commentary).

The expected value and variance of g^* are obtained by numerical integration, and are given in Table 1 for $\delta = 0.25$, 0.50, 0.75, 1.00, and 1.50 and $n = 10$, 20, 30, 40, and 50. As expected, the bias of g^* (as an estimator of δ) tends to be large for small n and small (but nonzero) values of δ. The bias decreases rapidly as δ increases, so that the bias is less than 10 percent for $\delta \geq 1.0$ and $n \geq 20$ per group. For smaller δ, the bias can be quite severe. For $\delta = 0.5$ and $n = 15$ the bias in g^* is nearly 100 percent, and for $\delta = 0.25$ the bias of g^* is over 200 percent even if $n = 40$ per group.

The absolute bias $|E(g^*) - \delta|$ tends to zero as δ approaches zero and also tends to zero for large δ. The absolute bias has a maximum between $\delta = 0.3$ (for $n = 10$) and $\delta = 0.2$ (for $n = 40$). Thus the absolute bias is largest for small but nonzero effect sizes. The relative bias $E(g^*)/\delta$ increases as δ approaches zero, but tends to unity as $|\delta|$ becomes large. The bias of g^* as a function of δ is illustrated in Fig. 1.

ABLE 1

xpected Value $E(g^*)$ and Variance $V(g^*)$ of the Sample Effect Size g^* When Only Significant Results at the 5-Percent Level of Significance) Are Observed[a]

	δ									
	0.25		0.50		0.75		1.00		1.50	
	$E(g^*)$	$V(g^*)$	$E(g^*)$	$V(g^*)$	$E(g^*)$	$V(g^*)$	$E(g^*)$	$V(g^*)$	$E(g^*)$	$V(g^*)$
Exact	1.01	0.44	1.22	0.11	1.30	0.10	1.39	0.13	1.67	0.24
Simulated	0.99	0.48	1.23	0.11	—	—	1.39	0.14	1.67	0.24
Exact	0.77	0.08	0.87	0.04	0.96	0.06	1.10	0.08	1.53	0.14
Simulated	0.77	0.08	0.87	0.04	—	—	1.09	0.08	1.54	0.14
Exact	0.65	0.03	0.73	0.03	0.85	0.05	1.03	0.07	1.52	0.09
Simulated	0.65	0.03	0.73	0.03	—	—	1.03	0.07	1.53	0.09
Exact	0.57	0.02	0.65	0.02	0.80	0.04	1.01	0.06	1.52	0.07
Simulated	0.57	0.02	0.65	0.02	—	—	1.01	0.06	1.52	0.07
Exact	0.51	0.01	0.60	0.02	0.77	0.04	1.00	0.05	1.50	0.05
Simulated	0.52	0.01	0.60	0.02	—	—	1.00	0.05	1.51	0.06

[a] Exact results were obtained by numerical integration; simulated results were obtained from 2000–10,000 lications for each combination of sample size and effect size.

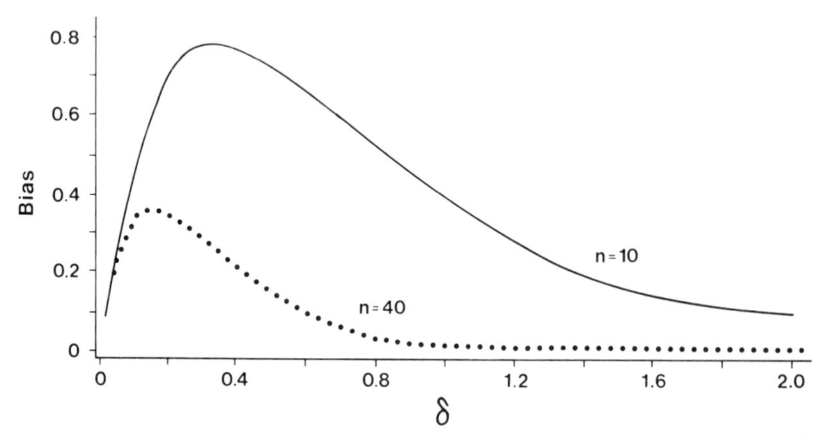

Fig. 1 The bias of the observed effect size g^* as a function of the population effect size δ when only effect sizes corresponding to mean differences that are statistically significant at the $\alpha = 0.05$ level may be observed.

Unfortunately, many quantitative research syntheses are likely to involve studies with sample sizes of less than 40 per group and effect sizes less than 1.00. Hence the severe bias of g^* for small n and small δ may lead to substantial overestimates of effect size if nonsignificant results are not reported. Lane and Dunlap (1978) found a similar severe bias in estimators of a different index of effect magnitude (Hays's ω^2) when nonsignificant results were not available.

Note that the bias of the observed effect size (or mean difference) has implications for the interpretation of research results from a single study. The results in this section show that journal editorial policies that require statistical significance for publication have the effect of inflating the values of treatment effects that appear in print. Similarly, the tendency to report and discuss only significant effects in a research study inflates the size of experimental effects that are reported in the literature. Such inflation of observed effects may seriously distort our perception of the magnitude of experimental effects.

C. ESTIMATION OF EFFECT SIZE FROM A SINGLE STUDY WHEN ONLY SIGNIFICANT RESULTS ARE OBSERVED

The results reported in the previous section suggest that the use of g^* (the sample estimate of effect size when only statistically significant outcomes are observed) as an estimator of δ can lead to serious bias. When sample sizes and effect sizes are both moderate, the bias of g^* may be substantial enough to influence substantive conclusions and could lead to problems in the interpretation of analyses based on effect sizes.

In this section, we consider a method for estimating the effect size from a single study when only statistically significant mean differences are observed. The method involves maximum likelihood estimation of δ based on the distribution of g^*. This procedure implicitly (through the distribution of g^*) corrects for the censoring of g^* values corresponding to statistically nonsignificant mean differences. First we derive the maximum likelihood estimator of δ based on a g^* value from a single experiment; then we obtain its exact distribution numerically, and use this distribution to study the bias of the estimator (see Hedges, 1984).

C.1 ESTIMATION OF EFFECT SIZE

The maximum likelihood estimator $\hat{\delta}$ of the effect size δ is obtained by maximizing the likelihood (or log likelihood) of g^* as a function of δ. When this is done we find that the same observed value of the standardized mean difference leads to a smaller maximum likelihood estimate of δ in the model where only significant results are observed than in the model where all results are observed. When all results are observed, the maximum likelihood estimate of δ is approximately the observed value of the standardized mean difference. Therefore, we would expect the maximum likelihood estimator of δ based on g^* to be smaller than g^*, especially when g^* is itself reasonably small.

The maximum likelihood estimator $\hat{\delta}$ of δ based on g^* cannot be obtained in closed form, but the likelihood can easily be maximized numerically. Figure 2 is a graphic representation of $\hat{\delta}$ as a function of g^* for $n = 10$, 20, and 40. In each case, the reference line corresponding to $g^* = \hat{\delta}$ also appears. Only positive values of g^* are plotted, because of symmetry. That is, if g^* corresponds to the estimate $\hat{\delta}_0$, then $-g^*$ corresponds to the estimate $-\hat{\delta}_0$.

As expected, small values of g^* lead to much smaller values of $\hat{\delta}$. This is illustrated in Fig. 2, where the function relating g^* to $\hat{\delta}$ lies below the reference line for small g^*. One interpretation is that barely significant values of g^* tend to be associated with small values of δ, and thus, since more values are censored in this case, the maximum likelihood estimator is small. Large values of g^* lead to values of $\hat{\delta}$ that are almost identical to g^*. This is illustrated in Fig. 2, where the function relating g^* to $\hat{\delta}$ does not deviate greatly from the reference line for large values of g^*. An interpretation of this finding is that large values of g^* arise because of large values of δ and hence the censoring rarely occurs, so that g^* is a good estimate of δ.

One additional feature of the curves in Fig. 2 is important. Note that $\hat{\delta}$ decreases as g^* tends to the nonzero minimum observable value, but the

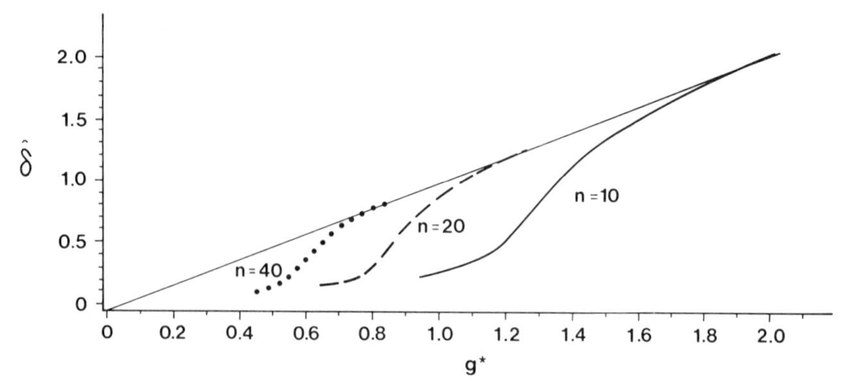

Fig. 2 The maximum likelihood estimator $\hat{\delta}$ of effect size as a function of the observed effect size g^*.

minimum $\hat{\delta}$ (for positive g^*) is not zero. For $n = 40$ this minimum value is $\hat{\delta} = 0.111$, whereas for $n = 10$ this minimum value is $\hat{\delta} = 0.242$.

Although the maximum likelihood estimator $\hat{\delta}$ based on g^* is not difficult to obtain numerically, the computations require a specialized computer program. Therefore we include Table 2 (from Hedges, 1984), which gives $\hat{\delta}$ as a function of g^* for values of n between 10 and 100. It also gives the maximum likelihood estimator of δ for each value of g^* as well as the minimum observable value of g^* (at the $\alpha = 0.05$ significance level) and the minimum value of $\hat{\delta}$ for a given n. Values of $\hat{\delta}$ can be determined without extensive computation by linear interpolation between tabulated values. Note that only positive values of g^* and $\hat{\delta}$ are tabulated. Negative values of $\hat{\delta}$ are obtained by symmetry. That is, if g_0 corresponds to $\hat{\delta}_0$, then $-g_0$ corresponds to $-\hat{\delta}_0$. We find $\hat{\delta}$ for negative values of g^* by finding the $\hat{\delta}$ that corresponds to $|g^*|$ and reversing the sign. Also note that values of g^* larger than 2.00 are not tabulated, because in that case $\hat{\delta} \doteq g^*$.

In this chapter we deal exclusively with the case $n^E = n^C = n$. If the experimental and control group sample sizes are unequal, an approximate procedure is to set $n = (n^E + n^C)/2$ and proceed as if the sample sizes were equal to this value.

C.2 THE DISTRIBUTION OF THE MAXIMUM LIKELIHOOD ESTIMATOR

The maximum likelihood estimator $\hat{\delta}$ of δ is an increasing function of g^*, so that from the distribution of g^* we can obtain, in principle, the distribution of $\hat{\delta}$. The practical problem is that the explicit mathematical form of the function that carries $\hat{\delta}$ to g^* is not known. One approach to this problem

TABLE 2

Maximum Likelihood Estimator $\hat{\delta}$ as a Function of Sample Size per Group n and Observed Effect Size g^{*a}

	n																
	10	12	14	16	18	20	25	30	35	40	45	50	60	70	80	90	100
g_{Min}	0.940	0.847	0.777	0.722	0.677	0.640	0.569	0.517	0.477	0.445	0.419	0.397	0.362	0.334	0.312	0.294	0.279
$\hat{\delta}_{Min}$	0.242	0.216	0.197	0.183	0.171	0.161	0.142	0.129	0.120	0.111	0.108	0.101	0.092	0.085	0.079	0.074	0.070
g^*																	
0.30	—	—	—	—	—	—	—	—	—	—	—	—	—	—	—	0.075	0.080
0.35	—	—	—	—	—	—	—	—	—	—	—	—	—	0.090	0.100	0.116	0.145
0.40	—	—	—	—	—	—	—	—	—	—	—	0.100	0.112	0.135	0.186	0.253	0.299
0.45	—	—	—	—	—	—	—	—	—	0.113	0.122	0.133	0.179	0.273	0.339	0.378	0.401
0.50	—	—	—	—	—	—	—	—	0.132	0.146	0.168	0.210	0.337	0.405	0.442	0.463	0.476
0.55	—	—	—	—	—	—	—	0.147	0.169	0.210	0.291	0.370	0.458	0.498	0.519	0.531	0.539
0.60	—	—	—	—	—	—	0.159	0.187	0.249	0.361	0.445	0.495	0.546	0.571	0.584	0.591	0.595
0.65	—	—	—	—	—	0.166	0.196	0.269	0.411	0.504	0.556	0.588	0.619	0.634	0.642	0.646	0.648
0.70	—	—	—	—	0.182	0.195	0.261	0.430	0.548	0.608	0.643	0.663	0.682	0.692	0.696	0.698	0.699
0.75	—	—	—	0.196	0.212	0.239	0.401	0.573	0.652	0.693	0.715	0.729	0.740	0.746	0.748	0.749	0.750
0.80	—	—	0.209	0.228	0.259	0.317	0.563	0.682	0.737	0.764	0.780	0.789	0.795	0.798	0.800	0.800	0.800
0.85	—	0.218	0.237	0.273	0.342	0.469	0.687	0.771	0.810	0.829	0.839	0.845	0.847	0.849	0.850	0.850	0.850
0.90	—	0.243	0.279	0.349	0.498	0.631	0.788	0.847	0.875	0.888	0.895	0.898	0.899	0.900	0.900	0.900	0.900
0.95	0.246	0.277	0.340	0.494	0.659	0.757	0.872	0.916	0.935	0.945	0.948	0.950	0.949	0.950	0.950	0.950	0.950
1.00	0.272	0.324	0.455	0.662	0.789	0.862	0.947	0.979	0.991	0.998	1.000	1.001	1.000	1.000	1.000	1.000	1.000
1.05	0.305	0.394	0.626	0.801	0.896	0.951	1.015	1.038	1.047	1.050	1.051	1.052	1.050	1.050	1.050	1.050	1.050
1.10	0.347	0.523	0.782	0.916	0.987	1.030	1.077	1.094	1.100	1.101	1.102	1.102	1.100	1.100	1.100	1.100	1.000
1.15	0.409	0.695	0.910	1.013	1.068	1.100	1.136	1.148	1.151	1.153	1.153	1.152	1.150	1.150	1.150	1.150	1.150
1.20	0.512	0.852	1.018	1.098	1.141	1.166	1.193	1.201	1.202	1.203	1.203	1.203	1.200	1.200	1.200	1.200	1.200
1.25	0.672	0.981	1.111	1.175	1.209	1.228	1.247	1.252	1.253	1.254	1.253	1.253	1.250	1.250	1.250	1.250	1.250
1.30	0.844	1.092	1.196	1.246	1.272	1.286	1.300	1.304	1.304	1.303	1.303	1.303	1.300	1.300	1.300	1.300	1.300

(Continued)

TABLE 2 (*Continued*)

	n																
	10	12	14	16	18	20	25	30	35	40	45	50	60	70	80	90	100
1.35	0.990	1.189	1.272	1.311	1.332	1.342	1.353	1.354	1.354	1.353	1.353	1.353	1.350	1.350	1.350	1.350	1.350
1.40	1.116	1.276	1.343	1.374	1.389	1.398	1.403	1.404	1.404	1.403	1.403	1.403	1.400	1.400	1.400	1.400	1.400
1.45	1.225	1.355	1.409	1.433	1.445	1.450	1.455	1.454	1.454	1.453	1.454	1.453	1.450	1.450	1.450	1.450	1.450
1.50	1.322	1.429	1.471	1.490	1.499	1.504	1.505	1.505	1.504	1.504	1.504	1.503	1.500	1.500	1.500	1.500	1.500
1.55	1.409	1.497	1.531	1.545	1.552	1.555	1.556	1.555	1.554	1.554	1.553	1.552	1.500	1.550	1.550	1.550	1.550
1.60	1.490	1.562	1.589	1.600	1.604	1.606	1.606	1.605	1.604	1.604	1.603	1.603	1.600	1.600	1.600	1.600	1.600
1.65	1.565	1.622	1.645	1.654	1.656	1.656	1.656	1.655	1.655	1.654	1.654	1.653	1.650	1.650	1.650	1.650	1.650
1.70	1.635	1.683	1.699	1.705	1.707	1.707	1.706	1.706	1.704	1.705	1.704	1.704	1.700	1.700	1.700	1.700	1.700
1.75	1.701	1.741	1.753	1.757	1.758	1.758	1.756	1.755	1.755	1.755	1.754	1.753	1.750	1.750	1.750	1.750	1.750
1.80	1.764	1.796	1.805	1.808	1.808	1.809	1.807	1.806	1.805	1.804	1.804	1.803	1.800	1.800	1.800	1.800	1.800
1.85	1.825	1.850	1.857	1.859	1.859	1.858	1.857	1.855	1.855	1.854	1.854	1.853	1.850	1.850	1.850	1.850	1.850
1.90	1.884	1.905	1.909	1.910	1.910	1.908	1.906	1.905	1.905	1.905	1.904	1.903	1.900	1.900	1.900	1.900	1.900
1.95	1.942	1.956	1.960	1.960	1.959	1.959	1.956	1.956	1.955	1.954	1.954	1.954	1.950	1.950	1.950	1.950	1.950
2.00	1.998	2.010	2.011	2.011	2.009	2.009	2.007	2.006	2.005	2.004	2.004	2.003	2.000	2.000	2.000	2.000	2.000

[a] The minimum observable g^* value is g_{Min} when all observable mean differences are significant at the $\alpha = 0.05$ level and $\hat{\delta}_{\text{Min}}$ is the maximum likelihood estimator corresponding to g_{Min}.

is to use a smooth approximation to this function, one that can be used for computations. Cubic spline approximations were found to give useful approximations to the density function of $\hat{\delta}$ (Ahlberg, Nilson, & Walsh, 1967).

Plots of the probability density function of $\hat{\delta}$ for $n = 10$ and 40 and $\delta = 0.0$, 0.50, and 1.00 are given in Fig. 3. Note that the distributions are essentially bimodal, with one peak near the smallest possible estimate and another peak near the actual value of δ. These distributions are highly nonnormal for $\delta = 0.0$ and $\delta = 0.50$. Note also the broad, flat tails of the

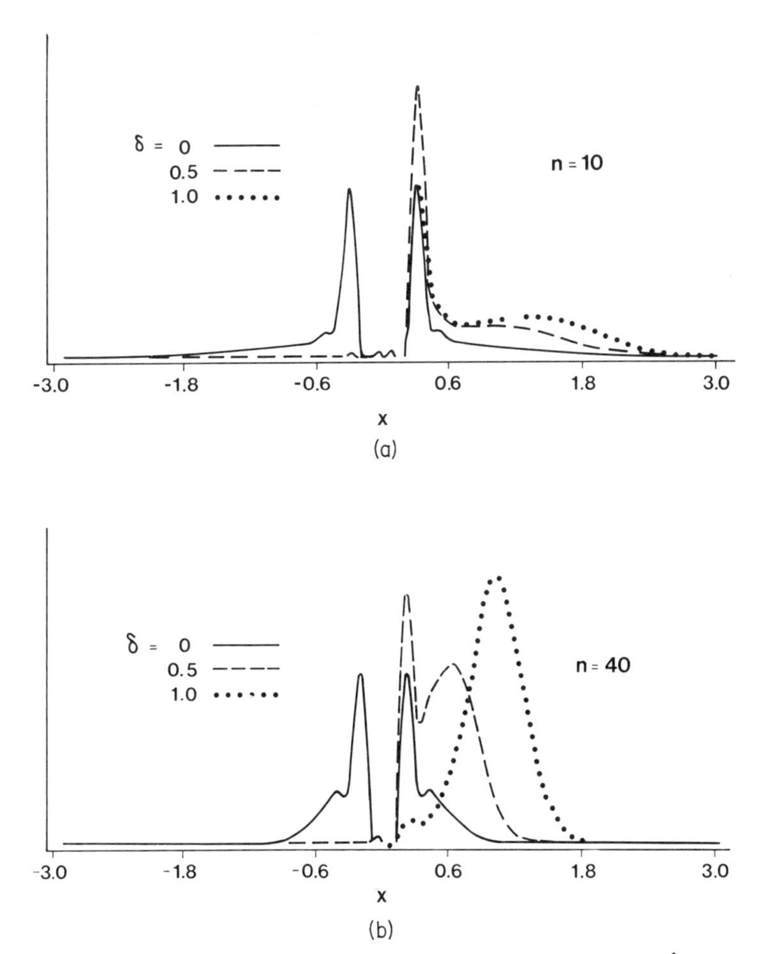

Fig. 3 Probability density function of the maximum likelihood estimator $\hat{\delta}$ of effect size for sample sizes (a) $n = 10$, and (b) $n = 40$.

density function, which suggests that the estimate of δ under censoring is not as precise as when all studies are observed.

The density function obtained by cubic spline approximations can be integrated numerically to give the expectation and variance of $\hat{\delta}$. The mean and variance of $\hat{\delta}$ for various values of δ are given in Table 3. The expected value of $\hat{\delta}$ is not always equal to δ. Therefore $\hat{\delta}$ is a biased estimator, and the extent of the bias depends on n and $\hat{\delta}$. Figure 4 is a plot of the bias $E(\hat{\delta}) - \delta$ as a function of δ for $n = 10$, 20, and 40.

Examination of the figure shows that $\hat{\delta}$ overestimates δ for small values of δ. For moderate values of δ the estimator $\hat{\delta}$ underestimates the true value, and $\hat{\delta}$ becomes almost unbiased for large values of δ. In most practical situations, the sample size is moderate, in which case the absolute magnitude of the bias is small. For example, if $n = 20$, the bias never exceeds 0.19, and if $n = 40$, the maximum bias does not exceed 0.11. Even for an n as small as 10, the bias does not exceed 0.27. Comparing the expected value of $\hat{\delta}$ with the expected values of g^* given in Table 1, we see that the bias of $\hat{\delta}$ is substantially smaller than the bias of g^*.

In many practical situations, the bias can be considerably smaller than the maximum bias. For effect sizes in the range of 0.5 to 0.8, the bias never exceeds 0.18 even if n is as small as 10. For example, if $n = 10$ and $\delta = 0.75$, the bias is only 0.06. Similarly, when $n = 20$ and $\delta = 0.50$, the bias is only 0.02.

TABLE 3

Mean $E(\hat{\delta})$ and Variance $V(\hat{\delta})$ of the Maximum Likelihood Estimator $\hat{\delta}$ of Effect Size When Only Significant Results Are Observed[a]

		δ									
		0.25		0.50		0.75		1.00		1.50	
n		$E(\hat{\delta})$	$V(\hat{\delta})$	$E(\hat{\delta})$	$V(\hat{\delta})$	$E(\hat{\delta})$	$V(\hat{\delta})$	$E(\hat{\delta})$	$V(\hat{\delta})$	$E(\hat{\delta})$	$V(\hat{\delta})$
10	Exact	0.52	0.28	0.68	0.27	0.81	0.33	0.96	0.40	1.40	0.52
	Simulated	0.51	0.30	0.70	0.27	—	—	0.97	0.42	1.44	0.54
20	Exact	0.39	0.10	0.52	0.13	0.69	0.18	0.92	0.21	1.48	0.20
	Simulated	0.39	0.10	0.52	0.13	—	—	0.92	0.21	1.51	0.18
30	Exact	0.34	0.07	0.48	0.10	0.68	0.13	0.95	0.13	1.49	0.11
	Simulated	0.34	0.07	0.48	0.10	—	—	0.95	0.13	1.53	0.10
40	Exact	0.31	0.05	0.46	0.08	0.70	0.10	0.98	0.08	1.50	0.08
	Simulated	0.31	0.05	0.46	0.08	—	—	0.98	0.08	1.52	0.07
50	Exact	0.28	0.04	0.44	0.07	0.71	0.07	0.99	0.06	1.49	0.07
	Simulated	0.28	0.04	0.45	0.07	—	—	0.99	0.06	1.51	0.06

[a] Exact results were obtained by numerical integration; simulated results were obtained from 2000–10,000 replications for each combination of sample size and effect size.

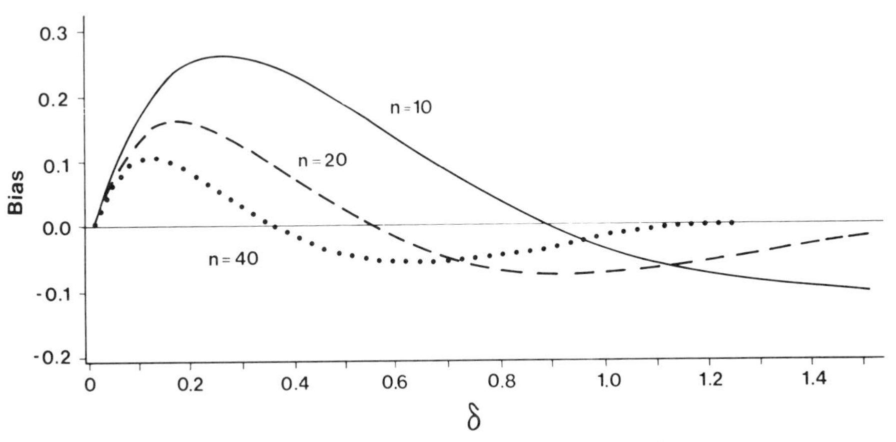

Fig. 4 Bias $E(\hat{\delta}) - \delta$ of the maximum likelihood estimator $\hat{\delta}$ of effect size.

Because substantive interpretations are based on the overall magnitude of the effect size, the relatively small absolute bias of $\hat{\delta}$ does not seem serious enough to affect substantive conclusions unless both sample sizes and effect sizes are quite small.

It is also interesting to compare the efficiency of $\hat{\delta}$ to that of the sample effect size g when the results of all studies are observed. Using expression (8) of Chapter 5 to calculate the variance of g, we see that for $\delta = 0.25$ the estimator $\hat{\delta}$ has approximately the same variance as does g. When both n and δ are large, the variance of $\hat{\delta}$ is also quite close to the variance of g. For intermediate situations, when δ is moderate to large but n is not large, $\hat{\delta}$ is considerably less efficient than g. For example, when $n = 20$ and $\delta = 1.0$, the variance of $\hat{\delta}$ is nearly twice the variance of g.

D. ESTIMATION OF EFFECT SIZE FROM A SERIES OF INDEPENDENT EXPERIMENTS WHEN ONLY SIGNIFICANT RESULTS ARE OBSERVED

The previous discussion dealt with estimating an effect size from a single experiment when only statistically significant results are observed. However, most practical situations involve the combination of estimates of effect size from several experiments. We now discuss methods for estimating an effect size from a series of independent experiments when only statistically significant mean differences are observed. We first present a simple counting procedure for estimating an effect size from a series of studies with homogeneous sample sizes, and then discuss maximum likelihood estimation of effect size. Finally, we consider a linear combination of estimates derived from each of the studies.

D.1 ESTIMATION OF EFFECT SIZE USING COUNTING PROCEDURES

Our starting point is a single study consisting of an experimental and a control group with population means μ^E and μ^C. The probability $p^+(\delta)$ that the study yields a positive mean difference large enough to be statistically significant at the 100α-percent level is given by the right tail of a noncentral t-variate with $2n - 2$ degrees of freedom and noncentrality parameter $\delta\sqrt{n/2}$. Similarly, the probability $p^-(\delta)$ that the study yields a negative mean difference that is large enough to be statistically significant at the 100α-percent level is given by the left tail of the same distribution. These probabilities depend on the true effect size δ, the sample size n, and the level of significance.

We define the probability $p(\delta)$ to be the ratio of the area in the right tail to that in the sum of the right and left tails:

$$p(\delta) = \frac{p^+(\delta)}{p^+(\delta) + p^-(\delta)}. \tag{1}$$

For a sample of k studies, each of which has a statistically significant mean difference, we wish to estimate δ. If the studies have a common sample size n, then we can consider each of the k studies as a Bernoulli trial, in which a trial is defined as a "success" if the sign of the mean difference $(\overline{Y}^E - \overline{Y}^C)$ is positive. It is well known that the sample number U of successes from a series of k trials has the binomial distribution with parameter $p(\delta)$.

The maximum likelihood estimator of the parameter $p(\delta)$ of the binomial distribution with k trials is U/k, the sample proportion of "successes." For a given sample size n per group, significance level α, and number k of trials, the probability $p(\delta)$ is a strictly monotonic function of δ. Hence, the maximum likelihood estimator of δ, based on the sample number of positive significant results, is $\hat{\delta}$ defined by

$$p(\hat{\delta}) = U/k. \tag{2}$$

Given U/k, we wish to find the corresponding value of $\hat{\delta}$. This can be accomplished with the use of Table 4, which gives values of $\hat{\delta}$ from values of U/k for various sample sizes n and for the $\alpha = 0.05$ significance level. A confidence interval for δ can be obtained by first obtaining a confidence interval

TABLE 4

Conditional Probability $p(\delta, n) = P\{t > 0 \mid |t| > c_t\}$ That a Two-Sample t-Statistic with n Subjects per Group Is Positive Given That the Absolute Value of the t-Statistic Exceeds the $\alpha = 0.05$ Critical Value for Effect Size δ

	δ												
n	0.00	0.02	0.04	0.06	0.08	0.10	0.15	0.20	0.25	0.30	0.40	0.50	0.70
2	0.500	0.516	0.531	0.547	0.562	0.577	0.615	0.651	0.685	0.718	0.777	0.827	0.900
4	0.500	0.528	0.556	0.584	0.611	0.638	0.700	0.756	0.805	0.845	0.906	0.945	0.982
6	0.500	0.537	0.573	0.609	0.643	0.676	0.751	0.814	0.863	0.901	0.950	0.976	0.994
8	0.500	0.544	0.586	0.628	0.668	0.706	0.788	0.852	0.899	0.932	0.971	0.988	0.998
10	0.500	0.549	0.598	0.644	0.689	0.729	0.816	0.879	0.923	0.952	0.982	0.993	0.999
12	0.500	0.555	0.608	0.659	0.706	0.750	0.838	0.900	0.940	0.964	0.988	0.996	1.000
14	0.500	0.559	0.617	0.672	0.722	0.767	0.857	0.916	0.952	0.973	0.992	0.998	1.000
16	0.500	0.564	0.625	0.683	0.736	0.783	0.872	0.929	0.961	0.979	0.994	0.998	1.000
18	0.500	0.568	0.633	0.694	0.749	0.796	0.886	0.939	0.968	0.984	0.996	0.999	1.000
20	0.500	0.572	0.640	0.704	0.760	0.809	0.897	0.947	0.974	0.987	0.997	0.999	1.000
22	0.500	0.575	0.647	0.713	0.771	0.820	0.907	0.954	0.978	0.990	0.998	1.000	1.000
24	0.500	0.579	0.654	0.721	0.781	0.830	0.915	0.960	0.982	0.992	0.998	1.000	1.000
50	0.500	0.614	0.716	0.800	0.864	0.910	0.970	0.991	0.997	0.999	1.000	1.000	1.000
100	0.500	0.659	0.789	0.878	0.933	0.964	0.993	0.999	1.000	1.000	1.000	1.000	1.000

$[p_L, p_U]$ for $p(\delta)$, from which the confidence limits δ_L and δ_U for δ are calculated. This involves solving the equations

$$p(\delta_L) = p_L \qquad \text{and} \qquad p(\delta_U) = p_U \tag{3}$$

for δ_L and δ_U.

Confidence intervals for $p(\delta)$ are obtained by any of the usual methods for finding a confidence interval for the parameter of the binomial distribution. For large k, the normal approximation to the binomial distribution can be used to obtain asymptotic confidence intervals for $p(\delta)$. This approximation says that U/k is normally distributed with mean $p(\delta)$ and variance $p(\delta)[1 - p(\delta)]/k$. Any of the other procedures for finding confidence intervals for $p(\delta)$ given in Chapter 4 can also be used here.

D.1.a Vote-Counting Estimates When All Significant Mean Differences Are in the Same Direction

Occasionally, all the statistically significant mean differences may have the same sign. This is likely to occur when sample sizes and effect sizes are large, as in the case of the Lane and Dunlap (1978) simulation. When all the significant mean differences have the same sign, the counting method given in the previous section cannot be applied directly, since the maximum likelihood estimate of $p(\delta, \alpha, n)$ is unity and therefore does not correspond to a unique value of $\hat{\delta}$. Instead we use other estimation procedures to obtain an estimate of $p(\delta)$.

When the significant mean differences are all in the same direction the true value of $p(\delta)$ is likely to be very small or very large. We can exploit knowledge about $p(\delta)$ by assuming a prior distribution for the magnitude of $p(\delta)$ to obtain a Bayes estimate of $p(\delta)$. For example, we can assume that $p(\delta)$ has a truncated uniform prior distribution; that is, there is a p_0 such that any value of $p(\delta)$ in the interval $[p_0, 1]$ is equally likely.

The Bayes estimate $\hat{p}$ of $p(\delta)$ is

$$\hat{p} = (k + 1)(1 - p_0^{k+2})/(k + 2)(1 - p_0^{k+1}). \tag{4}$$

For example, if $p_0 = 0.4$ and there are $k = 10$ trials, all of which are successes, then the Bayes estimate is $\hat{p} = 0.9167$. For $n = 20$, $\delta = 0.5$, transforming $\hat{p}$ to a value of $\hat{\delta}$ using Table 4 gives $\hat{\delta} = 0.17$. A more complete discussion of Bayes estimation for the parameter of the binomial distribution when all the trials are successes is given in Chew (1971).

D.1.b Limitations of Vote-Counting Estimates

The counting estimates described are simple to apply and have general appeal. In particular, they can be applied to provide an estimate of effect size based on very minimal data from a series of studies. However, these estimators are subject to the same limitations as the vote-counting estimators described in Chapter 4—limitations that restrict their applicability in practice. One limitation is that counting estimators of effect size require a rather large number of studies to obtain an accurate estimate of $p(\delta)$. Moreover, the estimate $\hat{\delta}$ of δ obtained by these methods can be sensitive to variations in $p(\delta)$, especially when $p(\delta)$ is close to zero or unity. Thus a large sample of studies is even more important when $p(\delta)$ is close to zero or unity. Formally, the asymptotic theory that underlies vote-counting estimators holds as the number k of studies becomes large.

Another limitation stems from the assumption of a common sample size for the k studies, a situation that occurs infrequently. This problem was discussed in some detail in Section G of Chapter 4, and the methods described there can be used here.

D.2 THE MAXIMUM LIKELIHOOD ESTIMATOR OF EFFECT SIZE

The method of maximum likelihood provides an alternative to simple vote-counting methods, and has the advantage that it can be applied to any collection of studies, including those in which the studies do not have identical sample sizes. The method of maximum likelihood also uses more of the information from each study (e.g., the magnitude of the g^* values as well as just the sign) than do vote-counting methods. The biggest disadvantage of maximum likelihood estimation is that it requires a special computer program to compute the likelihood and its maximum.

If $\mathbf{g}^* = (g_1^*, \ldots, g_k^*)$ is the vector of observed effect sizes from k independent experiments with sample sizes $n_1, \ldots, n_k$ per group and if δ is the population effect size, then the log likelihood $L(\mathbf{g}^* \mid \delta)$ is the sum of the log likelihoods for each g_i^*. Explicit expressions for the likelihood are given in Section E.

Although the value $\hat{\delta}$ that maximizes the log likelihood cannot be obtained in closed form, it can be obtained numerically. One simple procedure to find $\hat{\delta}$ is to evaluate the log likelihood for a grid of trial values and then to select $\hat{\delta}$ as the value of δ that yields the largest value of $L(\mathbf{g}^* \mid \delta)$. If a reasonably accurate "first guess" about the value of δ is available (e.g., from a vote-counting estimate or the method given in the next section), the evaluation of a relatively small grid of trial values may give $\hat{\delta}$ to an accuracy of two decimal places or more.

More elaborate numerical procedures can also be developed. For example, the Newton–Raphson method can be used to iterate rapidly to the value of $\hat{\delta}$. Such procedures involve the evaluation of fairly complicated derivatives, however, and are warranted only if rather extensive use of the resulting computer programs is envisioned. (For further details on the Newton–Raphson method see, e.g., Householder, 1953.)

D.3 WEIGHTED ESTIMATORS OF EFFECT SIZE

An alternative to finding the maximum likelihood estimate of δ based on all k experiments is to obtain the maximum likelihood estimates $\hat{\delta}_1, \ldots, \hat{\delta}_k$ from each experiment separately, and then to pool these estimates. This procedure has several advantages. First, there is no need for a specialized computer program, because Table 2 can be used to obtain the explicit maximum likelihood estimate of δ from each experiment. In addition, the estimates $\hat{\delta}_1, \ldots, \hat{\delta}_k$ can be examined individually to detect potential outliers or values that deviate greatly from the other estimates. Finally, the analysis for a linear combination of $\hat{\delta}_i$ values is similar to the analyses usually used in quantitative research syntheses.

The simplest combination of $\hat{\delta}_1, \ldots, \hat{\delta}_k$ is the unweighted mean. However, when the experiments have different sample sizes, the unweighted mean will not be optimal, since experiments with large sample sizes produce more precise estimates of δ and should be given more weight. For a weighted estimator of the form

$$\sum_{i=1}^{k} w_i \hat{\delta}_i \Big/ \sum_{i=1}^{k} w_i, \tag{5}$$

the weights that minimize the variance of (5) are proportional to the reciprocals of the sampling variances of the $\hat{\delta}_i$, so that

$$w_i = 1/\mathrm{Var}(\hat{\delta}_i), \qquad i = 1, \ldots, k. \tag{6}$$

Because the sampling variance of $\hat{\delta}_i$ is a function of the unknown parameter δ, the optimal weights cannot be calculated exactly in most applications.

However, we can use weights that are close to optimal, but are selected in a manner that does not involve δ. Because the sampling variance of $\hat{\delta}_i$ is approximately proportional to $1/n_i$, the use of $w_i = n_i$ results in a weighted estimator that is reasonably close to optimal. The weighted estimator will also tend to be less biased than the unweighted mean, since estimates from larger experiments (with correspondingly smaller bias) will be given more weight.

Example

The techniques described in this chapter are applied to some studies collected by Hedges, Giaconia, and Gage (1981). Some data from 10 studies that examined the effects of open and traditional education on student creativity are given in Table 5. The sample size, the effect size estimate, and the two-sample t-statistic are given for each study. The difference between the open and traditional group means is statistically significant at the $\alpha = 0.05$ level in four of the 10 studies. We illustrate the techniques described in this chapter by applying them to the effect size estimates derived from these four studies.

The unweighted average of all 10 effect size estimates is 0.29. The unweighted average of the g_i values obtained in the first four studies is 0.45, somewhat larger than the average for all studies. The unweighted average of the $\hat{\delta}_i$ is 0.33. In these data, the sample sizes vary considerably, so one might expect that weighted averages would differ from unweighted averages of estimates. The (sample size) weighted average of the effect size estimates for all studies is $g = 0.07$, and the average of the $\hat{\delta}_i$ is 0.01. It is interesting to note that the maximum likelihood estimator of δ based on the first four studies is $\hat{\delta} = 0.01$, which is the same as the weighted average of the $\hat{\delta}_i$.

D.4 APPLICATIONS OF THE METHODS THAT ASSUME CENSORING

The statistical procedures described in this chapter provide sharp estimates of effect size under nonrandom sampling when the censoring rule is known precisely. One such application is for the "shrinkage" of effect size estimates

TABLE 5
Studies of the Effects of Open versus Traditional Education on Student Creativity

Study	Common sample size $n^E = n^C$	Standardized mean difference g	t	$\hat{\delta}$	$n\hat{\delta}$	ng
1	90	−0.583	−3.91*	−0.57	−51.3	−52.47
2	40	0.535	2.39*	0.19	7.6	21.40
3	36	0.779	3.31*	0.71	25.6	28.04
4	20	1.052	3.33*	0.98	19.6	21.04
5	22	0.563	1.87	—	—	12.39
6	10	0.308	0.69	—	—	3.08
7	10	0.081	0.18	—	—	0.81
8	10	0.598	1.34	—	—	5.98
9	39	−0.178	−0.79	—	—	−6.94
10	50	−0.234	−1.17	—	—	−11.70
Mean	237	0.29		0.33	0.38	

that can only be calculated if the mean difference is statistically significant. Books or journal articles that report only "not significant" for results that do not attain the $\alpha = 0.05$ level of significance are an example. In such cases statistics are only reported (and hence an effect size estimate can only be calculated) if the mean difference is significant. The results developed herein correct effect size estimates for the effects of censoring nonsignificant results.

Another potential application is when the censoring rule is unknown, but is believed to be less extreme than complete censoring of all nonsignificant results. This occurs, for example, when some, but not all, nonsignificant results are censored. The reviewer can impose the more stringent censoring model considered in this chapter by using only effect size estimates that are statistically significant. The effect size can then be estimated from only statistically significant effect size estimates by using the methods described herein.

A third application of the methods described in this chapter is to examine the potential effects on conclusions of an unknown censoring rule. Here the question is whether a tendency not to sample (publish) studies with statistically nonsignificant results could have dramatically inflated the observed effect size estimates. By comparing the estimate of effect size from the observed effect sizes that are statistically significant with that obtained by a simple or weighted average of the observed effect sizes (including the ones that are not statistically significant), the reviewer can evaluate the potential bias due to censoring. If the overall estimates of effect size do not differ greatly, then it is difficult to argue that the overall observed effect size estimate has been inflated by the censoring of nonsignificant results without proposing a more extreme and less plausible censoring model. This application of checking the plausibility of censoring as a potential explanation of average effect magnitudes may be one of the most important applications of the methods presented herein.

E. OTHER METHODS FOR ESTIMATING EFFECT SIZES WHEN NOT ALL STUDY OUTCOMES ARE OBSERVED

Estimation of effect size is the most direct method of dealing with biases caused by censoring of effect size estimates from yielding statistically nonsignificant mean differences. Two other approaches to dealing with this problem have also been developed. The first addresses the question of how many (censored) null results would be necessary to overturn a conclusion obtained from available data. If the number of studies needed to overturn the result (the "fail-safe N") is large, then the plausibility that the conclusion is an artifact of censoring is small. The other approach uses an elaboration of the random effects model for effect sizes.

E.1 ASSESSING THE NUMBER OF STUDIES (WITH NULL RESULTS) NEEDED TO OVERTURN A CONCLUSION

In the models studied, we assume that we observe some, but not all, effect size estimates. Thus, it is possible to imagine that the k effect size estimates that are observed are a censored sample from K effect size estimates that are actually derived in research studies. The $K - k$ effect size estimates corresponding to nonsignificant mean differences are not available, because of the failure of authors to report the results or the failure of publication outlets (such as academic journals) to publish the results. What happens to the nonsignificant results? Rosenthal (1979) colorfully described the unpublished nonsignificant results as residing in the "file drawers" of the researchers who conducted the studies. He argues that the problem of assessing the importance of censored samples in research synthesis might therefore be called the file drawer problem.

Rosenthal (1979) did not address the impact of the file drawer problem on effect size estimates. Instead he examined the effect of unpublished studies with null results on tests of the significance of combined results (see Chapter 3). If a combined significance test yields a significant result, then the question is how many studies with null results must be in the file drawers to overturn the results of the combined significance test. Using the inverse normal method for testing the statistical significance of combined results, the combined significance test statistic based on the k observed studies is

$$z_0 = \sum_{i=1}^{k} \frac{\Phi^{-1}(p_i)}{\sqrt{k}},$$

where p_i is the one-tailed p-value for the ith study, and $\Phi(x)$ is the standard normal cumulative distribution function. Assume that k_0 additional studies have null results, and that $\sum \Phi^{-1}(p_i) = 0$ for these additional studies. Then the combined test statistic for the k observed studies and the k_0 unobserved studies is

$$z = \sum_{i=1}^{k} \frac{\Phi^{-1}(p_i)}{\sqrt{k + k_0}} = \frac{\sqrt{k}\, z_0}{\sqrt{k + k_0}}. \tag{7}$$

Setting z in (7) equal to a critical value C_α of the standard normal distribution and solving for k_0 yields

$$k_0 = k(z_0^2 - C_\alpha^2)/C_\alpha^2. \tag{8}$$

Thus (8) is a formula for the minimum number k_0 of null results needed so that the combined significance test statistic (7) will not be significant. The number k_0 is the so-called fail-safe number of studies.

If k_0 is extremely large (as is often the case) then the reviewer can argue that it is unlikely that so many unpublished studies exist. Consequently, the outcome of the combined significance test is unlikely to be an artifact of selective sampling of studies. If, on the other hand, k_0 is small, then the outcome of the combined significance test could easily be overturned by the discovery of a few additional studies in the file drawers. When k_0 is small, the outcome of the combined significance test could easily be an artifact of selective sampling.

The weakness in this argument is its reliance on the combined test statistic. We argued in Chapter 3 that combined significance tests seldom answer the questions of interest to research reviewers. Rejection of the combined null hypothesis implies that at least one study has a nonzero effect. Because this alternative hypothesis is usually not substantively meaningful, methods that increase the plausibility of the hypothesis are not particularly useful.

A variation of the fail-safe procedure that takes effect size into account was proposed by Orwin (1983). He proposes calculating the number of studies with null results necessary to reduce the average effect size estimate to a negligible level. Suppose that the average effect size estimate in k studies is $\bar{d}$, and assume that k_0 additional studies have an average effect size estimate of zero. A direct argument shows that the number k_0 of studies necessary to reduce the average effect size estimate of the $k + k_0$ studies to the critical level d_c is

$$k_0 = k(\bar{d} - d_c)/d_c. \tag{9}$$

By choosing d_c to be an effect size that is small enough to be negligible, the fail-safe number k_0 is the number of additional studies with null results that would be needed to reduce the observed average effect size $\bar{d}$ to a negligible size d_c.

E.2 RANDOM EFFECTS MODELS WHEN NOT ALL STUDY OUTCOMES ARE OBSERVED

In Sections C and D we examined fixed effects models for estimation of effect size when not all study outcomes are observed. That is, we assumed the effect sizes $\delta_1, \ldots, \delta_k$ are fixed but unknown constants. Random effects models treat $\delta_1, \ldots, \delta_k$ as realizations of a random variable Δ. In random effects models the object of the statistical analysis is to estimate the variance $\sigma^2(\Delta)$ of Δ and the mean $\bar{\Delta}$ of Δ. Champney (1983) considered a random effects model for effect sizes when not all study outcomes are observed. In his formulation the parameters $\delta_1, \ldots, \delta_k$ have a normal prior distribution with mean δ_* and variance σ_*^2. Champney used empirical Bayes methods to estimate these hyperparameters. His approach used the EM algorithm (Dempster, Laird, & Rubin, 1977) in a manner analogous to that described in Section G of Chapter 9. A complete explanation of the details of the procedure is given in Champney (1983).

F. TECHNICAL COMMENTARY*

Section B.2: The probability density function $f(x \mid \delta, n, \alpha)$ of g^* is

$$f(x \mid \delta, n, \alpha) = \begin{cases} h(x \mid \delta, n)/A(\delta, n, \alpha) & \text{if} \quad x^2 > 2F(\alpha, n)/n \\ 0 & \text{otherwise,} \end{cases}$$

where $h(x \mid \delta, n)$ is the probability density function of $\sqrt{2/n}$ times a noncentral t-variate with $2n - 2$ degrees of freedom and noncentrality parameter $\delta\sqrt{n/2}$, and $A(\delta, n, \alpha)$ is the probability that a noncentral F-variate with 1 and $2n - 2$ degrees of freedom and noncentrality parameter $n\delta^2/2$ exceeds the critical value $F(\alpha, n)$. The complete expression for $h(x \mid \delta, n)$ is given in (11) below.

Section C.1: The problem is to maximize

$$\log f(x \mid \delta, n, \alpha) = \log h(x \mid \delta, n) - \log A(\delta, n, \alpha)$$

for $x^2 > 2F(\alpha, n)/n$. Because $A(\delta, n, \alpha)$ is the probability of a statistically significant mean difference, it is an increasing function of δ for nonnegative δ, and it tends to unity as $\delta \to \infty$. Hence the value of δ that maximizes $\log f(x \mid \delta, n, \alpha)$ will tend to be smaller in absolute magnitude than the value of δ that maximizes $\log h(x \mid \delta, n)$.

Section C.2: If q denotes the function transforming $\hat{\delta}$ to g^*., i.e.,

$$g^* = q(\hat{\delta}),$$

the probability density function of $\hat{\delta}$ is

$$f(x \mid \delta, n, \alpha) = \left|\frac{dq}{dx}\right| h(q(x) \mid \delta, n).$$

Section D.2: The log likelihood is given by

$$L(g^* \mid \delta) = \sum_{i=1}^{k} \log h(g_i^* \mid \delta, n_i) - \sum_{i=1}^{k} \log A(\delta, n_i, \alpha). \tag{10}$$

Sections B and C. Derivations and Computations: Computations for the distribution of g^* were based on $\sqrt{n/2}g^*$ and on the corresponding noncentral t-distribution with noncentrality parameter $\lambda = \sqrt{n/2}\delta$ and $m = 2n - 2$ degrees of freedom. The density function for a noncentral t-variate with m degrees of freedom and noncentrality parameter λ is given (see, e.g., Resnikoff & Lieberman, 1957) by

$$h(x \mid \lambda, n) = \frac{\Gamma(m+1)}{2^{(m-1)/2}\Gamma(m/2)\sqrt{\pi m}} \exp\left(-\frac{1}{2}\frac{m\lambda^2}{m+x^2}\right) \times \left[\frac{m}{m+x^2}\right]^{(m+1)/2} Hh_m\left[\frac{-x\lambda}{\sqrt{m+x^2}}\right], \tag{11}$$

where

$$Hh_m(y) = \int_0^\infty \frac{v^m}{\Gamma(m+1)} \exp\left(-\frac{1}{2}(v+y)^2\right) dv.$$

For $m < 20$, values of $Hh_m(y)$ can be obtained as

$$Hh_m(y) = P_m(y)Hh_0(y) + Q_m(y)Hh_{-1}(y),$$

where $P_m(y)$ and $Q_m(y)$ are polynomials (see Resnikoff & Lieberman, 1957), and

$$Hh_0(y) = \frac{1}{2\pi}\int_0^y \exp\left(-\frac{1}{2}t^2\right) dt,$$

$$Hh_{-1}(y) = [\exp(-\tfrac{1}{2}y^2)]/2\pi.$$

A recurrence relationship among the polynomials $P_m(y)$ and $Q_m(y)$ simplifies their computation. For large m it is easier to compute $Hh_m(y)$ by using an asymptotic expansion given by Resnikoff and Lieberman (1957) which is accurate to five decimal places when $m \geq 20$:

$$Hh_m(y) = \frac{1}{\Gamma(m+1)} t^m \exp\left(-\frac{1}{2}(t+y)^2\right)\sqrt{\frac{2t^2}{m+t^2}} \times \left(1 - \frac{3m}{4(m+t^2)^2} + \frac{5m^2}{6(m+t^2)^3}\right),$$

where

$$t = -y\sqrt{y^2 + 4m}/2.$$

The IMSL (1977) subroutine MDTN was used to obtain integrals of the noncentral t-distribution. All of the numerical integrations were computed using IMSL subroutines DCADRE and DCSQDU as in Hedges (1984).

The exact density function of $\hat{\delta}$ was computed by obtaining a table of approximately 40 values of g^* and $\hat{\delta}$ for each n. The tabulated values were then used as knot points for cubic spline interpolation between tabulated values (e.g., approximation of the function $g^* = q[\hat{\delta}]$) using IMSL subroutines ICSCC and ICSEVU. The same splines were used to evaluate the Jacobian $dq/d\hat{\delta}$. Slightly different knot points were used for each value of n to obtain the most accurate approximation, and the final results were checked in several ways, including the simulation study.

The simulation studies described in this chapter were conducted using a file of g_i values that were generated for the simulation study reported in Chapter 5. The g_i values were derived using standard normal deviates and chi-square random numbers generated by the IMSL subroutines GGNML and GGCHS. For each value of δ (0.25, 0.50, 1.00, or 1.50) and each value of n (10, 20, 30, 40, or 50), 10,000 g_i values were generated using the identity

$$g = X/\sqrt{S/m},$$

where X is a normal variate with variance $2/n$ and mean δ, and S is an independent chi-square variate with $m = 2n - 2$ degrees of freedom. In the present study, values of g^* were obtained by selecting observations with absolute values that exceeded the $\alpha = 0.05$ critical value of the null distribution of g. Additional values of g were generated when necessary to yield at least 2000 values of g^* for each combination of n and δ. The g^* values were transformed to $\hat{\delta}_i$ values using cubic spline interpolation on a table of values of g^* and $\hat{\delta}$.

CHAPTER 15

Meta-Analysis in the Physical and Biological Sciences

Because human behavior is generally very complex, we normally expect that replicated studies with many variations will be required in order to understand particular phenomena in the social, behavioral, and educational sciences. Perhaps surprisingly, replication in the physical and biological sciences also has a long history. However, the source and form of problems in the physical sciences are quite different from those in the social sciences.

We illustrate experimental replication in the physical sciences by noting the history of measurements of gravitational force. Table 1 provides a list of results of 24 studies (see Speake, 1983), each providing a measure of gravitational force, but by different methods and apparatus. For such a set of studies we would not consider combining the results because presumably more accuracy is achieved in recent experiments.

The study by Heyl (1930) offers a good example of experimentation in the physical sciences for which some integration is required, and we review this study in greater detail. Three sets of experiments to determine gravitational force were conducted in which gold, platinum, and glass balls were used. The results of Heyl are given in Table 2.

TABLE 1

Measurements of *G* Gravitational Constant

Author	Date	(10^{-11})[in (meters)^3/(kilogram)(seconds)^2]	Method[a]
H. Cavendish	1798	6.75 ± 0.05	TB
F. Reich	1838	6.70 ± 0.04	TB
	1852	6.59 ± 0.04	TP
F. Baily	1843	6.63 ± 0.07	TB
A. Cornu & J. B. Baille	1873	6.64 ± 0.017	TB
Ph. von Jolly	1878	6.47 ± 0.11	B
J. Wilsing	1889	6.594 ± 0.015	VP
J. H. Poynting	1891	6.70 ± 0.04	B
C. V. Boys	1895	6.658 ± 0.007	TB
R. von Eötvös	1896	6.657 ± 0.013	TP
C. Braun	1897	6.649 ± 0.002	TP
F. Richarz & O. Krigar-Menzel	1898	6.685 ± 0.011	B
G. K. Burgess	1902	6.64	TB
P. R. Heyl	1930	6.670 ± 0.005	TP
J. Zahradnicek	1933	6.66 ± 0.04	RP
P. R. Heyl & P. Chzanowski	1942	6.673 ± 0.003	TP
J. Renner	1970	6.670 ± 0.008	TP
C. Pontikis	1972	6.6714 ± 0.0006	RP
G. G. Luther, W. R. Towler, R. D. Deslattes, R. A. Lowry, & J. W. Beams	1976	6.6699 ± 0.0014	TBR
O. V. Karagioz, V. P. Izmaylov, N. I. Agafoner, E. G. Kocheryan, & Yu. A. Tarakanov	1976	6.668 ± 0.002	TP
W. A. Koldewyn & J. Faller	1976	6.57 ± 0.17	TP
M. U. Sagitov, V. R. Miliukov, E. A. Monakhov V. S. Nazarenko, & Kh. G. Tadzhidinov	1978	6.6745 ± 0.0008	TP
G. G. Luther & W. R. Towler	1982	6.6726 ± 0.0005	TP
Current CODATA value	1973	6.6720 ± 0.0041	TP

[a] RP = resonant pendulum, TP = torsion pendulum, TB = torsion balance, B = beam balance, VP = vertical pendulum, TBR = torsion balance, rotating.

It is interesting to note Heyl's analysis. He argues that the range for the gold balls is too large, and in effect discards the results of that set of experiments. Since the range for platinum is equal to that for glass, he averages the means with equal weights:

$$\tfrac{1}{2}(6.664) + \tfrac{1}{2}(6.674) = 6.669,$$

with a mean deviation of 0.005.

TABLE 2

Gravitational Constant G ($\times 10^{-8}$) under Three Experimental Conditions

	Gold	Platinum	Glass
	6.683	6.661	6.678
	6.681	6.661	6.671
	6.676	6.667	6.675
	6.678	6.667	6.672
	6.679	6.664	6.674
	6.672		
Mean	6.678	6.664	6.674
Mean absolute deviation	0.003	0.002	0.002
Range	0.011	0.006	0.006

The author then reconsiders the results of the experiments with gold balls and takes another weighted average in which gold receives a weight one-third the weight of platinum and glass. This leads to the weighted average

$$\tfrac{3}{7}(6.664) + \tfrac{3}{7}(6.674) + \tfrac{1}{7}(6.678) = 6.670,$$

with a mean deviation of 0.005. This is the value quoted for the Heyl study (see Table 1).

In accordance with our discussion, we might also use a weighted average, with weights depending on the precision (variance) of the observations in each experiment. The variances for the three experiments are 15, 9, 7.5 (times the same scale factor), so that the weights are proportional to $\frac{1}{15} = 0.067$, $\frac{1}{9} = 0.111$, $1/7.5 = 0.133$. This yields a weighted average of

$$\frac{0.067}{0.311}(6.664) + \frac{0.111}{0.311}(6.674) + \frac{0.133}{0.311}(6.678) = 6.674.$$

This example has some intrinsic interest, but it serves to point out that combining results in the physical sciences is not devoid of problems. The characteristic being measured may be better defined, but imprecisions still remain.

In another context, in the calibration of instruments or in the analysis of chemicals, it is customary to send samples to several laboratories. Although these laboratories measure the same ingredients, there are variations in apparatus and methodology. The problem is to provide an estimate of the underlying parameters of concern. An example of this (Eberhardt, 1983) is given in Fig. 1. Note that in this example each laboratory measures 3 analytes and that laboratories B and F have some unusually large discrepancies. Should these laboratories be discounted? How should one combine these results?

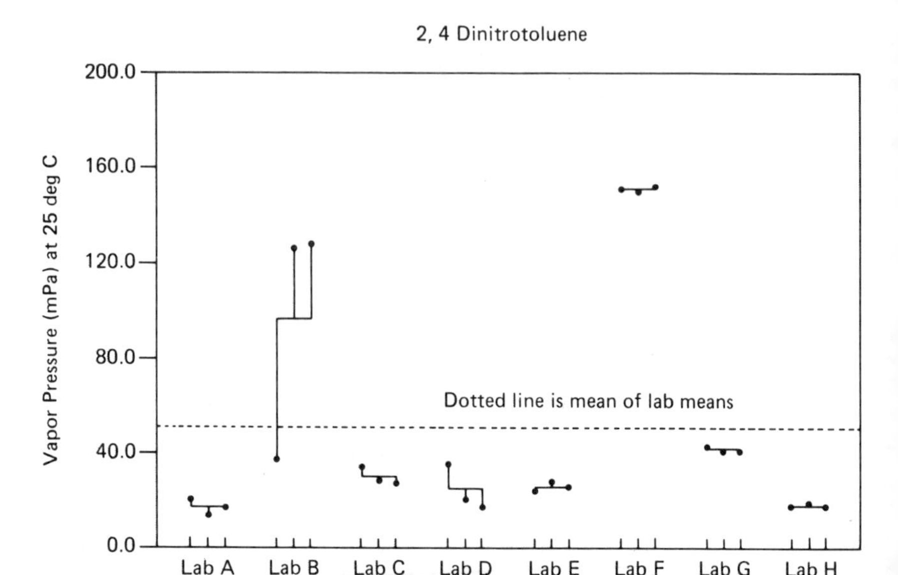

Fig. 1 Three measurements from each of eight laboratories.

The difference, then, between replications in the social and behavioral sciences and those in the physical and biological sciences is that in the former the relevant statistics to carry out a synthesis may be elusive and difficult to obtain. The models are rather amorphous, so specific assumptions are tenuous at best. At most one would assume a normal distribution—or a large sample normal approximation. But more assumptions than this are probably unwarranted.

On the other hand, in the physical and biological sciences there is an appearance of precision mainly because the goal is well defined and the models are specific. However, the analyses are frequently complicated, and often require numerical solutions or simulations. Sometimes, as in the case of tracking experiments, there is an excess of observations, whereas in other experiments the cost for each observation is high (either in time or in money) and the number of observations is relatively small. In the latter case the statistical methodology has to be as finely tuned as possible in order to provide maximum efficiency.

In this chapter we focus on several models that might be appropriate in some contexts in the physical and medical sciences. Obviously, no attempt at completeness could be achieved in one chapter, and our choice is by taste and experience.

In Sections A and B two multiplicative models are discussed, and in Sections C and D two parallel additive models are presented and analyzed. Experiments that have a treatment and a control are compared in Section E.

A. A MULTIPLICATIVE MODEL*

In this prototype model we are presented with the problem of determining whether two drugs are "equivalent." This determination is to be made on the basis of the results on p diagnostic psychological tests.

The first and central issue is to decide on a definition of equivalence. In some instances two drugs differ in strength, but by adjusting the strength the consequences may be quite similar. This is the case in comparing the effects on motor coordination of the effects of drinking beer versus wine, say, for which it is known that prescribed amounts of beer and wine will produce specified blood-alcohol levels.

Suppose

$$\mathbf{x} = (x_1, \ldots, x_p)', \qquad \mathbf{y} = (y_1, \ldots, y_p)'$$

represent the results on p diagnostic tests for the two drugs. One statistical model for this experiment is to assume that

$$\mathbf{x} = \boldsymbol{\mu} + \mathbf{e}_x, \qquad \mathbf{y} = \boldsymbol{\nu} + \mathbf{e}_y, \tag{1}$$

where $\mathbf{e}_x$ and $\mathbf{e}_y$ are independent, normally distributed errors with zero means and a common covariance matrix $\boldsymbol{\Sigma} = (\sigma_{ij})$.

If indeed the two drugs are equivalent in the context described, then we expect the means in the p diagnostic tests to differ by a multiplicative constant. That is,

$$\begin{aligned} \nu_1 &= \gamma\mu_1, \\ \nu_2 &= \gamma\mu_2, \\ &\vdots \\ \nu_p &= \gamma\mu_p. \end{aligned}$$

The multiplicative constant γ is called the *potency factor* and represents the alteration in strength of one drug to yield a profile (on the diagnostic tests) similar to that of the other drug.

In the experiment n_1 patients are given drug A and n_2 patients are given drug B. Each patient completes a set of p diagnostic tests, yielding a set of observations as in Table 3.

In addition, we observe a pooled sample covariance matrix $\mathbf{S}$ made up from the two sample covariance matrices $\mathbf{S}_1$ and $\mathbf{S}_2$;

$$\mathbf{S} = \frac{(n_1 - 1)\mathbf{S}_1 + (n_2 - 1)\mathbf{S}_2}{n_1 + n_2 - 2}.$$

Our goal is to obtain an estimate (and confidence interval) of the potency factor. The maximum likelihood estimator depends on the quantities

$$t_{11} = \bar{\mathbf{x}}'\mathbf{S}^{-1}\bar{\mathbf{x}}, \qquad t_{12} = \bar{\mathbf{x}}'\mathbf{S}^{-1}\bar{\mathbf{y}}, \qquad t_{22} = \bar{\mathbf{y}}'\mathbf{S}^{-1}\bar{\mathbf{y}}, \tag{2}$$

TABLE 3
Observations Obtained from the Experiment

Drug A					Drug B				
	Test					Test			
Patient	1	2	$\cdots$	p	Patient	1	2	$\cdots$	p
1	x_{11}	x_{12}	$\cdots$	x_{1p}	1	y_{11}	y_{12}	$\cdots$	y_{1p}
2	x_{21}	x_{22}	$\cdots$	x_{2p}	2	y_{21}	y_{22}	$\cdots$	y_{2p}
$\vdots$		$\vdots$		$\vdots$	$\vdots$		$\vdots$		$\vdots$
n_1	$x_{n_1 1}$	$x_{n_1 2}$	$\cdots$	$x_{n_1 p}$	n_2	$y_{n_2 1}$	$y_{n_2 2}$	$\cdots$	$y_{n_2 p}$
Means	$\bar{x}_1$	$\bar{x}_2$	$\cdots$	$\bar{x}_p$	Means	$\bar{y}_1$	$\bar{y}_2$	$\cdots$	$\bar{y}_p$

where $\bar{\mathbf{x}} = (\bar{x}_1, \ldots, \bar{x}_p)$ and $\bar{\mathbf{y}} = (\bar{y}_1, \ldots, \bar{y}_p)$ are the vectors of means on the p diagnostic tests. The maximum likelihood estimator $\hat{\gamma}$ of γ is given by

$$\hat{\gamma} = \frac{n_1 t_{11} - n_2 t_{22} + \sqrt{(n_1 t_{11} - n_2 t_{22})^2 + 4 n_1 n_2 t_{12}^2}}{2 n_1 t_{12}}. \tag{3}$$

Example

In a study described in Morrison (1976, p. 138) there are 49 elderly men separated into two groups (senile factor and no senile factor). Each of the individuals is scored on four variables (information, similarities, arithmetic, and picture completion) from the Wechsler Adult Intelligence Scale.

The means for the groups are as follows:

Group	Information	Similarities	Arithmetic	Picture completion	Sample size
No senile factor	12.57	9.57	11.49	7.97	37
Senile factor	8.75	5.33	8.50	4.75	12

The pooled sample covariance matrix is

$$\mathbf{S} = \begin{bmatrix} 11.2553 & 9.4042 & 7.1489 & 3.3830 \\ & 13.5318 & 7.3830 & 2.5532 \\ & & 11.5744 & 2.6170 \\ & & & 5.8085 \end{bmatrix}.$$

On the assumption that there is a multiplicative difference between the means of the two groups, we estimate the potency factor γ.

Using (2), we compute $\mathbf{S}^{-1}$ and t_{11}, t_{12}, t_{22}:

$$\mathbf{S}^{-1} = \begin{bmatrix} 0.259064 & -0.135783 & -0.058797 & -0.064719 \\ & 0.186449 & -0.038305 & 0.014382 \\ & & 0.150964 & -0.016920 \\ & & & 0.211171 \end{bmatrix},$$

$$t_{11} = 19.41, \qquad t_{12} = 13.44, \qquad t_{22} = 9.90,$$

so that

$$n_1 t_{11} - n_2 t_{22} = (37)(19.41) - (12)(9.90) = 599.37,$$
$$4n_1 n_2 t_{12}^2 = 4(37)(12)(13.44)^2 = 320{,}805.27,$$

and hence, from (3),

$$\hat{\gamma} = \frac{599.37 + 824.65}{2(37)(13.44)} = 1.43.$$

Thus, if indeed the senile group and no senile group are related by a multiplicative model, the maximum likelihood estimate of the multiplicative factor is 1.43. This is an estimate of the common ratio of means. In our example, the ratio of means is

$$\frac{12.57}{8.75} = 1.44, \qquad \frac{9.57}{5.33} = 1.80, \qquad \frac{11.49}{8.50} = 1.35, \qquad \frac{7.97}{4.75} = 1.68,$$

so that the estimate of γ appears quite reasonable.

When n_1 and n_2 are large the distribution of $\hat{\gamma}$ is approximately normal with mean γ and variance $(n_1\gamma^2 + n_2)/n_1 n_2 \tau_{22}$, where $\tau_{22} = \mathbf{v}\mathbf{\Sigma}^{-1}\mathbf{v}'$. Unfortunately, τ_{22} is a nuisance parameter in the asymptotic variance. In addition, the asymptotic variance depends on the unknown parameter γ. Both difficulties can be resolved when the sample sizes are large. First, make the variance-stabilizing transformation

$$A(\hat{\gamma}) = n_1^{-1/2} \operatorname{arcsinh}(\hat{\gamma}\sqrt{n_1/n_2}), \tag{4}$$

so that $A(\hat{\gamma})$ is approximately normal with mean $A(\gamma)$ and variance $1/\tau_{22}$. Note the relation

$$\operatorname{arcsinh} x = \log(x + \sqrt{x^2 + 1}), \tag{5}$$

which may be easier to use for hand computations.

A $100(1 - \alpha)$-percent confidence interval (A_L, A_U) for $A(\gamma)$ is

$$A_L = A(\hat{\gamma}) - C\sqrt{\frac{n_1 + n_2 - 2}{n_1 n_2 t_{22}}}, \qquad A_U = A(\hat{\gamma}) + C\sqrt{\frac{n_1 + n_2 - 2}{n_1 n_2 t_{22}}}, \tag{6}$$

where C is the critical value obtained from the standard normal tables, and t_{22} given by (2) is an estimate of τ_{22}. Given a confidence interval for $A(\gamma)$, we can convert this to a confidence interval (γ_L, γ_U) for γ by solving (4) for γ:

$$\gamma_L = \sqrt{\frac{n_2}{n_1}} \frac{\exp(2A_L\sqrt{n_1}) - 1}{2\exp(A_L\sqrt{n_1})}, \qquad \gamma_U = \sqrt{\frac{n_2}{n_1}} \frac{\exp(2A_U\sqrt{n_1}) - 1}{2\exp(A_U\sqrt{n_1})}. \tag{7}$$

Example

Continuing with the previous example, the variance-stabilizing transformation (4) gives the value

$$\begin{aligned} A(\hat{\gamma}) &= (\tfrac{1}{37}) \operatorname{arcsinh}[(1.43)(\tfrac{37}{12})] \\ &= (0.027) \operatorname{arcsinh} 4.41 = (0.027)(2.190) \\ &= 0.0591. \end{aligned}$$

A 95-percent confidence interval for $A(\gamma)$ is

$$A_L = 0.0591 - 1.96\sqrt{\frac{37 + 12 - 2}{(37)(12)(9.90)}} = -0.1436,$$

$$A_U = 0.0591 + 1.96\sqrt{\frac{37 + 12 - 2}{(37)(12)(9.90)}} = 0.2618.$$

To find a confidence interval for γ we use (7):

$$\gamma_L = \sqrt{\frac{12}{37}} \frac{\exp[2(-0.1436)\sqrt{37}] - 1}{2\exp[(-0.1436)\sqrt{37}]} = -0.56,$$

$$\gamma_U = \sqrt{\frac{12}{37}} \frac{\exp[2(0.2618)\sqrt{37}] - 1}{2\exp[(0.2618)\sqrt{37}]} = 1.34.$$

Since the potency factor is positive, we use the interval (0, 1.34).

B. ESTIMATING DISPLACEMENT AND POTENCY FACTOR*

Because some drugs contain "filler," the definition of equivalence might include a displacement factor. One interpretation of this is that the means of the diagnostic tests with drugs A and B satisfy the relations

$$\begin{aligned} \nu_1 &= \gamma\mu_1 + \eta, \\ \nu_2 &= \gamma\mu_2 + \eta, \\ &\vdots \\ \nu_p &= \gamma\mu_p + \eta, \end{aligned} \tag{8}$$

where γ is still the potency factor, but now η is a common displacement.

For example, if $\boldsymbol{\mu} = (1, 3, 4, 6, 10)'$, $\gamma = 2$, and $\eta = 5$, then $\mathbf{v} = (7, 11, 13, 17, 25)'$, so that the v values become more spread out—but by the same factor. The difference between two sets of v, say, $v_2 - v_1 = 4$ and $v_5 - v_4 = 8$, relates to the differences $\mu_2 - \mu_1 = 2$ and $\mu_5 - \mu_4 = 4$ by a common factor of 2.

The maximum likelihood estimate $\hat{\gamma}$ of γ has the same form as that in Section A, except that the definitions of t_{11}, t_{12}, t_{22} are modified. Define

$$\begin{aligned} t_{11}^* &= \bar{\mathbf{x}}'\mathbf{S}^{-1}\bar{\mathbf{x}} - (\bar{\mathbf{x}}'\mathbf{S}^{-1}\mathbf{e})^2/\mathbf{e}'\mathbf{S}^{-1}\mathbf{e}, \\ t_{12}^* &= \bar{\mathbf{x}}'\mathbf{S}^{-1}\bar{\mathbf{y}} - (\bar{\mathbf{x}}'\mathbf{S}^{-1}\mathbf{e})(\bar{\mathbf{y}}'\mathbf{S}^{-1}\mathbf{e})/\mathbf{e}'\mathbf{S}^{-1}\mathbf{e}, \\ t_{22}^* &= \bar{\mathbf{y}}'\mathbf{S}^{-1}\bar{\mathbf{y}} - (\bar{\mathbf{y}}'\mathbf{S}^{-1}\mathbf{e})^2/\mathbf{e}'\mathbf{S}^{-1}\mathbf{e}, \end{aligned} \tag{9}$$

where $\mathbf{e}$ is the column vector consisting of ones. The formula for $\hat{\gamma}$ is now

$$\hat{\gamma} = \frac{n_{11}t_{11}^* - n_2 t_{22}^* + \sqrt{(n_1 t_{11}^* - n_2 t_{22}^*)^2 + 4n_1 n_2 t_{12}^{*2}}}{2n_1 t_{12}^*}. \tag{10}$$

The maximum likelihood estimate $\hat{\eta}$ of η is

$$\hat{\eta} = (\bar{\mathbf{x}}'\mathbf{S}^{-1}\mathbf{e} - \hat{\gamma}\bar{\mathbf{y}}'\mathbf{S}^{-1}\mathbf{e})/\mathbf{e}'\mathbf{S}^{-1}\mathbf{e}. \tag{11}$$

Example

For the data of the example in Section A, we now suppose that there is a displacement in addition to a multiplication factor. To obtain the maximum likelihood estimate of γ we require the computations

$$\bar{\mathbf{x}}'\mathbf{S}^{-1}\mathbf{e} = 1.82449, \qquad \bar{\mathbf{y}}'\mathbf{S}^{-1}\mathbf{e} = 1.13814, \qquad \mathbf{e}'\mathbf{S}^{-1}\mathbf{e} = 0.207357,$$

which together with the previous computations yield values of t_{11}^*, t_{12}^*, t_{22}^* from (9):

$$\begin{aligned} t_{11}^* &= 19.41 - (1.82449)^2/0.207357 = 3.36, \\ t_{12}^* &= 13.44 - (1.82449)(1.13814)/0.207357 = 3.43, \\ t_{22}^* &= 9.90 - (1.13814)^2/0.207357 = 3.65. \end{aligned}$$

Inserting these values in (10) yields

$$\hat{\gamma} = \frac{(37)(3.36) - (12)(3.65) + \sqrt{[(37)(3.36) - (12)(3.65)]^2 + 4(37)(12)(3.43)^2}}{2(37)(3.43)}$$

$$= 0.97.$$

Insertion of this value in (11) yields the maximum likelihood estimate of the displacement parameter,

$$\hat{\eta} = \frac{1.82449 - (0.97)(1.13814)}{0.207357} = 3.47.$$

Approximate large sample confidence intervals for (γ_L, γ_U) and (η_L, η_U) are given by

$$\gamma_L = \hat{\gamma} - CB\sqrt{\mathbf{e}'\mathbf{S}^{-1}\mathbf{e}}\sqrt{\frac{1}{n_1} + \frac{\hat{\gamma}^2}{n_2}},$$

$$\gamma_U = \hat{\gamma} + CB\sqrt{\mathbf{e}'\mathbf{S}^{-1}\mathbf{e}}\sqrt{\frac{1}{n_1} + \frac{\hat{\gamma}^2}{n_2}} \tag{12}$$

and

$$\eta_L = \hat{\eta} - CB\sqrt{\bar{\mathbf{y}}'\mathbf{S}^{-1}\bar{\mathbf{y}}}\sqrt{\frac{1}{n_1} + \frac{\hat{\gamma}^2}{n_2}},$$

$$\eta_U = \hat{\eta} + CB\sqrt{\bar{\mathbf{y}}'\mathbf{S}^{-1}\bar{\mathbf{y}}}\sqrt{\frac{1}{n_1} + \frac{\hat{\gamma}^2}{n_2}}, \tag{13}$$

respectively, where C is the critical value of the standard normal distribution, and

$$B = 1/[(\mathbf{e}'\mathbf{S}^{-1}\mathbf{e})(\bar{\mathbf{y}}'\mathbf{S}^{-1}\bar{\mathbf{y}}) - (\bar{\mathbf{y}}'\mathbf{S}^{-1}\mathbf{e})^2] \tag{14}$$

is determined from the sample.

Example

We obtain approximate 95-percent confidence intervals for γ and η directly from (12) and (13). However, we first need the value of B from (14):

$$B = [(0.207357)(9.90489) - (1.13814)^2]^{-1} = 1.318.$$

Now

$$\gamma_L = 0.97 - (1.96)(1.318)\sqrt{0.207357}\sqrt{\frac{1}{37} + \frac{(0.97)^2}{12}} = 0.59,$$

$$\gamma_U = 0.97 + (1.96)(1.318)\sqrt{0.207357}\sqrt{\frac{1}{37} + \frac{(0.97)^2}{12}} = 1.35,$$

and

$$\eta_L = 3.47 - (1.96)(1.318)\sqrt{9.90489}\sqrt{\frac{1}{37} + \frac{(0.97)^2}{12}} = 0.83,$$

$$\eta_U = 3.47 + (1.96)(1.318)\sqrt{9.90489}\sqrt{\frac{1}{37} + \frac{(0.97)^2}{12}} = 6.11.$$

Remark: Because $\hat{\gamma}$ and $\hat{\eta}$ are correlated it is advisable to use the Bonferroni inequality when obtaining simultaneous confidence intervals for γ and η.

The results of Sections A and B were obtained by Kraft, Olkin, and van Eeden (1972), who also provide counterpart tests of hypotheses. This model was also studied by Guttman, Menzefricke, and Tyler (1983), who give a Bayesian estimate of the potency factor.

C. A MULTIPLICATIVE MODEL WITH SCALING*

Although the potency factor is multiplicative, the underlying model $\mathbf{x} = \boldsymbol{\mu} + \mathbf{e}_x$, $\mathbf{y} = \boldsymbol{\nu} + \mathbf{e}_y$ is additive. However, a multiplicative model would be more appropriate when the observations themselves, and not only the parameters, are related as in a ratio. That is, suppose p measurements are made on each of two machines, say, $\mathbf{x} = (x_1, \ldots, x_p)'$ and $\mathbf{y} = (y_1, \ldots, y_p)'$. In the multiplicative model the y machine might operate in such a way that each y measurement is related to an x measurement, that is, $y_i = \gamma x_i$, $i = 1, \ldots, p$, where γ is a multiplicative factor.

If we assume that $\mathbf{x}$ is normally distributed with mean $\boldsymbol{\mu}$ and covariance matrix $\boldsymbol{\Sigma}$, then $\mathbf{y}$ is normally distributed with mean $\gamma\boldsymbol{\mu}$ and covariance matrix $\gamma^2\boldsymbol{\Sigma}$, so that not only the means but also the variances undergo a scaling.

For the sake of simplicity of notation, write $\gamma = 1/\Gamma$. Further, if $\hat{\Gamma}$ is the maximum likelihood estimate of Γ, then $\hat{\gamma}$ is the maximum likelihood estimate of γ, so that nothing is lost in this simple transformation.

The maximum likelihood estimate $\hat{\Gamma}$ of Γ is obtained as the solution of an equation of degree $2p$:

$$mc_0 + (m-1)c_1\Gamma + (m-2)c_2\Gamma^2 + \cdots + (m-2p)c_{2p}\Gamma^{2p} = 0, \qquad (15)$$

where $m = 2pn_2/(n_1 + n_2)$, the constants $c_0, c_1, \ldots, c_{2p}$ are obtained from the determinantal expansion

$$|\mathbf{A} - \mathbf{B}\Gamma + \mathbf{C}\Gamma^2| = c_0 + c_1\Gamma + c_2\Gamma^2 + \cdots + c_{2p}\Gamma^{2p}, \qquad (16)$$

and the matrices **A**, **B**, **C** are

$$\begin{aligned} \mathbf{A} &= \mathbf{S}_1 + (n_1n_2/n)\bar{\mathbf{x}}\bar{\mathbf{x}}', \\ \mathbf{B} &= [n_1n_2/(n_1 + n_2)](\bar{\mathbf{x}}\bar{\mathbf{y}}' + \bar{\mathbf{y}}\bar{\mathbf{x}}'), \\ \mathbf{C} &= \mathbf{S}_2 + [n_1n_2(n_1 + n_2)]\bar{\mathbf{y}}\bar{\mathbf{y}}'. \end{aligned} \qquad (17)$$

Here $\mathbf{x} = (\bar{x}_1, \ldots, \bar{x}_p)$ and $\bar{\mathbf{y}} = (\bar{y}_1, \ldots, \bar{y}_p)$ are the mean vectors for the two diagnostic test, and $\mathbf{S}_1$ and $\mathbf{S}_2$ are the sample covariance matrices for the two diagnostic tests.

Although the determination of the maximum likelihood estimator $\hat{\gamma}$ requires considerable computation, its large sample distribution is relatively simple; namely, $\sqrt{n_1 + n_2}(\hat{\gamma} - \gamma)$ is normally distributed with mean zero and asymptotic variance

$$\sigma_\infty^2(\hat{\gamma}) = \gamma^2(n_1 + n_2)^2/[n_1 n_2(2p + \boldsymbol{\mu}'\boldsymbol{\mu})]. \tag{18}$$

Consequently, an approximate confidence interval (γ_L, γ_U) for γ is given by

$$\gamma_L = \frac{\hat{\gamma}}{1 + C(2p + \hat{\boldsymbol{\mu}}'\hat{\boldsymbol{\mu}})n_1 n_2/(n_1 + n_2)},$$

$$\gamma_U = \frac{\hat{\gamma}}{1 - C(2p + \hat{\boldsymbol{\mu}}'\hat{\boldsymbol{\mu}})n_1 n_2/(n_1 + n_2)},$$

where

$$\hat{\boldsymbol{\mu}} = (n_1\bar{\mathbf{x}} + n_2\bar{\mathbf{y}}\hat{\Gamma})/(n_1 + n_2), \tag{19}$$

and C is the critical value of the standard normal distribution.

This model is studied in some detail by Guttman and Olkin (1985), who also provide a Bayesian estimator.

D. AN ADDITIVE MODEL FOR INTERLABORATORY DIFFERENCES*

Different laboratories may have different errors of measurements. Thus, even though each laboratory is measuring the same element, we cannot simply average measurements from more and less precise laboratories. Different models have been used to describe interlaboratory discrepancies.

In one model we assume that there is a measurement error σ_0^2 common to all laboratories and that, in addition, there is a measurement error σ_j^2 unique to the jth laboratory. Frequently the common measurement error σ_0^2 is known.

Thus, initially we assume that there are k laboratories each represented by a normal distribution with a common mean μ and respective variances

$$\begin{aligned} \tau_1 &= \sigma_0^2 + \sigma_1^2, \\ \tau_2 &= \sigma_0^2 + \sigma_2^2, \\ &\vdots \\ \tau_k &= \sigma_0^2 + \sigma_k^2. \end{aligned} \tag{20}$$

Samples of size $n_1, n_2, \ldots, n_k$ are taken from the k laboratories, yielding sample means $x_1, \ldots, x_k$ and sample variances $s_1^2, \ldots, s_k^2$.

The maximum likelihood estimator $\hat{\mu}$ of μ cannot be obtained in closed form, but can be obtained numerically as the solution of

$$\frac{n_1^2(\bar{x}_1 - \mu)}{s_1^2 + n_1(\bar{x}_1 - \mu)^2} + \cdots + \frac{n_k^2(\bar{x}_k - \mu)}{s_k^2 + n_k(\bar{x}_k - \mu)^2} = 0. \tag{21}$$

For large samples an approximate confidence interval (μ_L, μ_U) for μ is given by

$$\mu_L = \hat{\mu} - C\sqrt{\sum \frac{n_i}{t_i}\Big/\sum n_i}, \qquad \mu_U = \hat{\mu} + C\sqrt{\sum \frac{n_i}{t_i}\Big/\sum n_i}, \tag{22}$$

where

$$t_i = [s_i^2 + n_i(\bar{x}_i - \hat{\mu})^2]/n_i, \tag{23}$$

and C is a percentage point of the standard normal distribution.

E. AN ADDITIVE MODEL WITH A CONTROL*

An experiment consists of the administration of three drugs, after which the patient takes p psychological tests. The first drug is a control drug and leads to measurements $\mathbf{w} = (w_1, \ldots, w_p)$. The other two drugs have different strengths and lead to measurements $\mathbf{x} = (x_1, \ldots, x_p)$ and $\mathbf{y} = (y_1, \ldots, y_p)$.

The additive model assumed is

$$\mathbf{w} = \boldsymbol{\eta} + \mathbf{e}_w, \qquad \mathbf{x} = \boldsymbol{\eta} + \boldsymbol{\mu} + \mathbf{e}_x, \qquad \mathbf{y} = \boldsymbol{\eta} + \gamma\boldsymbol{\mu} + \mathbf{e}_y, \tag{24}$$

where the errors $\mathbf{e}_w, \mathbf{e}_x, \mathbf{e}_y$ are independent, each with zero mean and common covariance matrix $\boldsymbol{\Sigma}$. The parameter $\boldsymbol{\eta}$ is the baseline effect resulting from the control (placebo) drug. The parameter $\boldsymbol{\mu}$ is the additive effect from the first drug, and the parameter γ is the potency factor that distinguishes the two drugs.

Samples of size n_0, n_1, n_2 are taken from the placebo experiment and the experiments with the first and second drugs, yielding vectors $\mathbf{w}, \bar{\mathbf{x}}, \bar{\mathbf{y}}$ of the means on the p psychological tests. Because the population covariance matrices for the three experiments are the same, we obtain a pooled sample covariance matrix $\mathbf{S}$.

The maximum likelihood estimate of the parameter γ is obtained as one of the two roots of the quadratic equation

$$c_0\gamma^2 - c_1\gamma + c_2 = 0, \tag{25}$$

where

$$
\begin{aligned}
c_0 &= a_{12}b_{22} - a_{22}b_{12}, \\
c_1 &= a_{11}b_{22} - a_{22}b_{11}, \\
c_2 &= a_{11}b_{12} - b_{11}a_{12},
\end{aligned} \tag{26}
$$

$$
\begin{aligned}
a_{11} &= (\mathbf{w} - \overline{\mathbf{x}})'\mathbf{S}^{-1}(\mathbf{w} - \overline{\mathbf{x}}), \quad a_{12} = (\mathbf{w} - \overline{\mathbf{x}})'\mathbf{S}^{-1}(\mathbf{w} - \overline{\mathbf{y}}), \\
a_{22} &= (\mathbf{w} - \overline{\mathbf{y}})'\mathbf{S}^{-1}(\mathbf{w} - \overline{\mathbf{y}}), \\
b_{11} &= n_0 n_1 + n_1 n_2, \qquad b_{12} = n_1 n_2, \\
b_{22} &= n_0 n_2.
\end{aligned} \tag{27}
$$

To determine which of the two roots of (9) is the maximum likelihood estimator, we compute the expression

$$
\frac{a_{11} - 2\gamma a_{12} + \gamma^2 a_{22}}{b_{11} - 2\gamma b_{12} + \gamma^2 b_{22}} \tag{28}
$$

for each of the two roots and then choose the root that yields the maximum. Once the maximum likelihood estimator $\hat{\gamma}$ is obtained, then

$$
\hat{\boldsymbol{\eta}} = \frac{n_0 \bar{w} + c(\bar{x}\hat{\gamma} - \bar{y})}{n_0 + c(\hat{\gamma} - 1)}, \tag{29}
$$

where

$$
c = n_1 n_2(\hat{\gamma} - 1)/(n_1 + n_2\hat{\gamma}^2), \tag{30}
$$

and

$$
\boldsymbol{\mu} = \frac{n_1(\bar{x} - \hat{\eta}) + \hat{\gamma} n_2(\bar{y} - \hat{\eta})}{n_1 + n_2\hat{\gamma}^2}. \tag{31}
$$

The asymptotic distribution of $\hat{\gamma}$ is

$$
\sqrt{n_0 + n_1 + n_2}(\hat{\gamma} - \gamma) \sim \mathcal{N}(0, \sigma_\infty^2(\hat{\gamma})), \tag{32}
$$

where

$$
\sigma_\infty^2(\hat{\gamma}) = n_2 \boldsymbol{\mu}\boldsymbol{\mu}' \frac{n_1(n_0 + n_2) + (n_0 + n_1)\gamma^2}{n_1(n_0 + n_2) + (n_0 + n_1)\gamma^2 + n_2^2(1 - \gamma)^2}. \tag{33}
$$

Consequently a confidence interval (γ_L, γ_U) for γ is given by

$$
\gamma_L = \hat{\gamma} - C\hat{\sigma}_\infty(\hat{\gamma})/\sqrt{n_0 + n_1 + n_2}, \quad \gamma_U = \hat{\gamma} + C\hat{\sigma}_\infty(\hat{\gamma})/\sqrt{n_0 + n_1 + n_2}, \tag{34}
$$

where C is the critical value obtained from the standard normal distribution, and $\hat{\sigma}_{\infty}(\hat{\gamma})$ is obtained from (33) with $\hat{\gamma}$ and $\hat{\mu}$ replacing γ and μ, respectively.

Many other physical examples can be given, but the above may suffice to illustrate the differences between combining measurements in the physical versus the social sciences. In essence, the models are crisper and better defined in the physical sciences, but the estimation procedures can often be quite complicated. Also, because the distribution theory is difficult, large sample theory or simulations are used to approximate the distributional results.

APPENDIX

APPENDIX A

Table of the Standard Normal Cumulative Distribution Function $\Phi(z)$

	z									
	0.00	0.01	0.02	0.03	0.04	0.05	0.06	0.07	0.08	0.09
0.0	0.50000	0.50399	0.50798	0.51197	0.51595	0.51994	0.52392	0.52790	0.53188	0.53586
0.1	0.53983	0.54380	0.54776	0.55172	0.55567	0.55962	0.56362	0.56749	0.57142	0.57535
0.2	0.57926	0.58317	0.58706	0.59095	0.59483	0.59871	0.60257	0.60642	0.61026	0.61409
0.3	0.61791	0.62172	0.62552	0.62930	0.63307	0.63683	0.64058	0.64431	0.64803	0.65173
0.4	0.65542	0.65910	0.66276	0.66640	0.67003	0.67364	0.67724	0.68082	0.68439	0.68793
0.5	0.69146	0.69497	0.69847	0.70194	0.70540	0.70884	0.71226	0.71566	0.71904	0.72240
0.6	0.72575	0.72907	0.73237	0.73565	0.73891	0.74215	0.74537	0.74857	0.75175	0.75490
0.7	0.75804	0.76115	0.76424	0.76730	0.77035	0.77337	0.77637	0.77935	0.78230	0.78524
0.8	0.78814	0.79103	0.79389	0.79673	0.79955	0.80234	0.80511	0.80785	0.81057	0.81327
0.9	0.81594	0.81859	0.82121	0.82381	0.82639	0.82894	0.83147	0.83398	0.83646	0.83891
1.0	0.84134	0.84375	0.84614	0.84849	0.85083	0.85314	0.85543	0.85769	0.85993	0.86214
1.1	0.86433	0.86650	0.86864	0.87076	0.87286	0.87493	0.87698	0.87900	0.88100	0.88298
1.2	0.88493	0.88686	0.88877	0.89065	0.89251	0.89435	0.89617	0.89796	0.89973	0.90147
1.3	0.90320	0.90490	0.90658	0.90824	0.90988	0.91149	0.91309	0.91466	0.91621	0.91774
1.4	0.91924	0.92073	0.92220	0.92364	0.92507	0.92647	0.92785	0.92922	0.93056	0.93189
1.5	0.93319	0.93448	0.93574	0.93699	0.93822	0.93943	0.94062	0.94179	0.94295	0.94408
1.6	0.94520	0.94630	0.94738	0.94845	0.94950	0.95053	0.95154	0.95254	0.95352	0.95449
1.7	0.95543	0.95637	0.95728	0.95818	0.95907	0.95994	0.96080	0.96164	0.96246	0.96327
1.8	0.96407	0.94685	0.96562	0.96638	0.96712	0.96784	0.96856	0.96926	0.96995	0.97062
1.9	0.97128	0.97193	0.97257	0.97320	0.97381	0.97441	0.97500	0.97558	0.97615	0.97670

2.0	0.97725	0.97778	0.97831	0.97882	0.97932	0.97982	0.98030	0.98077	0.98124	0.98169
2.1	0.98214	0.98257	0.98300	0.98341	0.98382	0.98422	0.98422	0.98461	0.98500	0.98574
2.2	0.98610	0.98645	0.98679	0.98713	0.98745	0.98778	0.98809	0.98840	0.98870	0.98899
2.3	0.98928	0.98956	0.98983	0.99010	0.99036	0.99061	0.99086	0.99111	0.99134	0.99158
2.4	0.99180	0.99202	0.99224	0.99245	0.99266	0.99286	0.99305	0.99324	0.99343	0.99361
2.5	0.99379	0.99396	0.99413	0.99430	0.99446	0.99461	0.99477	0.99492	0.99506	0.99520
2.6	0.99534	0.99547	0.99560	0.99573	0.99585	0.99598	0.99609	0.99621	0.99632	0.99643
2.7	0.99653	0.99664	0.99674	0.99683	0.99693	0.99702	0.99711	0.99720	0.99728	0.99736
2.8	0.99744	0.99752	0.99760	0.99767	0.99774	0.99781	0.99788	0.99795	0.99801	0.99807
2.9	0.99813	0.99819	0.99825	0.99831	0.99836	0.99841	0.99846	0.99851	0.99856	0.99861
3.0	0.99865	0.99869	0.99874	0.99878	0.99882	0.99886	0.99889	0.99893	0.99896	0.99900
3.1	0.99903	0.99906	0.99910	0.99913	0.99916	0.99918	0.99921	0.99924	0.99926	0.99929
3.2	0.99931	0.99934	0.99936	0.99938	0.99940	0.99942	0.99944	0.99946	0.99948	0.99950
3.3	0.99952	0.99953	0.99955	0.99957	0.99958	0.99960	0.99961	0.99962	0.99964	0.99965
3.4	0.99966	0.99968	0.99969	0.99970	0.99971	0.99972	0.99973	0.99974	0.99975	0.99976
3.5	0.99977	0.99978	0.99978	0.99979	0.99980	0.99981	0.99981	0.99982	0.99983	0.99983
3.6	0.99984	0.99985	0.99985	0.99986	0.99986	0.99987	0.99987	0.99988	0.99988	0.99989
3.7	0.99989	0.99990	0.99990	0.99990	0.99991	0.99991	0.99992	0.99992	0.99992	0.99992
3.8	0.99993	0.99993	0.99993	0.99994	0.99994	0.99994	0.99994	0.99995	0.99995	0.99995
3.9	0.99995	0.99995	0.99996	0.99996	0.99996	0.99996	0.99996	0.99996	0.99997	0.99997
4.0	0.99997	0.99997	0.99997	0.99997	0.99997	0.99997	0.99998	0.99998	0.99998	0.99998

APPENDIX B

Percentiles of Chi-square Distributions: Left and Right Tail Areas

Degrees of freedom	Left tail area					
	0.005	0.01	0.025	0.05	0.10	0.25
1	$0.0^{4}3927$	$0.0^{3}1571$	$0.0^{3}9821$	$0.0^{2}3932$	0.01579	0.1015
2	0.01003	0.02010	0.05064	0.1026	0.2107	0.5754
3	0.07172	0.1148	0.2158	0.3518	0.5844	1.213
4	0.2070	0.2971	0.4844	0.7107	1.064	1.923
5	0.4117	0.5543	0.8312	1.145	1.610	2.675
6	0.6757	0.8721	1.237	1.635	2.204	3.455
7	0.9893	1.239	1.690	2.167	2.833	4.255
8	1.344	1.646	2.180	2.733	3.490	5.071
9	1.735	2.088	2.700	3.325	4.168	5.899
10	2.156	2.558	3.247	3.940	4.865	6.737
11	2.603	3.053	3.816	4.575	5.578	7.584
12	3.074	3.571	4.404	5.226	6.304	8.438
13	3.565	4.107	5.009	5.892	7.042	9.299
14	4.075	4.660	5.629	6.571	7.790	10.17
15	4.601	5.229	6.262	7.261	8.547	11.04
16	5.142	5.812	6.908	7.962	9.312	11.91
17	5.697	6.408	7.564	8.672	10.09	12.79
18	6.265	7.015	8.231	9.390	10.86	13.68
19	6.844	7.633	8.907	10.12	11.65	14.56
20	7.434	8.260	9.591	10.85	12.44	15.45
21	8.034	8.897	10.28	11.59	13.24	16.34
22	8.643	9.542	10.98	12.34	14.04	17.24
23	9.260	10.20	11.69	13.09	14.85	18.14
24	9.886	10.86	12.40	13.85	15.66	19.04
25	10.52	11.52	13.12	14.61	16.47	19.94
26	11.16	12.20	13.84	15.38	17.29	20.84
27	11.81	12.88	14.57	16.15	18.11	21.75
28	12.46	13.56	15.31	16.93	18.94	22.66
29	13.12	14.26	16.05	17.71	19.77	23.57
30	13.79	14.95	16.79	18.49	20.60	24.48
31	14.46	15.66	17.54	19.28	21.43	25.39
32	15.13	16.36	18.29	20.07	22.27	26.30
33	15.82	17.07	19.05	20.87	23.11	27.22
34	16.50	17.79	19.81	21.66	23.95	28.14
35	17.19	18.51	20.57	22.47	24.80	29.05
36	17.89	19.23	21.34	23.27	25.64	29.97
37	18.59	19.96	22.11	24.07	26.49	30.89
38	19.29	20.69	22.88	24.88	27.34	31.81
39	20.00	21.43	23.65	25.70	28.20	32.74
40	20.71	22.16	24.43	26.51	29.05	33.66

APPENDIX B ***(Continued)***

Degrees of freedom	Left tail area					
	0.005	0.01	0.025	0.05	0.10	0.025
50	27.99	29.71	32.36	34.76	37.69	42.94
60	35.53	37.48	40.48	43.19	46.46	52.29
70	43.28	45.44	48.76	51.74	55.33	61.70
80	51.17	53.54	57.15	60.39	64.28	71.14
90	59.20	61.75	65.65	69.13	73.29	80.62
100	67.33	70.06	74.22	77.93	82.36	90.13
110	75.55	78.46	82.87	86.79	91.47	99.67
120	83.85	86.92	91.57	95.70	100.6	109.2
130	92.22	95.45	100.3	104.7	109.8	118.8
140	100.7	104.0	109.1	113.7	119.0	128.4
150	109.1	112.7	118.0	122.7	128.3	138.0
160	117.7	121.3	126.9	131.8	137.5	147.6
170	126.3	130.1	135.8	140.8	146.8	157.2
180	134.9	138.8	144.7	150.0	156.2	166.9
190	143.5	147.6	153.7	159.1	165.5	176.5
200	152.2	156.4	162.7	168.3	174.8	186.2

Degrees of freedom	Right tail area						
	0.50	0.75	0.90	0.95	0.975	0.99	0.995
1	0.4549	1.323	2.706	3.841	5.025	6.635	7.879
2	1.386	2.773	4.605	5.991	7.378	9.210	10.60
3	2.366	4.108	6.251	7.815	9.348	11.34	12.84
4	3.357	5.385	7.779	9.488	11.14	13.28	14.86
5	4.351	6.626	9.236	11.07	12.83	15.09	16.75
6	5.348	7.841	10.64	12.59	14.45	16.81	18.55
7	6.346	9.037	12.02	14.07	16.01	18.48	20.28
8	7.344	10.22	13.36	15.51	17.53	20.09	21.95
9	8.343	11.39	14.68	16.92	19.02	21.67	23.59
10	9.342	12.55	15.99	18.31	20.48	23.21	25.19
11	10.34	13.70	17.28	19.68	21.92	24.72	26.76
12	11.34	14.85	18.55	21.03	23.34	26.22	28.30
13	12.34	15.98	19.81	22.36	24.74	27.69	29.82
14	13.34	17.12	21.06	23.68	26.12	29.14	31.32
15	14.34	18.25	22.31	25.00	27.49	30.58	32.80
16	15.34	19.37	23.54	26.30	28.85	32.00	34.27
17	16.34	20.49	24.77	27.59	30.19	33.41	35.72
18	17.34	21.60	25.99	28.87	31.53	34.81	37.16
19	18.34	22.72	27.20	30.14	32.85	36.19	38.58
20	19.34	23.83	28.41	31.41	34.17	37.57	40.00

(Continued)

APPENDIX B (*Continued*)

Degrees of freedom	Right tail area						
	0.50	0.75	0.90	0.95	0.975	0.99	0.99
21	20.34	24.93	29.62	32.67	35.48	38.93	41.40
22	21.34	26.04	30.81	33.92	36.78	40.29	42.80
23	22.34	27.14	32.01	35.17	38.08	41.64	44.18
24	23.34	28.24	33.20	36.42	39.36	42.98	45.56
25	24.34	29.34	34.38	37.65	40.65	44.31	46.93
26	25.34	30.43	35.56	38.89	41.92	45.64	48.29
27	26.34	31.53	36.74	40.11	43.19	46.96	49.64
28	27.34	32.62	37.92	41.34	44.46	48.28	50.99
29	28.34	33.71	39.09	42.56	45.72	49.59	52.34
30	29.34	34.80	40.26	43.77	46.98	50.89	53.67
31	30.34	35.89	41.42	44.99	48.23	52.19	55.00
32	31.34	36.97	42.58	46.19	49.48	53.49	56.33
33	32.34	38.06	43.75	47.40	50.73	54.78	57.65
34	33.34	39.14	44.90	48.60	51.97	56.06	58.96
35	34.34	40.22	46.06	49.80	53.20	57.34	60.27
36	35.34	41.30	47.21	51.00	54.44	58.62	61.58
			6	52.19	55.67	59.89	62.88
38	37.34	43.46	49.51	53.38	56.90	61.16	64.18
39	38.34	44.54	50.66	54.57	58.12	62.43	65.48
40	39.34	45.62	51.81	55.76	59.34	63.69	66.77
50	49.33	56.33	63.17	67.50	71.42	76.15	79.49
60	59.33	66.98	74.40	79.08	83.30	88.38	91.95
70	69.33	77.58	85.53	90.53	95.02	100.4	104.2
80	79.33	88.13	96.58	101.9	106.6	112.3	116.3
90	89.33	98.65	107.6	113.1	118.1	124.1	128.3
100	99.33	109.1	118.5	124.3	129.6	135.8	140.2
110	109.3	119.6	129.4	135.5	140.9	147.4	151.9
120	119.3	130.1	140.2	146.6	152.2	159.0	163.6
130	129.3	140.5	151.0	157.6	163.5	170.4	175.3
140	139.3	150.9	161.8	168.6	174.6	181.8	186.8
150	149.3	161.3	172.6	179.6	185.8	193.2	198.4
160	159.3	171.7	183.3	190.5	196.9	204.5	209.8
170	169.3	182.0	194.0	201.4	208.0	215.8	221.2
180	179.3	192.4	204.7	212.3	219.0	227.1	232.6
190	189.3	202.8	215.4	223.2	230.1	238.3	244.0
200	199.3	213.1	226.0	234.0	241.1	249.4	255.3

APPENDIX C

(a) Values of $z = \frac{1}{2}\log[(1 + r)/(1 - r)]$

	r									
	0.0	0.1	0.2	0.3	0.4	0.5	0.6	0.7	0.8	0.9
0.00	0.000	0.100	0.203	0.310	0.424	0.549	0.693	0.867	1.099	1.472
0.01	0.010	0.110	0.213	0.321	0.436	0.563	0.709	0.887	1.127	1.528
0.02	0.020	0.121	0.224	0.332	0.448	0.576	0.725	0.908	1.157	1.589
0.03	0.030	0.131	0.234	0.343	0.460	0.590	0.741	0.929	1.188	1.658
0.04	0.040	0.141	0.245	0.354	0.472	0.604	0.758	0.950	1.221	1.738
0.05	0.050	0.151	0.255	0.365	0.485	0.618	0.775	0.973	1.256	1.832
0.06	0.060	0.161	0.266	0.377	0.497	0.633	0.793	0.996	1.293	1.946
0.07	0.070	0.172	0.277	0.388	0.510	0.648	0.811	1.020	1.333	2.092
0.08	0.080	0.182	0.288	0.400	0.523	0.662	0.829	1.045	1.376	2.298
0.09	0.090	0.192	0.299	0.412	0.536	0.678	0.848	1.071	1.422	2.647

(b) Values of $r = (e^{2z} - 1)/(e^{2z} + 1)$

	z									
	0.0	0.1	0.2	0.3	0.4	0.5	0.6	0.7	0.8	0.9
0.00	0.000	0.100	0.197	0.291	0.380	0.462	0.537	0.604	0.664	0.716
0.01	0.010	0.110	0.207	0.300	0.388	0.470	0.544	0.611	0.670	0.721
0.02	0.020	0.119	0.217	0.310	0.397	0.478	0.551	0.617	0.675	0.726
0.03	0.030	0.129	0.226	0.319	0.405	0.485	0.558	0.623	0.680	0.731
0.04	0.040	0.139	0.235	0.327	0.414	0.493	0.565	0.629	0.686	0.735
0.05	0.050	0.149	0.245	0.336	0.422	0.501	0.572	0.635	0.691	0.740
0.06	0.060	0.159	0.254	0.345	0.430	0.508	0.578	0.641	0.696	0.744
0.07	0.070	0.168	0.264	0.354	0.438	0.515	0.585	0.647	0.701	0.749
0.08	0.080	0.178	0.273	0.363	0.446	0.523	0.592	0.653	0.706	0.753
0.09	0.090	0.188	0.282	0.371	0.454	0.530	0.598	0.658	0.711	0.757

	z									
	1.0	1.1	1.2	1.3	1.4	1.5	1.6	1.7	1.8	1.9
0.00	0.762	0.800	0.834	0.862	0.885	0.905	0.922	0.935	0.947	0.956
0.01	0.766	0.804	0.837	0.864	0.887	0.907	0.923	0.937	0.948	0.957
0.02	0.770	0.808	0.840	0.867	0.890	0.909	0.925	0.938	0.949	0.958
0.03	0.774	0.811	0.843	0.869	0.892	0.910	0.926	0.939	0.950	0.959
0.04	0.778	0.814	0.845	0.872	0.894	0.912	0.927	0.940	0.951	0.960
0.05	0.782	0.818	0.848	0.874	0.896	0.914	0.929	0.941	0.952	0.960
0.06	0.786	0.821	0.851	0.876	0.898	0.915	0.930	0.943	0.953	0.961
0.07	0.789	0.824	0.854	0.879	0.900	0.917	0.932	0.944	0.954	0.962
0.08	0.793	0.827	0.856	0.881	0.901	0.919	0.9 3	0.945	0.954	0.963
0.09	0.797	0.831	0.859	0.883	0.903	0.920	0.934	0.946	0.955	0.963

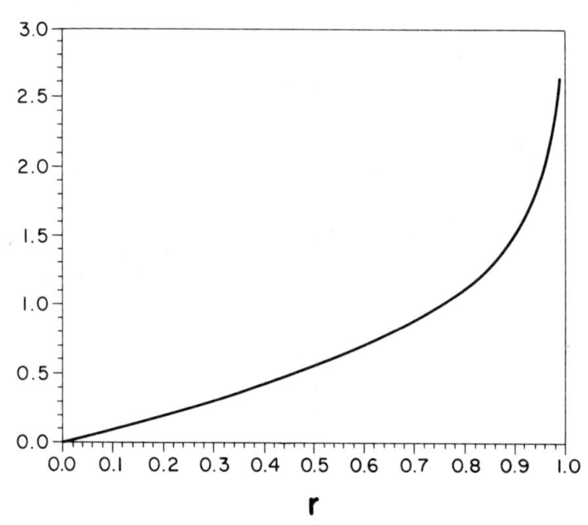

A graph of $z = \frac{1}{2} \log[(1+r)/(1-r)]$.

APPENDIX D

(a) Values of $\sqrt{2}\sinh^{-1} x$

	x									
	0.0	0.1	0.2	0.3	0.4	0.5	0.6	0.7	0.8	0.9
0.00	0.000	0.141	0.281	0.418	0.552	0.681	0.804	0.923	1.036	1.144
0.01	0.014	0.155	0.295	0.432	0.565	0.693	0.817	0.935	1.047	1.154
0.02	0.028	0.169	0.309	0.455	0.578	0.706	0.829	0.946	1.058	1.165
0.03	0.042	0.183	0.322	0.459	0.591	0.718	0.841	0.958	1.069	1.175
0.04	0.057	0.197	0.336	0.472	0.604	0.731	0.853	0.969	1.080	1.186
0.05	0.071	0.211	0.350	0.485	0.617	0.743	0.864	0.980	1.091	1.196
0.06	0.085	0.225	0.364	0.499	0.630	0.756	0.876	0.992	1.101	1.206
0.07	0.099	0.239	0.377	0.512	0.642	0.768	0.888	1.003	1.112	1.216
0.08	0.113	0.253	0.391	0.525	0.655	0.780	0.900	1.014	1.123	1.226
0.09	0.127	0.267	0.405	0.538	0.668	0.792	0.911	1.025	1.133	1.236

	x									
	1.0	1.1	1.2	1.3	1.4	1.5	1.6	1.7	1.8	1.9
0.00	1.246	1.344	1.437	1.525	1.609	1.690	1.766	1.840	1.910	1.977
0.01	1.256	1.353	1.446	1.534	1.618	1.697	1.774	1.847	1.917	1.984
0.02	1.266	1.363	1.455	1.542	1.626	1.705	1.781	1.854	1.923	1.990
0.03	1.276	1.372	1.464	1.551	1.634	1.713	1.789	1.861	1.930	1.997
0.04	1.286	1.382	1.473	1.559	1.642	1.721	1.796	1.868	1.937	2.003
0.05	1.296	1.391	1.482	1.568	1.650	1.728	1.803	1.875	1.944	2.010
0.06	1.306	1.400	1.490	1.576	1.658	1.736	1.811	1.882	1.951	2.016
0.07	1.315	1.409	1.499	1.585	1.666	1.744	1.818	1.889	1.957	2.023
0.08	1.325	1.419	1.508	1.593	1.674	1.751	1.825	1.896	1.964	2.029
0.09	1.334	1.428	1.517	1.601	1.682	1.759	1.832	1.903	1.970	2.035

(b) Values of $\sinh(x/\sqrt{2})$

	x									
	0.00	0.01	0.02	0.03	0.04	0.05	0.06	0.07	0.08	0.09
2.40	0.000	0.017	0.034	0.051	0.068	0.085	0.102	0.119	0.136	0.153
2.45	0.000	0.017	0.035	0.052	0.069	0.087	0.104	0.121	0.139	0.156
2.50	0.000	0.018	0.035	0.053	0.071	0.088	0.106	0.124	0.142	0.159
2.55	0.000	0.018	0.036	0.054	0.072	0.090	0.108	0.126	0.144	0.162
2.60	0.000	0.018	0.037	0.055	0.074	0.092	0.110	0.129	0.147	0.166
2.65	0.000	0.019	0.037	0.056	0.075	0.094	0.112	0.131	0.150	0.169
2.70	0.000	0.019	0.038	0.057	0.076	0.095	0.115	0.134	0.153	0.172
2.75	0.000	0.019	0.039	0.058	0.078	0.097	0.117	0.136	0.156	0.175
2.80	0.000	0.020	0.040	0.059	0.079	0.099	0.119	0.139	0.158	0.178
2.85	0.000	0.020	0.040	0.060	0.081	0.101	0.121	0.141	0.161	0.182

APPENDIX D (*Continued*)

	x									
	0.00	0.10	0.20	0.30	0.40	0.50	0.60	0.70	0.80	0.90
2.40	0.000	0.170	0.341	0.513	0.688	0.866	1.049	1.237	1.431	1.633
2.45	0.000	0.173	0.348	0.524	0.702	0.884	1.071	1.263	1.461	1.667
2.50	0.000	0.177	0.355	0.534	0.717	0.902	1.093	1.289	1.491	1.701
2.55	0.000	0.180	0.362	0.545	0.731	0.920	1.115	1.314	1.521	1.735
2.60	0.000	0.184	0.369	0.556	0.745	0.939	1.136	1.340	1.550	1.769
2.65	0.000	0.188	0.376	0.566	0.760	0.957	1.158	1.366	1.580	1.803
2.70	0.000	0.191	0.383	0.577	0.774	0.975	1.180	1.392	1.610	1.837
2.75	0.000	0.195	0.390	0.588	0.788	0.993	1.202	1.417	1.640	1.871
2.80	0.000	0.198	0.397	0.598	0.803	1.011	1.224	1.443	1.670	1.905
2.85	0.000	0.202	0.404	0.609	0.817	1.029	1.246	1.469	1.700	1.939

	x									
	0.00	1.00	2.00	3.00	4.00	5.00	6.00	7.00	8.00	9.00
2.40	0.000	1.842	4.644	9.867	20.232	41.141	83.492	169.359	343.492	696.648
2.45	0.000	1.880	4.741	10.072	20.653	41.998	85.232	172.887	350.648	711.161
2.50	0.000	1.919	4.838	10.278	21.075	42.855	86.971	176.415	357.804	725.675
2.55	0.000	1.957	4.934	10.483	21.496	43.712	88.711	179.944	364.960	740.188
2.60	0.000	1.996	5.031	10.689	21.918	44.569	90.450	183.472	372.116	754.701
2.65	0.000	2.034	5.128	10.895	22.339	45.427	92.190	187.000	379.272	769.215
2.70	0.000	2.072	5.225	11.100	22.761	46.284	93.929	190.529	386.428	783.728
2.75	0.000	2.111	5.321	11.306	23.182	47.141	95.668	194.057	393.584	798.242
2.80	0.000	2.149	5.418	11.511	23.604	47.998	97.408	197.585	400.741	812.755
2.85	0.000	2.187	5.515	11.717	24.025	48.855	99.147	201.113	407.897	827.269

APPENDIX E

Confidence Intervals of the Parameter p of the Binomial Distribution, for confidence coefficients of (a) 0.99, (b) 0.95, (c) 0.90, and (d) 0.80. The numbers on the curves indicate sample size.

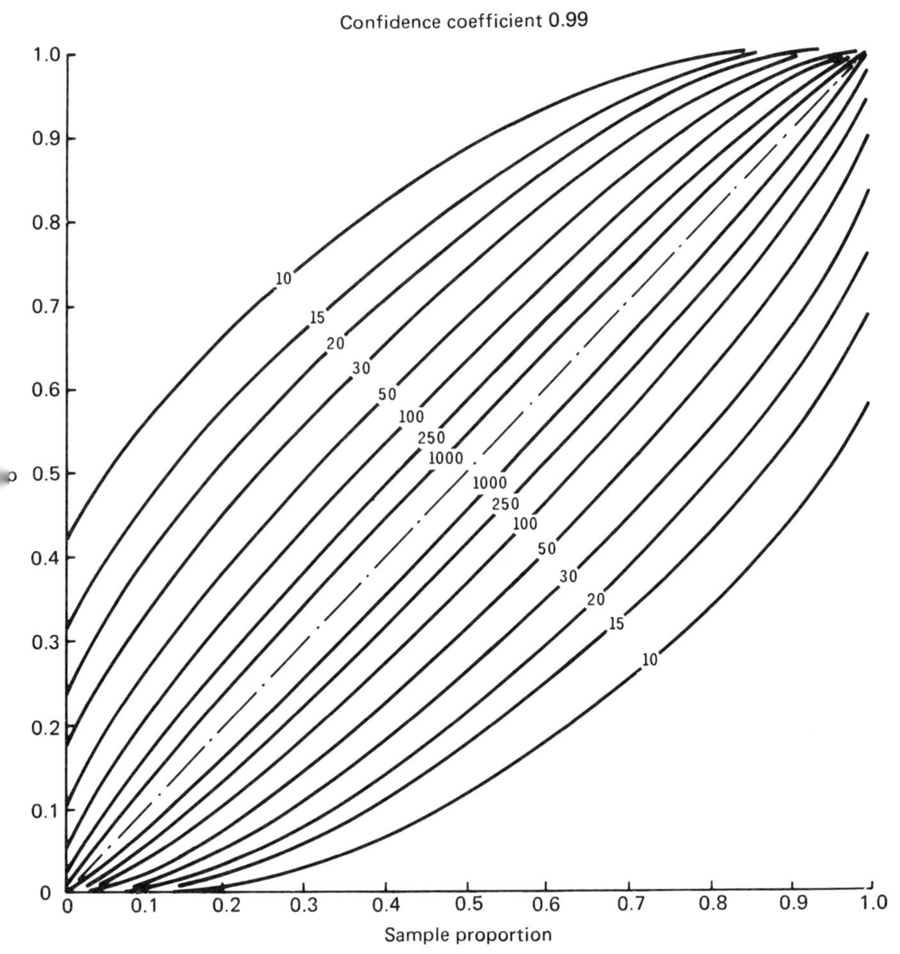

The numbers on the curves indicate sample size

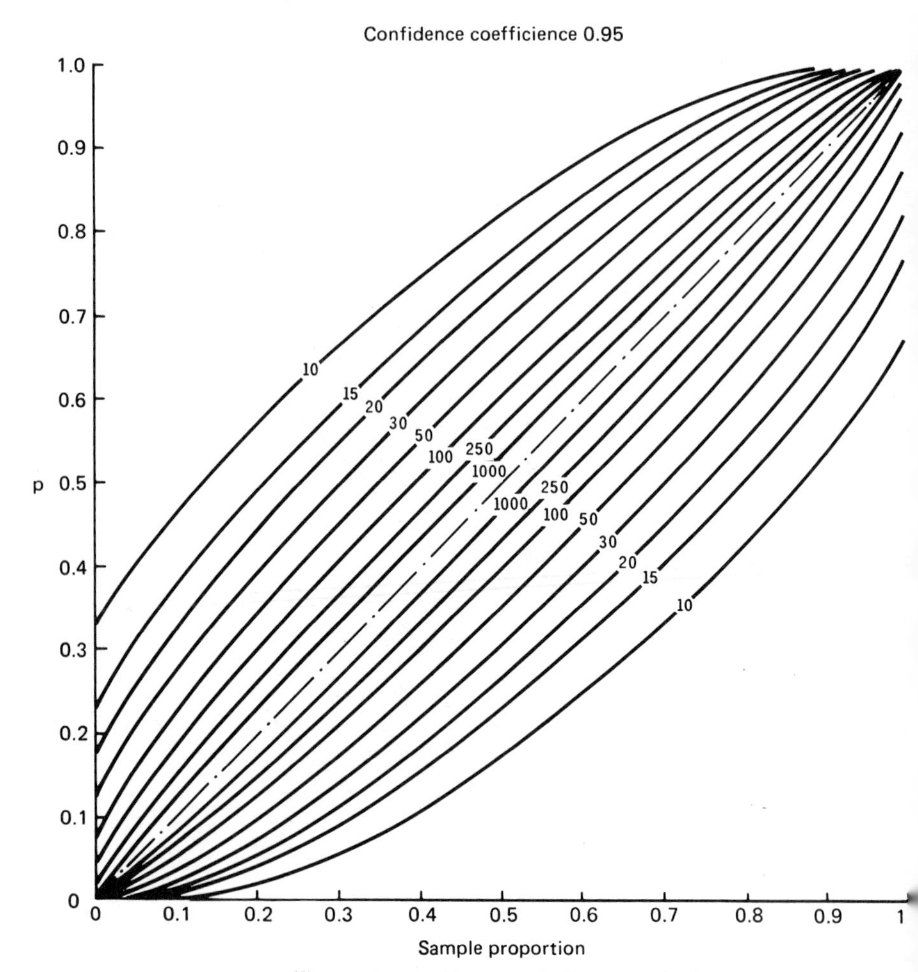

The numbers on the curves indicate sample size

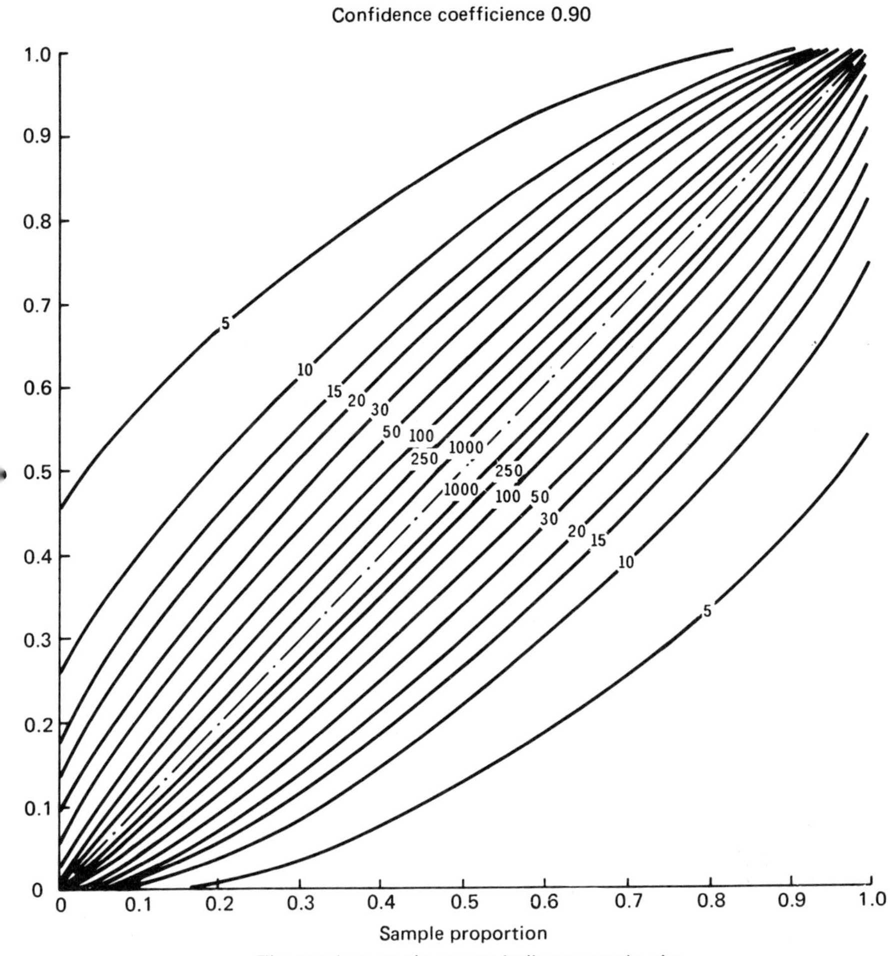

The numbers on the curves indicate sample size

Confidence coefficience 0.80

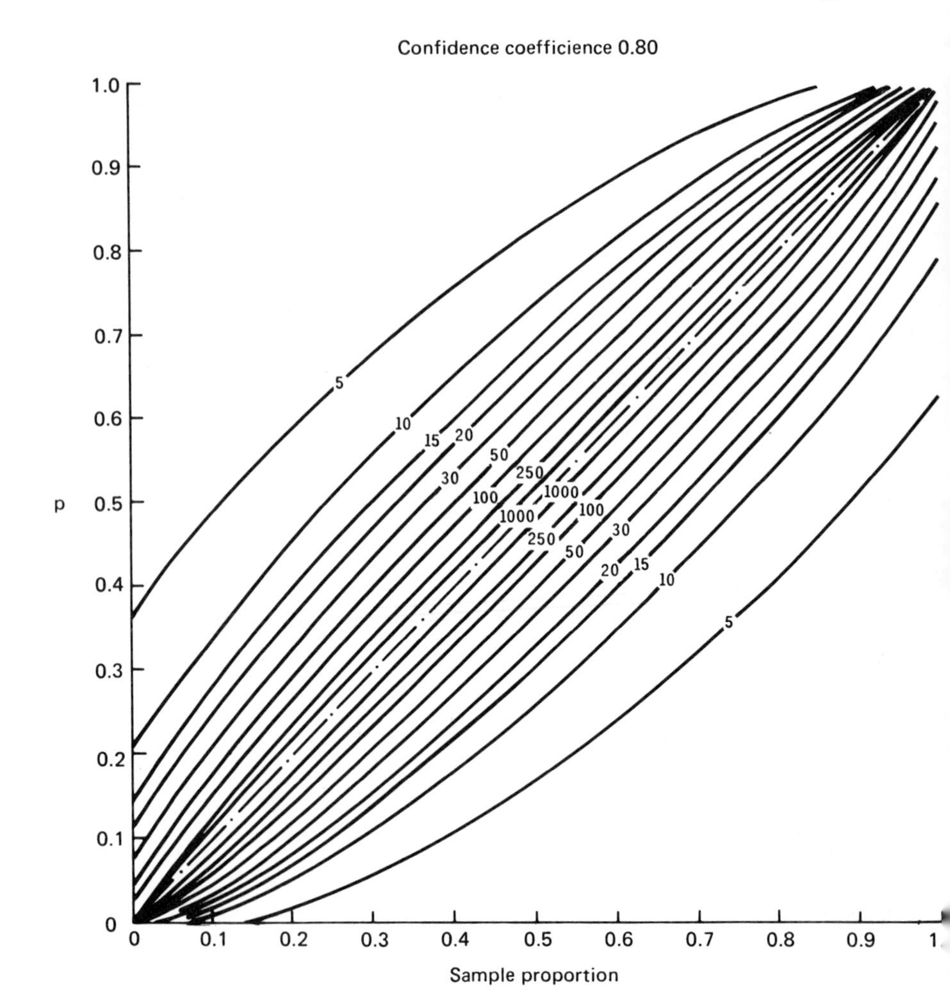

The numbers on the curves indicate sample size

APPENDIX F

Nomographs for Exact Confidence Intervals for δ when $2 \leq n \leq 10$, for (a) 90-, (b) 95-, and (c) 99-percent confidence intervals.

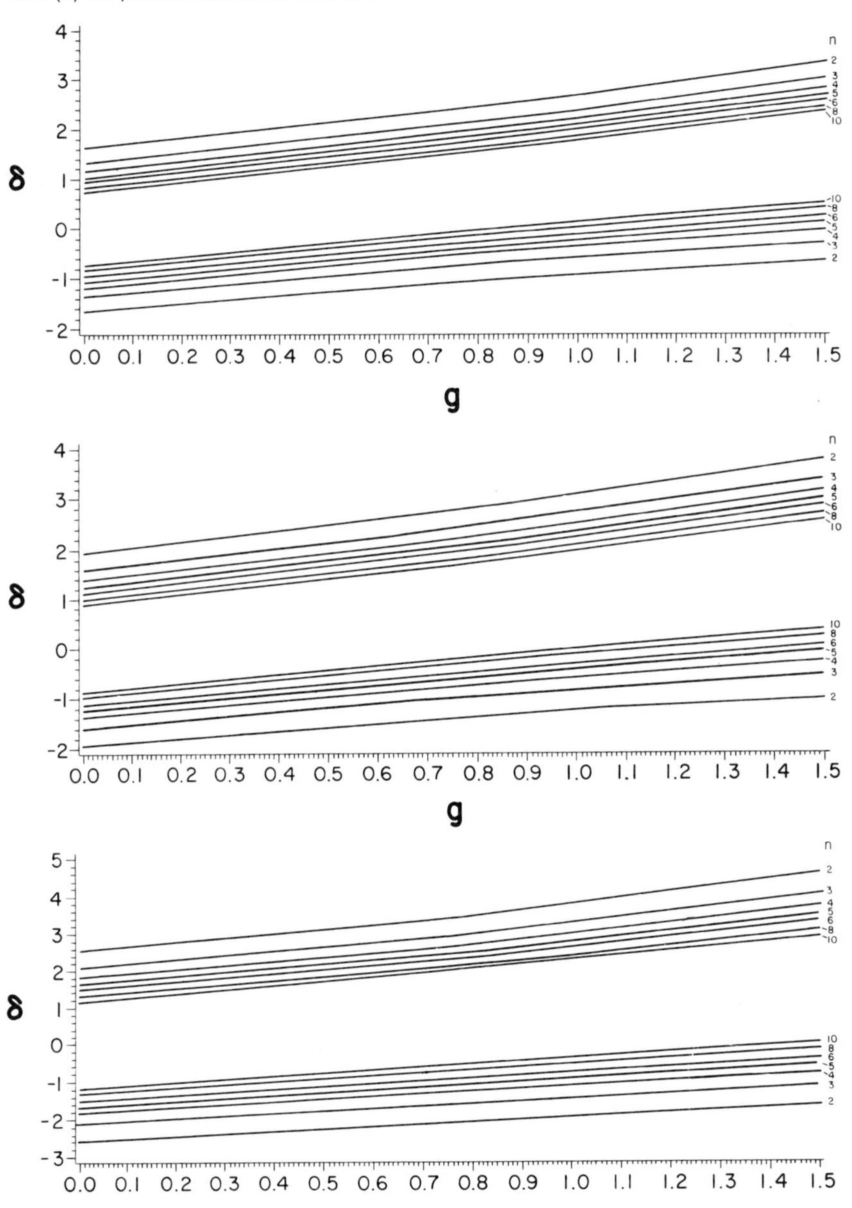

APPENDIX G

Confidence Intervals for the Correlation Coefficient for Different Sample Sizes, for confidence coefficients of (a) 0.90, (b) 0.95, (c) 0.98, and (d) 0.99. The numbers on the curves indicate sample size.

CONFIDENCE BELTS

CHART I. CHANCE OF REJECTING THE HYPOTHESIS WHEN TRUE = ·05 + ·05 = ·10

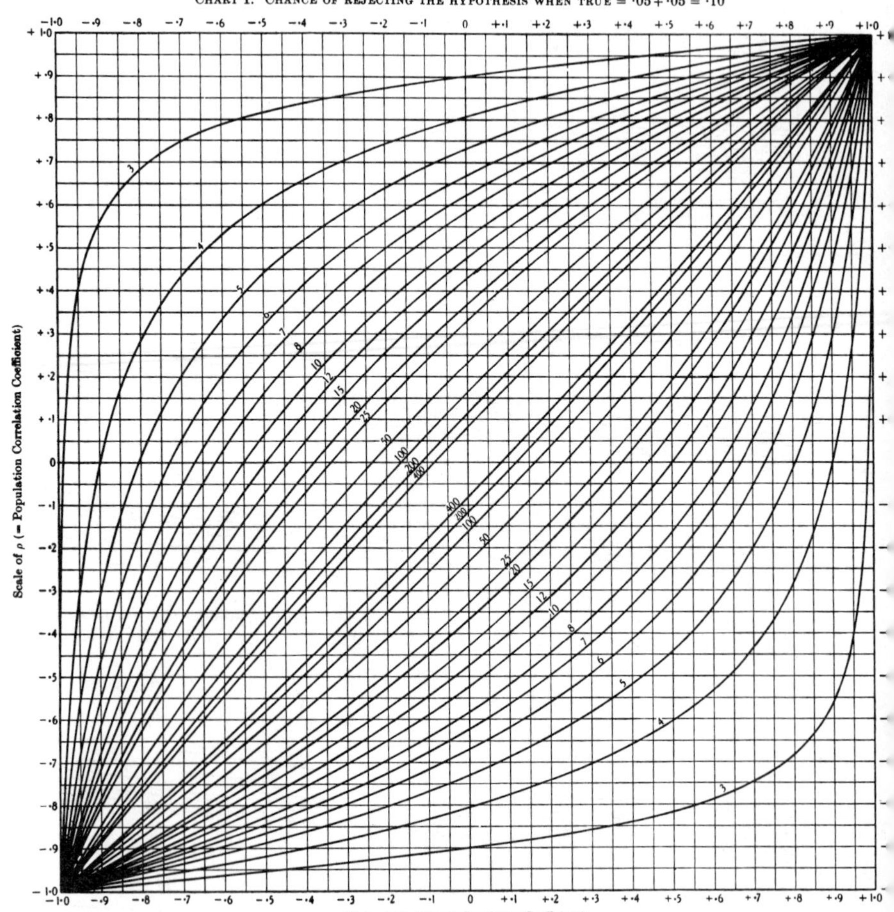

The numbers on the curves indicate sample size

CONFIDENCE BELTS

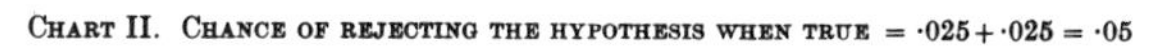

CHART II. CHANCE OF REJECTING THE HYPOTHESIS WHEN TRUE = ·025 + ·025 = ·05

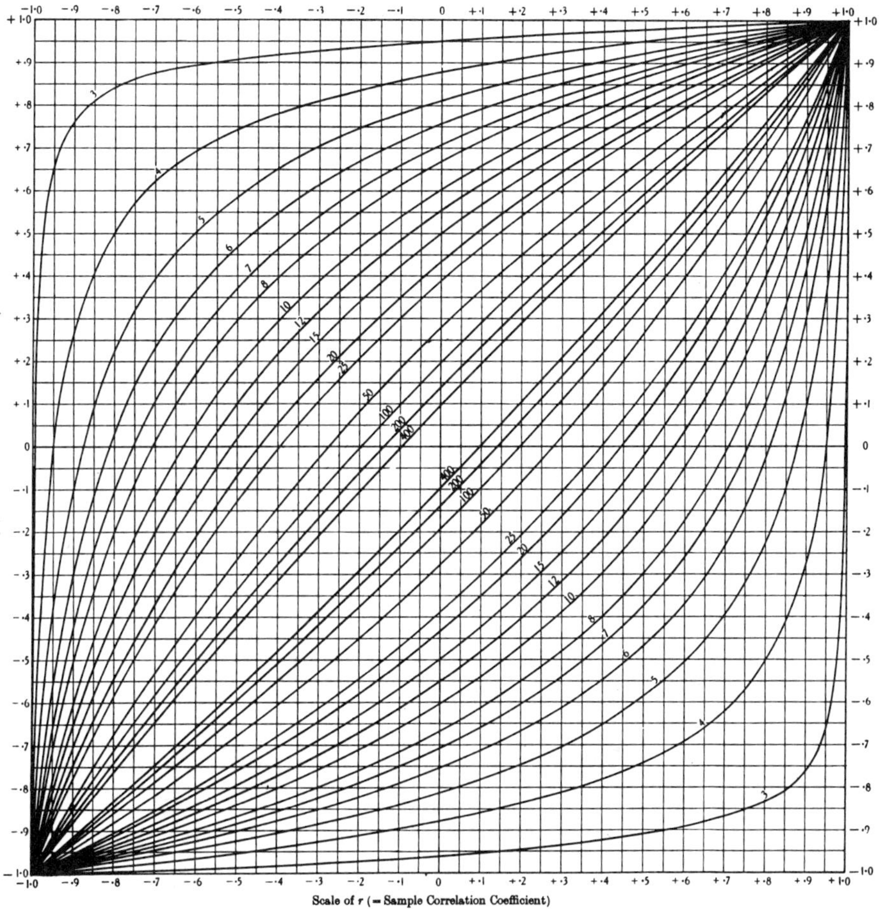

Scale of r (= Sample Correlation Coefficient)

The numbers on the curves indicate sample size

CONFIDENCE BELTS

CHART III. CHANCE OF REJECTING THE HYPOTHESIS WHEN TRUE = ·01 + ·01 = ·02

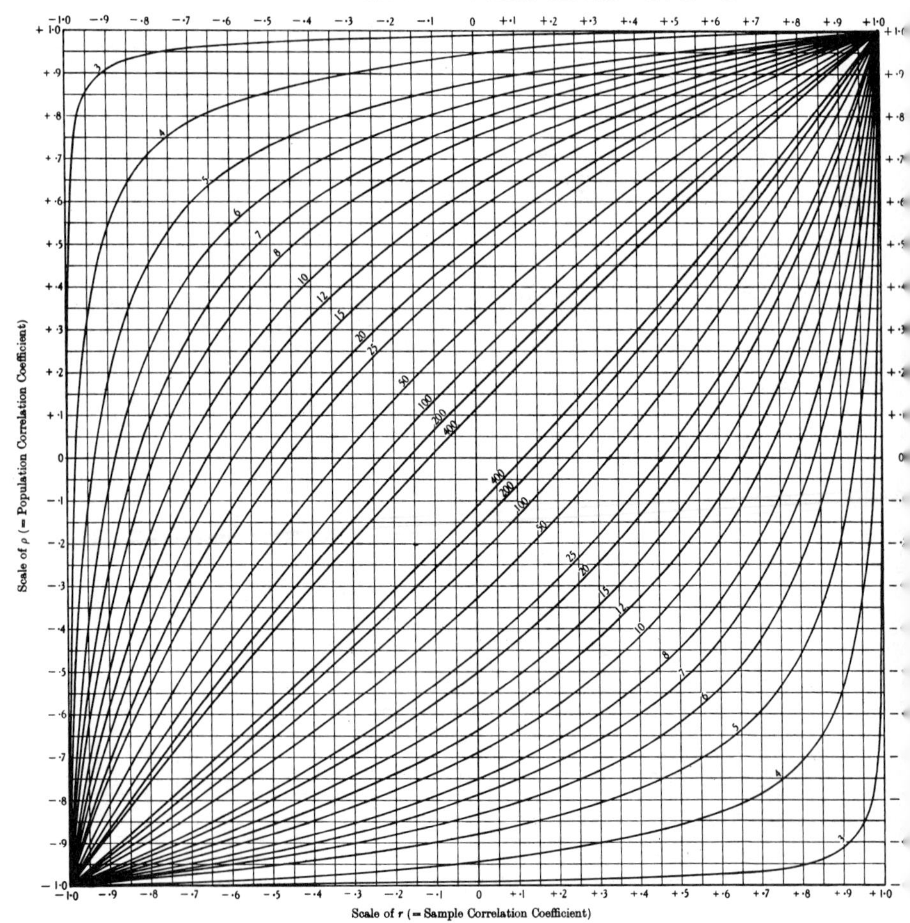

The numbers on the curves indicate sample size

CONFIDENCE BELTS

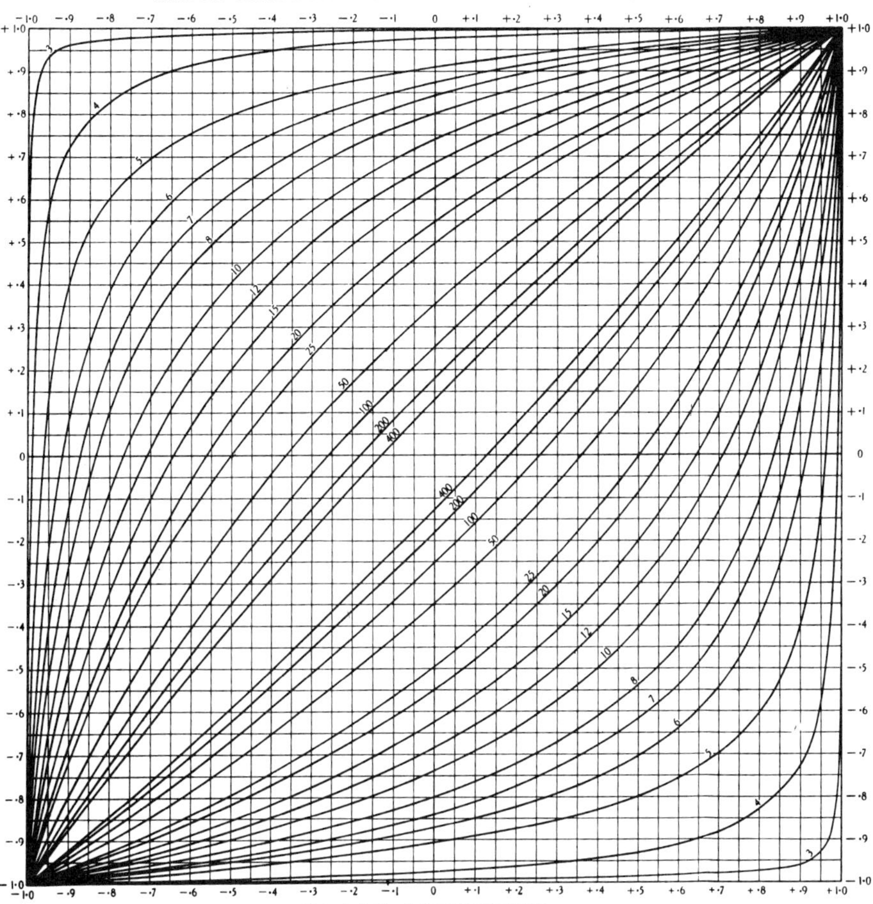

References

Abu Libdeh, O. (1984). *Strength of association in the simple general linear model: A comparative study of Hays' omega-squared.* Doctoral dissertation, The University of Chicago.

Adcock, C. J. (1960). A note on combining probabilities. *Psychometrika*, **25**, 303–305.

Ahlberg, J., Nilson, E., & Walsh, J. (1967). *The theory of splines and their application.* New York: Academic Press.

American College Testing Program. (1976–1977). *College student profiles: Norms for ACT assessment.* Iowa City, IA: ACT Publications.

Backman, M. E. (1972). Patterns in mental abilities: Ethnic, socioeconomic, and sex differences. *American Educational Research Journal*, **9**, 1–12.

Bahadur, R. R. (1965). An optimal property of the likelihood ratio statistic. In *Proceedings of the Fifth Berkeley Symposium on Mathematical Statistics and Probability.* Berkeley, CA: University of California Press.

Bahadur, R. R. (1967). Rates of convergence of estimates and test statistics. *Annals of Mathematical Statistics*, **38**, 303–324.

Bakan, D. (1966). The test of significance in psychological research. *Psychological Bulletin*, **66**, 432–437.

Barnett, V., & Lewis, T. (1978). *Outliers in statistical data.* Chichester, England: John Wiley & Sons.

Bayley, N., & Oden, M. (1955). The maintenance of intellectual ability in gifted adults. *Journal of Gerontology*, **10**, 91–107.

Becker, B. J. (1983, April). *Influence again: A comparison of methods for meta-analysis.* Paper presented at the annual meeting of the American Educational Research Association. Montreal.

Becker, B. J., & Hedges, L. V. (1984). Meta-analysis of cognitive gender differences: A comment on an analysis by Rosenthal and Rubin. *Journal of Educational Psychology*, **76**, 583–587.

Belsley, D. A., Kuh, E., & Welsch, R. E. (1980). *Regression diagnostics.* New York: John Wiley & Sons.

Berk, R. H., & Cohen, A. (1979). Asymptotically optimal methods of combining tests. *Journal of the American Statistical Association*, **74**, 812–814.

Bhattacharya, N. (1961). Sampling experiments on the combination of independent χ^2-tests. *Sankhyā*, **23A**, 191–196.

Bieri, J., Bradburn, W., & Galinsky, M. (1958). Sex differences in perceptual behavior. *Journal of Personality*, **26**, 1–12.

Birnbaum, A. (1954). Combining independent tests of significance. *Journal of the American Statistical Association*, **49**, 559–575.

Bloom, B. (1964). *Stability and change in human characteristics*. New York: John Wiley & Sons.

Blum, J. E., Fosshage, J. L., & Jarvik, L. F. (1972). Intellectual changes and sex differences in octogenarians: A twenty-year longitudinal study of aging. *Developmental Psychology*, **7**, 178–187.

Bogo, N., Winget, C., & Gleser, G. C. (1970). Ego defenses and perceptual styles. *Perceptual and Motor Skills*, **30**, 599–604.

Box, G. P., & Tiao, G. (1973). *Bayesian inference in statistical analysis*. Reading, MA: Addison-Wesley.

Bozarth, J. D., & Roberts, R. R. (1972). Signifying significant significance. *American Psychologist*, **27**, 774–775.

Campbell, D. T. (1969). Definitional versus multiple operationalism. *Et Al.*, **2**, 14–17.

Champney, T. F. (1983). *Adjustments for selection: Publication bias in quantitative research synthesis*. Doctoral dissertation, The University of Chicago.

Chan, T. F., Golub, G. H., & LeVeque, R. J. (1983). Algorithms for computing the sample variance: Analysis and recommendations. *The American Statistician*, **37**, 242–247.

Chase, L. J., & Chase, R. B. (1976). Statistical power analysis of applied psychological research. *Journal of Applied Psychology*, **61**, 234–237.

Chew, V. (1971). Point estimation of the parameter of the binomial distribution. *The American Statistician*, **25**, 47–50.

Clopper, C. J., & Pearson, E. S. (1934). The use of confidence or fiducial limits illustrated in the case of the binomial. *Biometrika*, **26**, 404–413.

Cochran, W. G. (1937). Problems arising in the analysis of a series of similar experiments. *Journal of the Royal Statistical Society* (*Suppl.*), **4**, 102–118.

Cochran, W. G. (1943). The comparison of different scales of measurement for experimental results. *Annals of Mathematical Statistics*, **14**, 205–216.

Cohen, J. (1962). The statistical power of abnormal-social psychological research: A review. *Journal of Abnormal and Social Psychology*, **65**, 145–153.

Cohen, J. (1969). *Statistical power analysis for the behavioral sciences*. New York: Academic Press.

Cohen, J. (1977). *Statistical power analysis for the behavioral sciences*, (2nd ed.). New York: Academic Press.

Cohen, P. A. (1981). Student ratings on instruction and student achievement: A meta-analysis of multisection validity studies. *Review of Educational Research*, **51**, 281–309.

Cook, T. D. (1974). The potential and limitations of secondary data analysis pp. 155–222. In *Educational evaluation: Analysis and responsibility* (Ed. by M. W. Apple, M. J. Subkoviak, & H. S. Lufter). Berkeley, CA: McCutchan.

Cook, T. D., & Weisberg, S. (1982). *Residuals and influence in regression*. London: Chapman & Hall.

Cooper, H. M. (1979). Statistically combining independent studies: Meta-analysis of sex differences in conformity research. *Journal of Personality and Social Psychology*, **37**, 131–146.

Cooper, H. M. (1982). Scientific guidelines for conducting integrative research reviews. *Review of Educational Research*, **52**, 291–302.

Cronbach, L. J. (1980). *Toward reform of program evaluation*. San Francisco: Jossey-Base.

David, F. N. (1934). On the P_{λ_η} test for randomness; remarks, further illustrations and table for P_{λ_η}. *Biometrika*, **26**, 1–11.

David, F. N. (1938). *Tables of the distribution of the correlation coefficient*. London: Biometrika Office.

Dempster, A. P., Laird, N. M., & Rubin, D. B. (1977). Maximum likelihood estimation from incomplete data via the EM algorithm. *Journal of the Royal Statistical Society, Series B*, **39**, 1–38.

Droege, R. C. (1967). Sex differences in aptitude maturation during high school. *Journal of Counseling Psychology*, **14**, 407–411.

Dunn, O. J. (1961). Multiple comparisons among means, *Journal of the American Statistical Association*, **56**, 52–64.

Eagly, A. H. (1978). Sex differences in influencability. *Psychological Bulletin*, **85**, 86–116.

Eagly, A. H., & Carli, L. L. (1981). Sex of researchers and sex typed communications as determinants of sex differences in influencability: A meta-analysis of social influence studies. *Psychological Bulletin*, **90**, 1–20.

Eberhardt, K. R. (1983), (Private communication.)

Edgeworth, F. Y. (1887). The choice of means. *Philosophical Magazine*, **24**, 268–271.

Edgington, E. S. (1972a). An additive method for combining probability values from independent experiments. *The Journal of Psychology*, **80**, 351–363.

Edgington, E. S. (1972b). A normal curve method for combining probability values from independent experiments. *The Journal of Psychology*, **82**, 85–89.

Erlenmeyer-Kimling, L., & Jarvik, L. F. (1963). Genetics and intelligence: A review. *Science*, **142**, 1477–1479.

Feldman-Summers, S., Montano, D. E., Kasprzyk, D., & Wagner, B. (1977). *Influence when competing views are gender related: Sex as credibility*. Unpublished manuscript, University of Washington, Seattle.

Fiebert, M. (1967). Cognitive styles in the deaf. *Perceptual and Motor Skills*, **24**, 319–329.

Fisher, R. A. (1915). Frequency distribution of the values of the correlation coefficient in samples from an indefinitely large population. *Biometrika*, **10**, 507–521.

Fisher, R. A. (1921). On the 'probable error' of a coefficient of correlation deduced from a small sample. *Metron*, **1**, 1–32.

Fisher, R. A. (1922). The goodness of fit of regression formulae and the distribution of regression coefficient. *Journal of the Royal Statistical Society*, **85**, 597–612. London: Oliver & Boyd.

Fisher, R. A. (1928). The general sampling distribution of the multiple correlation coefficient. *Proceedings of the Royal Statistical Society, Series A*, **121**, 654–673.

Fisher, R. A. (1932). *Statistical methods for research workers* (4th ed.). London: Oliver and Boyd.

Fisher, R. A., & Yates, F. (1957). *Statistical tables*. London: Oliver & Boyd.

Gage, N. L. (1978). *The scientific basis of the art of teaching*. New York: Teachers College Press.

Gates, A. I. (1961). Sex differences in reading ability. *Elementary School Journal*, **61**, 431–434.

George, E. O. (1977). *Combining independent one-sided and two-sided statistical tests—Some theory and applications*. Doctoral dissertation, University of Rochester.

George, E. O., & Mudholker, G. S. (1977). *On the convolution of logistic random variables*. Unpublished paper, University of Rochester, Rochester, NY.

Giaconia, R. M., & Hedges, L. V. (1982). Identifying features of effective open education. *Review of Educational Research*, **52**, 579–602.

Glass, G. V. (1976). Primary, secondary, and meta-analysis of research. *Educational Researcher*, **5**, 3–8.

Glass, G. V. (1978). Integrating findings: The meta-analysis of research. *Review of research in education*, (Ed. by L. S. Shulman) **5**, 351–379, Itasca, IL:F. E. Peacock.

Glass, G. V., & Hakstian, A. R. (1969). Measures of association in comparative experiments: Their development and interpretation. *Americal Educational Research Journal*, **6**, 403–414.

Glass, G. V., McGaw, B., & Smith, M. L. (1981). *Meta-analysis in social research*. Beverly Hills, CA: Sage Publications.

Glass, G. V., Peckham, P. D., & Sanders, J. R. (1972). Consequences of failure to meet assumptions underlying fixed effects analyses of variance and covariance. *Review of Educational Research*, **42**, 237–288.

Glass, G. V., & Smith, M. L. (1979). Meta-analysis of the relationship between class size and achievements. *Educational Evaluation and Policy Analysis*, **1**, 2–16.

Goldberger, A. S. (1964). *Econometric Theory*. New York: John Wiley & Sons.

Good, I. J. (1955). On the weighted combination of significance tests. *Journal of the Royal Statistical Society, Series B*, **17**, 264–265.

Greenwald, A. G. (1975). Consequences of prejudice against the null hypothesis. *Psychological Bulletin*, **82**, 1–20.

Gross, F. (1959). The role of set in perception of the upright. *Journal of Personality*, **27**, 95–103.

Gruen, A. (1955). The relation of dancing experience and personality to perception. *Psychological Monographs*, **69**.

Guttman, I., Manzefricke, U., & Tyler, D. (1984). Magnitudinal effects in the normal multivariate model. Unpublished manuscript.

Guttman, I., & Olkin, I. (1985). Statistical inference for constants of proportionality. In P. R. Krishnaiah (Ed.), *Multivariate Analysis VI*, Amsterdam: North-Holland.

Hall, J. A. (1978). Gender effects in decoding nonverbal cues. *Psychological Bulletin*, **85**, 845–857.

Harter, H. L. (1961). *Order statistics and their use in testing and estimation*. Washington, D.C.: U.S. Government Printing Office.

Hays, W. L. (1963). *Statistics*. New York: Holt, Rinehart, & Winston.

Hays, W. L. (1981). *Statistics* (3rd ed.). New York: Holt, Rinehart, and Winston.

Hedges, L. V. (1980). *Combining the results of experiments using different scales of measurement*. Doctoral dissertation, Stanford University, Stanford, CA.

Hedges, L. V. (1981). Distribution theory for Glass's estimator of effect size and related estimators. *Journal of Educational Statistics*, **6**, 107–128.

Hedges, L. V. (1982a). Fitting categorical models to effect sizes from a series of experiments. *Journal of Educational Statistics*, **7**, 119–137.

Hedges, L. V. (1982b). Estimating effect size from a series of independent experiments. *Phychological Bulletin*, **92**, 490–499.

Hedges, L. V. (1982c). Fitting continuous models to effect size data. *Journal of Educational Statistics*, **7**, 245–270.

Hedges, L. V. (1983a). Combining independent estimators in research synthesis. *British Journal of Mathematical and Statistical Psychology*, **36**, 123–131.

Hedges, L. V. (1983b). A random effects model for effect sizes. *Psychological Bulletin*, **93**, 388–395.

Hedges, L. V. (1984). Estimation of effect size under nonrandom sampling: The effects of censoring studies yielding statistically insignificant mean differences. *Journal of Educational Statistics*, **9**, 61–85.

Hedges, L. V., Giaconia, R. M., & Gage, N. L. (1981). *The empirical evidence on the effectiveness of open education*. Stanford, CA.: Stanford University School of Education.

Hedges, L. V., & Olkin, I. (1980). Vote-counting methods in research synthesis. *Psychological Bulletin*, **88**, 359–369.

Hedges, L. V., & Olkin, I. (1983a). Clustering estimates of effect magnitude from independent studies. *Psychological Bulletin*, **93**, 563–573.

Hedges, L. V., & Olkin, I. (1983b). Regression models in research synthesis. *The American Statistician*, **37**, 137–140.

Hedges, L. V., & Olkin, I. (1984). Nonparametric estimators of effect size in meta-analysis. *Psychological Bulletin*, **96**, 573–580.

Hedges, L. V., & Stock, W. (1983). The effects of class size: An examination of rival hypotheses. *American Educational Research Journal*, **20**, 63–85.

Heyl, P. R. (1930). Redetermination of the constant of gravitation. *National Bureau of Standards Journal of Research*, **5**, 1243–1290.

Hoaglin, D. C., & Welsch, R. E. (1978). The hat matrix in regression and ANOVA. *The American Statistician*, **32**, 17–22; Corrigenda. *The American Statistician*, **32**, 146.

Hotelling, H. (1925). The distribution of correlation ratios calculated from random data. *Proceedings of the National Academy of Sciences*, **11**, 657–662.

Hotelling, H. (1953). New light on the correlation coefficient and its transforms. *Journal of the Royal Statistical Society, Series B*, **15**, 193–232.

Householder, A. S. (1953). *Principles of numerical analysis*. New York: McGraw-Hill.

Humphreys, L. G. (1980). The statistics of failure to replicate: A comment on Buriels (1978) conclusions. *Journal of Educational Psychology*, **72**, 71–75.

Hunter, J. E., Schmidt, F. L., & Jackson, G. B. (1982). *Meta-analysis: Cumulating research findings across studies*. Beverly Hills, CA: Sage Publications.

Hyde, J. S. (1981). How large are cognitive gender differences? A meta-analysis using ω^2 and *d*. *American Psychologist*, **36**, 892–901.

International Mathematical and Statistical Libraries, Inc. (1977). *IMSL Library I*, 7th ed. Houston: Author.

Jackson, G. B. (1980). Methods for integrative reviews. *Review of Educational Research*, **50**, 438–460.

Jacobson, L. I., Berger, S. E., & Milham, J. (1970). Individual differences in cheating during a temptation period when confronting failure. *Journal of Personality and Social Psychology*, **15**, 48–56.

Johnson, D. W., Maruyama, G., Johnson, R., Nelson, D., & Skon, L. (1981). Effects of co-operative, competitive, and individualistic goal structures on achievement: A meta-analysis. *Psychological Bulletin*, **89**, 47–62.

Johnson, N. L., & Kotz, S. (1970). *Distributions in statistics: Continuous univariate distributions*—**2**. New York: John Wiley & Sons.

Johnson, N. L., & Welch, B. L. (1939). Applications of the noncentral *t*-distribution. *Biometrika*, **31**, 362–389.

Jones, L. V., & Fiske, D. (1953). Models for testing the significance of combined results. *Psychological Bulletin*, **50**, 375–382.

Kelley, T. L. (1935). An unbiased correlation ratio measure. *Proceedings of the National Academy of Science*, **21**, 554–559.

King, B. T. (1959). Relationships between susceptibility to opinion change and child rearing practices pp. 207–221. In *Personality and Persuasibility*, (Ed. by C. I. Hovland and I. L. Janis) New Haven CT: Yale University Press.

Koziol, J. A., & Perlman, M. D. (1978). Combining independent chi-square tests. *Journal of the American Statistical Association*, **73**, 753–763.

Kraemer, H. C. (1974). Improved approximation to the non-null distribution of the correlation coefficient. *Journal of the American Statistical Association*, **68**, 1004–1008.

Kraemer, H. C. (1975). On estimation and hypothesis testing problems for correlation coefficients. *Psychometrika*, **40**, 473–485.

Kraemer, H. C. (1979). Tests of homogeneity of independent correlation coefficients. *Psychometrika*, **44**, 329–335.

Kraemer, H. C. (1983). Theory of estimation and testing of effect sizes: Use in meta-analysis. *Journal of Educational Statistics*, **8**, 93–101.

Kraemer, H. C., & Andrews, G. (1982). A non-parametric technique for meta-analysis effect size calculation. *Psychological Bulletin*, **91**, 404–412.

Kraemer, H. C., & Paik, M. (1979). A central *t* approximation to the noncentral *t* distribution. *Technometrics*, **21**, 357–360.

Kraft, C. H., Olkin, I., & van Eeden, C. (1972). Estimation and testing for differences in magnitude or displacement in the mean vectors of two multivariate normal populations. *Annals of Mathematical Statistics*, **3**, 455–467.

Krauth, J. (1983). Nonparametric effect size estimation: A comment on Kraemer and Andrews. *Psychological Bulletin*, **94**, 190–192.

Krishnan, M. (1972). Series representations of a bivariate singly noncentral *t*-distribution. *Journal of the American Statistical Association*, **67**, 228–231.

Kulik, J. A., Kulik, C. C., & Cohen, P. A. (1979). A meta-analysis of outcome studies of Keller's personalized system of instruction. *American Psychologist*, **34**, 307–318.

Lancaster, H. O. (1949). The combination of probabilities arising from data in discrete distributions. *Biometrika*, **36**, 370–382.

Lancaster, H. O. (1961). The combination of probabilities: An application of orthonormal functions. *Australian Journal of Statistics*, **3**, 20–33.

Lane, D. M., & Dunlap, W. P. (1978). Estimating effect size: Bias resulting from the significance criterion in editorial decisions. *British Journal of Mathematical and Statistical Psychology*, **31**, 107–112.

Laubscher, N. F. (1960). Normalizing the noncentral *t* and *F* distributions. *Annals of Mathematical Statistics*, **31**, 1105–1112.

Laughlin, P. R., Branch, L. G., & Johnson, H. H. (1969). Individual versus triadic performance on a unidimensional complementary task as a function of initial ability level. *Journal of Personality and Social Psychology*, **12**, 144–150.

Light, R. J. (1983). *Evaluation studies review annual* (Vol. 8). Beverly Hills: Sage Publications.

Light, R. J. & Pillemer, D. B. (1982). Numbers and narrative: Combining their strengths in research reviews. *Harvard Educational Review*, **52**, 1–26.

Light, R. J., & Smith, P. V. (1971). Accumulating evidence: Procedures for resolving contradictions among different research studies. *Harvard Educational Review*, **41**, 429–471.

Lipták, T. (1958). On the combination of independent tests. *Magyar Tudományos Akadémia Matematikai Kutató Intezetenek Kozlemenyei*, **3**, 1971–1977.

Littell, R. C., & Folks, J. L. (1971). Asymptotic optimality of Fisher's method of combining independent tests. *Journal of the American Statistical Association*, **66**, 802–806.

Littell, R. C. & Folks, J. L. (1973). Asymptotic optimality of Fisher's method of combining independent tests II. *Journal of American Statistical Association*, **68**, 193–194.

Lord, F. M., & Novick, M. R. (1968). *Statistical theories of mental test scores*, with contributions by A. Birnbaum. Reading, MA: Addison-Wesley.

Maccoby, E. E., & Jacklin, C. N. (1974). *The psychology of sex differences*, Stanford, CA: Stanford University Press.

Matarazzo, J. D. (1972). *Wechsler's measurement and appraisal of adult intelligence* (5th ed.). Baltimore: Williams & Wilkins.

Melton, A. W. (1962.) Editorial. *Journal of Experimental Psychology*, **64**, 553–557.

Mendelsohn, G. A., & Griswold, B. B. (1966). Assessed creative potential, vocabulary level, and sex as predictors of the use of incidental cues in verbal problem solving. *Journal of Personality and Social Psychology*, 4, 423–431.

Miller, R. G., Jr. (1981). *Simultaneous statistical inference* (2nd ed.). New York: Springer-Verlag.

Molenaar, W. (1970). Approximations to the Poisson, binomial and hypergeometric distributions. Amsterdam: Mathematical Center Tracts.

Morf, M. E., & Howitt, R. (1970). Rod-and-frame test performance as a function of momentary arousal. *Perceptual and Motor Skills*, **31**, 703–733.

Morf, M. E., Kavanaugh, R. D., & McConville, M. (1971). Intratest and sex differences on a portable rod-and-frame test. *Perceptual and Motor Skills*, **32**, 727–733.

Morrison, D. F. (1976). *Multivariate statistical methods.* New York: McGraw-Hill.

Mosteller, F., & Bush, R. R. (1954). Selected quantitative techniques pp. 289–334. In *Handbook of social psychology* (Ed. by G. Lindzey). Cambridge, MA: Addison-Wesley.

Mudholkar, G. S., & George, E. O. (1979). The logit method for combining probabilities. In J. Rustagi (Ed.). *Symposium on optimizing methods in statistics* (pp. 345–366). New York: Academic Press.

Nash, S. C. (1975). The relationship among sex-role stereotyping, sex-role preference, and the sex difference in spatial visualization. *Sex Roles*, **1**, 15–32.

Neyman, J., & Scott, E. L. (1948). Consistent estimates based on partially consistent observations. *Econometrica*, **16**, 1–32.

Olkin, I. (1967). Correlations revisited, 102–128. In *Improving experiments: Design and statistical analysis*, J. Stanley (Ed.). Seventh Annual Phi Delta Kappa Symposium on Educational Research, Chicago: Rand McNally.

Olkin, I. (1973). *Do positive population correlation coefficients yield positive sample correlation coefficients*? Technical Report No. 73, Stanford University, Department of Statistics, Stanford, CA.

Olkin, I., & Pratt, J. W. (1958). Unbiased estimation of certain correlation coefficients. *Annals of Mathematical Statistics*, **29**, 201–211.

Olkin, I., & Siotani, M. (1965). Testing for the equality of correlation coefficients for various multivariate models. Technical Report, Stanford University, Stanford, CA.

Olkin, I., & Sylvan, M. (1977). Correlational analysis when some variances and covariances are known. pp. 175–191. In *Multivariate analysis—IV*, (Ed. by P. R. Krishnaiah) Amsterdam: North-Holland.

Oltman, P. R. (1968). A portable rod-and-frame apparatus. *Perceptual and Motor Skills*, **26**, 503–506.

Oosterhoff, J. (1969). *Combination of one-sided statistical tests.* Amsterdam: Mathematical Center Tracts.

Orwin, R. G. (1983). A fail-safe *N* for effect size in meta-analysis. *Journal of Educational Statistics*, **8**, 157–159.

Pape, E. S. (1972). A combination of *F*-statistics. *Technometrics*, **14**, 89–99.

Pearson, E. S. (1950). On questions raised by the combination of tests based on discontinuous distributions. *Biometrika*, **37**, 383–398.

Pearson, K. (1896). Mathematical contributions to the theory of evolution, III. Regression, heredity and panmixia. *Philosophical Transactions of the Royal Society of London, Series A*, **187**, 253–318. [Reprinted in Karl Pearson's Early Statistical Papers (1948), pp. 113–178. Cambridge: Cambridge University Press.]

Pearson, K. (1905). Mathematical contributions to the theory of evolution XIV. On the general theory of skew correlation and non-linear regression. *Drapers' Company Research Memoirs, Biometric Series, II*, **54**. [Reprinted, p. 477–528 in Karl Pearson's Early Statistical Papers (1948) pp. 477–528. Cambridge: Cambridge University Press.]

Pearson, K. (1933). On a method of determining whether a sample of given size *n* supposed to have been drawn from a parent population having a known probability integral has probably been drawn at random. *Biometrika*, **25**, 379–410.

Pillemer, D. B., & Light, R. J. (1980). Synthesizing outcomes: How to use research evidence from many studies. *Harvard Educational Review*, **50**, 176–195.

Presby, S. (1978). Overly broad categories obscure important differences. *American Psychologist*, **33**, 514–515.

Resnikoff, G. J., & Lieberman, G. J. (1957). *Tables of the noncentral t-distribution.* Stanford, CA: Stanford University Press.

Robbins, H. E. (1948). The distribution of a definite quadratic form. *Annals of Mathematical Statistics*, **19**, 266–270.

Rocke, D. M., Downs, G. W., & Rocke, A. J. (1982). Are robust estimators really necessary? *Technometrics*, **24**, 95–101.

Rosenthal, R. (1978). Combining results of independent studies. *Psychological Bulletin*, **85**, 185–193.

Rosenthal, R. (1979). The "file drawer problem" and tolerance for null results. *Psychological Bulletin*, **86**, 638–461.

Rosenthal, R., & Rubin, D. B. (1979a). Comparing significance levels of independent studies. *Psychological Bulletin*, **86**, 1165–1168.

Rosenthal, R., & Rubin, D. B. (1979b). A note on percent variance explained as a measure of the importance of effects. *Journal of Applied Social Psychology*, **9**, 395–396.

Rosenthal, R., & Rubin, D. B. (1982a). Further meta-analytic procedures for assessing cognitive gender differences. *Journal of Educational Psychology*, **74**, 708–712.

Rosenthal, R., & Rubin, D. B. (1982b). Comparing effect sizes of independent studies. *Psychological Bulletin*, **92**, 500–504.

Rosenthal, R., & Rubin, D. B. (1982c). A simple, general purpose display of magnitude of experimental effect. *Journal of Educational Psychology*, **74**, 166–169.

Sakoda, J. M., Cohen, B. H., & Beall, G. (1954). Test of significance for a series of statistical tests. *Psychological Bulletin*, **51**, 172–175.

Sampson, E. E., & Hancock, F. T. (1967). An examination of the relationship between ordinal position, personality, and conformity. *Journal of Personality and Social Psychology*, **5**, 398–407.

Sarhan, A. S., & Greenberg, B. G. (1962). *Contributions to order statistics*. New York: John Wiley & Sons.

Schaafsma, W. (1968). A comparison of the most stringent and the most stringent somewhere most powerful test for certain problems with restricted alternative. *Annals of Mathematical Statistics*, **39**, 531–546.

Scheffé, H. (1953). A method for judging all contrasts in the analysis of variance. *Biometrika*, **40**, 87–104.

Scheffé, H. (1959). *The analysis of variance*. New York: John Wiley & Sons.

Schwartz, D. W., & Karp, S. (1967). Field dependence in a geriatric population. *Perceptual and Motor Skills*, **24**, 495–504.

Sherman, J. (1978). *Sex related cognitive differences*. Springfield, Ill: Charles C. Thomas.

Shrout, P. E., & Fleiss, J. L. (1979). Intraclass correlation: Uses in assessing interrater reliability. *Psychological Bulletin*, **86**, 420–428.

Sidman, M. (1960). *Tactics of scientific research: Evaluating experimental data in psychology*. New York: Basic Books.

Sistrunk, F. (1971). Negro–white comparisons in social conformity. *Journal of Social Psychology*, **85**, 77–85.

Sistrunk, F. (1972). Masculinity–femininity and conformity. *Journal of Social Psychology*, **87**, 161–162.

Sistrunk, F., & McDavid, J. W. (1971). Sex variable in conformity behavior. *Journal of Personality and Social Psychology*, **17**, 200–207.

Smith, M. L., & Glass, G. V. (1977). Meta-analysis of psychotherapy outcome studies. *American Psychologist*, **32**, 752–760.

Smith, M. L., & Glass, G. V. (1980). Meta-analysis of class size and its relationship to attitudes and instruction. *American Educational Research Journal*, **17**, 419–433.

Speake, C. C. (1983). A beam balance method for determining the Newtonian constant of gravitation. Doctral dissertation, University of Cambridge.

Stafford, R. E. (1961). Sex differences in spatial visualization as evidence of sex-linked inheritance. *Perceptual and Motor Skills*, **13**, 428.

Stein, C. (1964). Inadmissibility of the usual estimator for the variance of a normal distribution with unknown mean. *Annals of the Institute of Statistical Mathematics*, **16**, 155–160.

Sterling, T. C. (1959). Publication decisions and their possible effects on inferences drawn from tests of significance—or vice versa. *Journal of the American Statistical Association*, **54**, 30–34.

Stigler, S. M. (1973). Simon Newcomb, Percy Daniel, and the history of robust estimation, 1885–1920. *Journal of the American Statistical Association*, **68**, 872–879.

Stigler, S. M. (1977). Do robust estimators work with *real* data? *Annals of Statistics*, **5**, 1055–1098.

Stouffer, S. A., Suchman, E. A., DeVinney, L. C., Star, S. A., & Williams, R. M., Jr. (1949). *The American soldier, Volume I. Adjustment during Army life*. Princeton, NJ: Princeton University Press.

Strube, M. J. (1981). Meta-analysis and cross-cultural comparison: Sex differences in child competitiveness. *Journal of Cross Cultural Psychology*, **12**, 3–20.

Tippett, L. H. C. (1931). *The method of statistics*. London: Williams & Norgate.

Tukey, J. W. (1949). Comparing individual means in the analysis of variance. *Biometrics*, **5**, 99–114.

Tukey, J. W. (1977). *Exploratory data analysis*. Reading, Mass.: Addison-Wesley.

Uguroglu, M. E., & Walberg, H. J. (1979). Motivation and achievement: A quantitative synthesis. *American Educational Research Journal*, **16**, 375–390.

Very, P. S. (1967). Differential factor structures in mathematical ability. *Genetic Psychology Monographs*, **75**, 169–207.

Viana, M. A. G. (1980). Statistical methods for summarizing independent correlational results. *Journal of Educational Statistics*, **5**, 83–104.

Walberg, H. J. (1969). Physics, femininity, and creativity. *Developmental Psychology*, **1**, 47–54.

Walker, G. A., & Saw, J. G. (1978). The distribution of linear combinations of *t*-variables. *Journal of the American Statistical Association*, **73**, 876–878.

Wallis, W. A. (1942). Compounding probabilities from independent significance tests. *Econometrica*, **10**, 229–248.

Wilkinson, B. (1951). A statistical consideration in psychological research. *Psychological Bulletin*, **48**, 156–158.

Williams, P. A., Haertel, E. H., Haertel, G. D., & Walberg, H. J. (1982). The impact of leisure-time television on school learning: A research synthesis. *American Educational Research Journal*, **19**, 19–50.

Willoughby, R. H. (1967). Field-dependence and locus of control. *Perceptual and Motor Skills*, **24**, 671–672.

Winer, B. J. (1971). *Statistical principles of experimental design* (2nd ed.). New York: McGraw-Hill.

Wishart, J. A. (1932). Note on the distribution of the correlation ratio. *Biometrika*, **24**, 441–456.

Wright, B. W., & Stone, M. H. (1979). *Best test design*. Chicago: Mesa Press.

Wyer, R. S., Jr. (1966). Effects of incentive to perform well, group attraction, and group acceptance on conformity in a judgmental task. *Journal of Personality and Social Psychology*, **4**, 21–26.

Wyer, R. S., Jr. (1967). Behavioral correlates of academic achievement: Conformity under achievement-affiliation-incentive conditions. *Journal of Personality and Social Psychology*, **6**, 255–263.

Yates, F., & Cochran, W. G. (1938). The analysis of groups of experiments. *Journal of Agricultural Science*, **28**, 556–580.

Zelen, M. (1957). The analysis of incomplete block designs. *Journal of the American Statistical Association*, **52**, 204–217.

Zelen, M., & Joel, L. S. (1959). On the weighted compounding of two independent significance tests. *Annals of Mathematical Statistics*, **30**, 885–895.

Selected Bibliography of Meta-Analytic Studies

Abrami, P. C., Leventhal, L., & Perry, R. P. (1982). Educational seduction. *Review of Educational Research*, **52**, 446–464.

Anderson, R. D., Kahl, S. R., Glass, G. V., & Smith, M. L. (1983). Science education: A meta-analysis of major questions. *Journal of Research in Science Teaching*, **20**, 379–385.

Andrews, C., Guitar, B., & Howie, P. (1980). Meta-analysis of the effects of stuttering treatment. *Journal of Speech & Hearing Disorders*, **45**, 287–307.

Baker, S. B., & Popowicz, C. L. (1983). Meta-analysis as a strategy for evaluating effects of career education interventions. *Vocational Guidance Quarterly*, **31**, 178–186.

Barclay, L. K. (1983). Using Spanish as the language of instruction with Mexican American Head Start children: A re-evaluation using meta-analysis. *Perceptual & Motor Skills*, **56**, 359–366.

Bassoff, E. S., & Glass, G. V. (1982). The relationship between sex roles and mental health: A meta-analysis of twenty-six studies. *Counseling Psychologist*, **10:4**, 105–112.

Beaman, A. L., Cole, C. M., Preston, M., Klentz, B., & Steblay, N. M. (1983). 15 years of foot in the door research: A meta-analysis. *Personality and Social Psychology Bulletin*, **9**, 181–196.

Bergin, A. E. (1983). Religiosity and mental health: A critical reevaluation and meta-analysis. *Professional Psychology: Research and Practice*, **14**, 170–184.

Blanchard, E. B., *et al.* (1980). Migraine and tension headache: A meta-analytic review. *Behavior Therapy*, **11**, 613–631.

Bucknam, R. B., & Brand, S. G. (1983). EBCE really works: A meta-analysis on experience based career education. *Educational Leadership*, **40:6**, 66–71.

Burger, J. M. (1981). Motivational biases in the attribution of responsibility for an accident: A meta-analysis of the defensive-attribution hypothesis. *Psychological Bulletin*, **90**, 496–512.

Carlberg, C., and Kàvale, K. (1980). The efficacy of special versus regular class placement for exceptional children: A meta-analysis. *Journal of Special Education*, **14**, 295–309.

Cohen, P. A. (1980). Effectiveness of student rating feedback for improving college instruction: A meta-analysis of findings. *Research in Higher Education*, **13**, 321–341.

Cohen, P. A. (1981). Student ratings of instruction and student achievement: A meta-analysis of multisection validity studies. *Review of Educational Research*, **51**, 281–309.

Cohen, P. A. (1982). Validity of student ratings in psychology courses: A research synthesis. *Teaching of Psychology*, **9**, 78–82.

Cohen, P. A., Ebeling, B. J., & Kulik, J. A. (1981). A meta-analysis of outcome studies of visual-based instruction. *Educational Communication & Technology*, **29**, 26–36.

Cohen, P. A., Kulik, J. A., and Kulik, C. C. (1982). Education outcomes of tutoring: A meta-analysis of findings. *American Educational Research Journal*, **19**, 237–248.

Cooper, H. M. (1979). Statistically combining independent studies: A meta-analysis of sex-differences in conformity research. *Journal of Personality and Social Psychology*, **37**, 131–146.

Crain, R. L., & Mahard, R. E. (1983). The effect of research methodology on desegregation-achievement studies: A meta-analysis. *American Journal of Sociology*, **88**, 839–854.

Dekkers, J., & Donatti, S. (1981). The integration of research studies on the use of simulation as an instructional strategy. *Journal of Educational Research*, **74**, 424–427.

DerSimonian, R., & Laird, N. M. (1983). Evaluating the effect of coaching on SAT scores: A meta-analysis. *Harvard Educational Review*, **53**, 1–15.

Druva, C. A., & Anderson, R. D. (1983). Science teacher characteristics by teacher behavior and by student outcome: A meta-analysis of research. *Journal of Research in Science Teaching*, **20**, 467–479.

Dusek, J. B., & Joseph, G. (1983). The bases of teacher expectancies: A meta-analysis. *Journal of Educational Psychology*, **75**, 327–346.

Eagly, A. H., & Carli, L. L. (1981). Sex of researchers and sex-typed communications as determinants of sex differences in influencability: A meta-analysis of social influence studies. *Psychological Bulletin*, **90**, 1–20.

Epstein, L. H., & Wing, R. R. (1980). Aerobic exercise and weight. *Addictive Behaviors*, **5**, 371–388.

Feltz, D. L., & Landers, D. M. (1983). The effects of mental practice on motor skill learning and performance: A meta-analysis. *Journal of Sport Psychology*, **5**, 25–57.

Fisher, C. D., & Gitelson, R. (1983). A meta-analysis of the correlates of role conflict and ambiguity. *Journal of Applied Psychology*, **68**, 320–333.

Fleming, M. L., & Malone, M. R. (1983). The relationship of student characteristics and student performance in science as viewed by meta-analysis research. *Journal of Research in Science Teaching*, **20**, 481–495.

Hansford, B. C., & Hattie, J. A. (1982). The relationship between self and achievement/performance measures. *Review of Educational Research*, **52**, 123–142.

Hattie, J. A., & Hansford, B. C. (1982). Self measures and achievement: Comparing a traditional review of literature with a meta-analysis. *Australian Journal of Education*, **26**, 71–75.

Horak, V. M. (1981). A meta-analysis of research findings on individualized instruction in mathematics. *Journal of Educational Research*, **74**, 249–253.

Hyde, J. S. (1981). How large are cognitive gender differences: A meta-analysis using omega and *d*. *American Psychologist*, **36**, 892–901.

Inglis, J., & Lawson, J. S. (1982). A meta-analysis of sex differences in the effects of unilateral brain damage on intelligence test results. *Canadian Journal of Psychology*, **36**, 670–683.

Iverson, B. K., & Levy, S. R. (1982). Using meta-analysis in health education research. *Journal of School Health*, **52**, 234–239.

Johnson, D. W., *et al.* (1981). Effects of cooperative, competitive, and individualistic goal structures on achievement: A meta-analysis. *Psychological Bulletin*, **89**, 47–62.

Johnson, D. W., & Johnson, R. T. (1983). Interdependence and interpersonal attraction among heterogeneous and homogeneous individuals: A theoretical formulation and a meta-analysis of the research. *Review of Educational Research*, **53**, 5–54.

Kavale, K. (1980). Auditory-visual integration and its relationship to reading achievement: A meta-analysis. *Perceptual & Motor Skills*, **51**, 947–955.

Kavale, K. (1981). The relationship between auditory perceptual skills and reading ability: A meta-analysis. *Journal of Learning Disabilities*, **14**, 539–546.

Kavale, K. (1981). Functions of the Illinois Test of Psycholinguistic Abilities (ITPA): Are they trainable? *Exceptional Children*, **47**, 496–510.

Kavale, K. (1982). Meta-analysis of the relationship between visual perceptual skills and reading achievement. *Journal of Learning Disabilities*, **15**, 42–51.

Kavale, K. (1982). Psycholinguistic training programs: Are there differential treatment effects? *Exceptional Child*, **29**, 21–30.

Kavale, K. (1982). The efficacy of stimulant drug treatment for hyperactivity: A meta-analysis. *Journal of Learning Disabilities*, **15**, 280–289.

Kavale, K. A., & Forness, S. R. (1983). Hyperactivity and diet treatment: A meta-analysis of the Feingold hypothesis. *Journal of Learning Disabilities*, **16**, 324–330.

Kavale, K. A. & Glass, G. V. (1982). The efficacy of special education interventions and practices: A compendium of meta-analysis findings. *Focus on Exceptional Children*, **15**, 1–14.

Kavale, K. A., & Mattson, P. D. (1983). One jumped off the balance beam: Meta-analysis of perceptual motor training. *Journal of Learning Disabilities*, **16**, 165–173.

Kulik, C. C., & Kulik, J. A. (1982). Effects of ability grouping on secondary school students: A meta-analysis of evaluation findings. *American Educational Research Journal*, **19**, 415–428.

Kulik, C. C., Kulik, J. A., & Cohen, P. A. (1979). A meta-analysis of outcome studies of Keller's personalized system of instruction. *American Psychologist*, **34**, 307–318.

Kulik, C. C., Kulik, J. A., & Cohen, P. A. (1980). Instructional technology and college teaching. *Teaching of Psychology*, **7**, 199–205.

Kulik, C. C., Kulik, J. A., & Shwalb, B. J. (1983). College programs for high-risk and disadvantaged-students: A meta-analysis of findings. *Review of Educational Research*, **53**, 397–414.

Kulik, C. C., Schwalb, B. J., & Kulik, J. A. (1982). Programmed instruction in secondary education: A meta-analysis of evaluation findings. *Journal of Educational Research*, **75**, 133–138.

Kulik, J. A., Kulik, C. C., & Cohen, P. A. (1980). Effectiveness of computer-based college teaching: A meta-analysis of findings. *Review of Educational Research*, **50**, 525–544.

Kulik, J. A., Bangert, R. L., & Williams, G. W. (1983). Effects of computer-based teaching on secondary school students. *Journal of Educational Psychology*, **75**, 19–26.

Luiten, J., Ames, W., & Ackerson, G. (1980). A meta-analysis of the effects of advance organizers on learning and retention. *American Educational Research Journal*, **17**, 211–218.

Mabe III, P. A., & West, S. G. (1982). Validity of self-evaluation of ability: A review and meta-analysis. *Journal of Applied Psychology*, **67**, 280–296.

Mullen, B., & Suls, J. (1982). The effectiveness of attention and rejection as coping styles: A meta-analysis of temporal differences. *Journal of Psychosomatic Research*, **26**, 43–49.

Parker, K. (1983). A meta-analysis of the reliability and validity of the Rorschach. *Journal of Personality Assessment*, **47**, 227–231.

Redfield, D. L., & Rousseau, E. W. (1981). A meta-analysis of experimental research on teacher questioning behavior. *Review of Educational Research*, **51**, 237–245.

Remmer, A. M., & Jernstedt, G. C. (1982). Comparative effectiveness of simulation games in secondary and college level instruction: A meta-analysis. *Psychological Reports*, **51**, 742.

Shapiro, D. A., & Shapiro, D. (1982). Meta-analysis of comparative therapy outcome studies: A replication and refinement. *Psychological Bulletin*, **92**, 581–604.

Shapiro, D. A., & Shapiro, D. (1982). Meta-analysis of comparative therapy outcome research A critical appraisal. *Behavioral Psychotherapy*, **10**, 4–25.

Shapiro, D. A., & Shapiro, D. (1983). Comparative therapy outcome research: Methodological implications of meta-analysis. *Journal of Consulting and Clinical Psychology*, **51**, 42–53.

Smith, M. L., & Glass, G. V. (1977). Meta-analysis of psychotherapy outcome studies. *American Psychologist*, **32**, 752–760.

Smith, M. L., & Glass, G. V. (1980). Meta-analysis of research on class size and its relationship to attitudes and instruction. *American Educational Research Journal*, **17**, 419–433.

Sparling, P. B. (1980). A meta-analysis of studies comparing maximal oxygen-uptake in men and women. *Research Quarterly for Exercise and Sport*, **51**, 542–552.

Steinbrueck, S. M., Maxwell, S. E., & Howard, G. S. (1983). A meta-analysis of psychotherapy and drug-therapy in the treatment of unipolar depression with adults. *Journal of Consulting and Clinical Psychology*, **51**, 856–863.

Stone, C. L. (1983). A meta-analysis of advance organizer studies. *Journal of Experimental Education*, **51**, 194–199.

Strube, M. J. (1981). Meta-analysis and cross-cultural comparison: Sex differences in child competitiveness. *Journal of Cross-Cultural Psychology*, **12**, 3–20.

Sweitzer, G. L., & Anderson, R. D. (1983). A meta-analysis of research on science teacher-education practices associated with inquiry strategy. *Journal of Research in Science Teaching*, **20**, 452–466.

Thurber, S., & Walker, C. E. (1983). Medication and hyperactivity: A meta-analysis, *Journal of General Psychology*, **108**, 79–86.

Trochim, W. M. (1982). Methodologically based discrepancies in compensatory education evaluations. *Evaluation Review*, **6**, 443–480.

Wampler, K. S. (1982). Bringing the review of literature into the age of quantification: Meta-analysis as a strategy for integrating research findings in family studies. *Journal of Marriage and the Family*, **44**, 1009–1023.

White, K. R. (1982). The relation between socioeconomic status and academic achievements. *Psychological Bulletin*, **91**, 461–481.

Whitley, B. E. (1983). Sex role orientation and self-esteem: A critical meta-analytic review. *Journal of Personality & Social Psychology*, **44**, 765–778.

Willett, J. B., Yamashita, J. J. M., & Anderson, R. D. (1983). A meta-analysis of instructional systems applied in science teaching. *Journal of Research in Science Teaching*, **20**, 405–417.

Willson, V. L. (1983). A meta-analysis of the relationship between science achievement and science attitude: Kindergarten through college. *Journal of Research in Science Teaching*, **20**, 839–850.

Willson, V. L., & Putnam, R. R. (1982). A meta-analysis of pretest sensitization effects in experimental design. *American Educational Research Journal*, **19**, 249–258.

Wise, K. C., & Okey, J. R. (1983). A meta-analysis of the effects of various science teaching strategies on achievement. *Journal of Research in Science Teaching*, **20**, 419–435.

Yeany, R. H., & Miller, P. A. (1983). Effects of diagnostic remedial instruction on science learnings: A meta-analysis. *Journal of Research in Science Teaching*, **20**, 19–26.

Author Index

Numbers in italics refer to the pages on which the complete references are listed.

A

Abu Libdeh, O., 103, *347*
Adcock, C. J., 45, *347*
Ahlberg, J., 295, *347*
Andrews, G., 92, 93, 94, 95, *351*

B

Backman, M. E., 17, 18, 19, 261, *347*
Bahadur, R. R., 33, *347*
Bakan, D., 287, *347*
Barnett, V., 257, *347*
Bayley, N., 18, *347*
Beall, G., 36, *354*
Becker, B. J., 16, 21, 171, 261, *347*
Belsley, D. A., 257, *347*
Berger, S. E., 17, *351*
Berk, R. H., 42, 43, *347*
Bhattacharya, N., 43, *347*
Bieri, J., 17, 18, *347*
Birnbaum, A., 32, *348*
Bloom, B., 6, 223, *348*
Blum, J. E., 18, 20, *348*
Bogo, N., 20, *348*
Box, G. P., 201, *348*
Bozarth, J. D., 286, *348*
Bradburn, W. M., 17, 18, *347*
Branch, L. G., 18, *352*
Bush, R. R., 27, 39, *353*

C

Campbell, D. T., 6, *348*
Carli, L. L., 21, 154, 286, *349*
Champney, T. F., 200, 202, 203, 307, *348*
Chan, T. F., 128, *348*
Chase, L. J., 286, *348*
Chase, R. B., 286, *348*
Chew, V., 69, 300, *348*
Clopper, C. J., 53, *348*
Cochran, W. G., 2, 327, *348, 355*
Cohen, A., 42, 43, *347*
Cohen, B. H., 36, *354*
Cohen, J., 7, 51, 52, 108, 286, *348*
Cohen, P. A., 148, 223, *348, 352*
Cook, T. D., 13, 258, *348*
Cooper, H. M., 14, 21, 45, *348*
Cronbach, L. J., 190, *348*

D

David, F. N., 42, 225, 228, *348*
Dempster, A. P., 200, 307, *349*
DeVinney, L. C., 39, *355*
Downs, G. W., 250, *354*
Droege, R. C., 17, 18, 19, *349*
Dunlap, W. P., 287, 290, 300, *352*
Dunn, O. J., 269, *349*

E

Eagly, A. H., 21, 154, 286, *349*
Eberhardt, K. R., 313, *349*
Edgeworth, F. Y., 250, *349*
Edgington, E. S., 27, 42, *349*
Erlenmeyer-Kimling, L., 272, *349*

F

Feldman-Summers, S., 22, *349*
Fiebert, M., 20, *349*
Fisher, R. A., 1, 27, 37, 44, 89, 102, 103, 225, 227, *349*
Fiske, D. W., 2, 27, *351*
Fleiss, J. L., 102, *354*
Folks, J. L., 33, *352*
Fosshage, J. L., 18, 203, *348*

G

Gage, N. L., 23, 26, 274, 302, *349, 350*
Galinsky, M. D., 17, 18, *347*
Gates, A. I., 18, *349*
George, E. O., 40, 41, 42, *349, 353*
Giaconia, R. M., 23, 148, 274, 302, *349, 350*
Glass, G. V., 6, 7, 9, 10, 11, 12, 13, 14, 57, 76, 78, 104, 108, 167, 189, 190, 205, *349, 350, 354*
Gleser, G. C., 20, *348*
Goldberger, A. S., 172, *350*
Golub, G. H., 128, *348*
Good, I. J., 38, *350*
Greenberg, B. G., 93, *354*
Greenwald, A. G., 285, 287, *350*
Griswold, B. B., 18, *352*
Gross, F., 20, *350*
Gruen, A., 20, *350*
Guttman, I., 321, 322, *350*

H

Haertel, E. H., 167, *355*
Haertel, G. D., 167, *355*
Hakstian, A. R., 104, *355*
Hall, J. A., 190, *350*
Hancock, F. T., 22, *354*
Harter, H. L., 283, *350*
Hays, W. L., 101, 103, *350*
Hedges, L. V., 16, 23, 50, 51, 67, 79, 81, 86, 89, 92, 112, 117, 118, 123, 124, 131, 148, 170, 175, 185, 190, 193, 224, 239, 261, 266, 274, 282, 287, 288, 291, 292, 302, 309, *347, 349, 350, 351*
Heyl, P. R., 311, 312, *351*
Hoaglin, D. C., 258, *351*
Hotelling, H., 102, 225, *351*
Householder, A. S., 302, *351*
Howitt, R., 20, *352*
Humphreys, L. G., 4, *351*
Hunter, J. E., 242, *351*
Hyde, J. S., 16, 262, *351*

J

Jacklin, C. N., 16, *352*
Jackson, G. B., 14, 242, *351*
Jacobson, L. I., 17, *351*
Jarvik, L. F., 18, 20, 272, *348, 349*
Joel, L. S., 38, *355*
Johnson, D. W., 190, *351*
Johnson, H. H., 18, *352*
Johnson, N. L., 39, 86, 91, *351*
Johnson, R., 190, *351*
Jones, L. V., 2, 27, *351*

K

Karp, S., 20, *354*
Kasprzyk, D., 22, *349*
Kavanaugh, R. D., 20, *353*
Kelley, T. L., 102, *351*
King, B. T., 22, *351*
Kotz, S., 39, 91, *351*
Koziol, J. A., 43, *351*
Kraemer, H. C., 89, 90, 92, 93, 94, 95, 121, 228, 234, 235, *351, 352*
Kraft, C. H., 321, *352*
Krauth, J., 92, *352*
Krishnan, M., 209, *352*

Subject Index

Numbers in italics refer to the pages on which a complete reference is listed.

A

Acceptance region, convex, 32
Accuracy, of large sample approximations, 174
Additive model
with a control, 323
for interlaboratory differences, 322
Admissibility, of a combined test, 31
Alternative hypotheses, 30
American College Testing Program, 17, 18, *347*
Analysis of variance, for effect sizes, 148
Attenuation, in effect sizes, 136

B

Bahadur efficiency, 33
Bayes estimate, of a proportion, 300
Bias, of effect size estimators, 83, 193
Bonferroni inequality, 162
in clustering, 279
for correlations, 239

C

Calibration, of instruments, 313
Categorical models
diagnostic procedures for, 257
fixed effects in, 150
Clustering procedures,
comparisons of, 281
for correlations, 271
disjoint, 266
effect of unequal sample sizes on, 276
effect sizes, 273
for normal variables, 266
overlapping, 266
Cognitive abilities, sex differences in, 16
Combined tests
admissibility of, 31
Edgington's method for, 42
Fisher's method for, 37
inverse normal method for, 39
Liptak's method for, 42
logit method for, 40
monotonicity of, 32
Pearson's method for, 32, 42

of significance, 2
summary of, 44
Tippett's method for, 32
Wilkinson's method for, 32
Combining estimates, of correlation coefficients, 262
Common factor, effect of validity on, 140
Comparisons
Scheffé's method for, 161
simultaneous, 160, 161
Computational formulas
for homogeneity test statistics, 127
for regressions, 262
Computations, using statistical packages, 173, 174, 241
Computer packages, 309
Confidence intervals
for effect size, 88, 91
for regression coefficients, 170
from vote counts, 53
Conformity, sex differences in, 21, 153–155, 158, 160, 161, 164, 171, 173
Consistency of results, 4
Convex acceptance region, 32
Correlated estimators, 210
pooling of, 221
Correlation coefficients,
as effect magnitude, 77, 101
bias of, 225
clustering procedures for, 271
combining, 223
combining estimates of, 229–234, 262
common, 229
confidence interval for, 227
distribution of, 225, 226
estimation of, 63
homogeneity of, 234
intraclass, 102
maximum likelihood estimator of, 232
product-moment, 7, 225
unbiased estimator of, 225
variance of, 225
weighted estimators of common, 230
z-transformation of, 120, 227
Correlation ratio, for effect magnitude, 101
Correlation variance component, unbiased estimate of, 244
Counting procedures, for estimating effect sizes, 298

D

Diagnostic procedures
for categorical models, 257
for homogeneous models, 251
for linear models, 257
Digamma function, 105
Displacement factor, 318
Distribution
of effect size, 288
of maximum likelihood estimator, 292

E

Edgington's method, for combined tests, 42
Effect magnitudes
consistency of, 4
correlation coefficient as, 77
defined by correlation coefficient, 101
defined by correlation ratio, 101
estimates of, 11
scale-free indices of, 6
Effect sizes, 7
analysis of variance for, 148
bias of estimators of, 83, 193, 286
clustering, 273
confidence intervals for, 88, 91
correlated, 210
counting procedures for estimating, 298
definition of sample, 9
disattenuated, 135, 136
distribution of, 288
estimates of, 57, 291
estimators, distribution of, 86
gain scores as estimates of, 97
homogeneity of, 122, 148, 213
interpretation of, 76
linear combinations of estimates of, 109
from a linear statistic, 129
maximum likelihood estimate of, 71, 81, 118, 131, 291, 301
mean, 198
mean-squared error of estimators of, 83
with missing data, 285
models, fitting of, 157–159
multivariate models for, 205
nonparametric estimators of, 93
proportions as estimates of, 95
random effects models for, 189
robust estimators of, 92

shrunken estimators of, 82, 303
significant, 287
from a single experiment, 76
with small sample sizes, 128
standardized mean difference for, 57, 78
testing for homogeneity of, 213
transformation of, 184, 274
transformed estimates of, 119
in truncated distributions, 290
unbiased estimators of, 81
variance components for, 195
variance of estimates of, 193
weighted estimators of, 302
Efficiency
Bahadur, 33
of tests, 33
of weighted estimators, 113
EM algorithm, 200
Empirical Bayes estimation, for random effects models, 200
Equivalence of drugs, 318
Errors of measurement, 134
Estimation
of a common correlation, 63
of effect sizes, 291
Estimators, pooling of correlated, 221
Expectation, of standardized mean difference, 79
Explanatory models, 14
Exponential family, 32

F

Fail-safe number, 306
File drawer problem, 305
Fisher's method, 2
for combined tests, 37
Fixed effects, categorical models, 150, 247

G

Gain scores, in estimating effect size, 97
Gamma function, convexity of, 105
Gaps, of order statistics, 267
Gender differences
in cognitive abilities, 16
in conformity, 21, 153–155, 158, 160, 161, 164, 171, 173
in field articulation ability, 20
in quantitative ability, 17
in verbal ability, 18
in visual-spatial ability, 19, 260, 261
Graphic methods, for detecting outliers, 251
Gravitational constant, measurements of, 312

H

Heterogeneity of variances, 11
Homogeneity
of effect sizes, 122, 148, 213
test for, 122, 153
test statistics, computation of, 127
Homogeneity statistics, for detecting outliers, 256
Hypotheses
alternative, 30
of no effects, 29

I

IMSL, 309, *351*
Intelligence, in twins, 272
Interaction, 8
Interlaboratory differences, additive model for, 322
Intraclass correlation coefficient, 102
Invalidity
corrections for, 141
effects of, 131
model for, 138
Inverse normal method, for combined tests, 39
Item response models, 70
Inverse probability distribution function, 42

J

Johnson–Welch approximation, 86

L

Large sample approximations, accuracy of, 86, 87, 114, 124, 125
Least squares procedures, 250
Likelihood ratio test, for homogeneity of correlations, 236
Linear models
diagnostic procedures for, 257
for fitting correlations, 237
for fitting effect sizes, 167
Linear statistic, for effect size, 129

Liptak's method, for combined tests, 42
Logit method, for combined tests, 40

M

Maximum likelihood estimates
  distribution of, 292
  of effect sizes, 71, 81, 118, 131, 291, 301
  of regression coefficients, 183
Mean difference, standardized, 57, 78
Mean effect size, estimation of, 198
Mean, of standardized mean differences, 79, 104
Mean-squared error
  of effect size estimators, 83
  of standardized mean differences, 104
Measurement error
  corrections for, 135
  effects of, 131, 134, 228
Meta-analysis, definition of, 13
Missing data, 200
Model specification
  testing for, 172, 240
Models, explanatory, 14
Monotonicity
  of a combined test, 32
  of Tippett's procedure, 46
Multiplicative models, 315
Multivariate models, for effect sizes, 205

N

Newton–Raphson procedures, 301
New York public schools, 19
Noncentral t-distribution, 62, 89, 209, 307
Nonparametric estimators
  comparison of, 99
  of effect sizes, 93
Nonparametric tests, 2, 28

O

Omega-squared index, 103
Omnibus tests, 2, 28
Open education
  effects of, 23
  effects on attitudes, 24, 194, 196, 197, 199, 217, 251, 254, 255, 256
  effects on student creativity, 25, 185, 303
  effects on student independence, 24, 274
  effects on student self-concept, 25
One-tailed tests, 29
Order statistics, in clustering, 266
Outliers, 248
  detection of, 253
  detection of, by specification statistics, 261
  in general linear models, 258
  graphic methods for, 251
  homogeneity statistics for, 256
  in measuring effect sizes, 93

P

P-values, 29
Pearson's method, 2
  for combined tests, 32, 42
Pooled data, 8
Pooled sample variances, 79
Pooled within-group variance, 62
Pooling, of correlated estimators, 221
Potency factor, 318
Power, of vote-counting procedures, 51
Primary analysis, definition of, 12
Project Talent, 17–19
Proportions
  Bayes estimate of, 300
  in estimating effect sizes, 95

R

Random effects models, 248
  for correlations, 242
  for effect sizes, 189
  empirical Bayes estimates in, 200
  for truncated distributions, 306
Regression coefficients
  confidence intervals for, 170
  estimating, 238
  maximum likelihood estimators of, 183
Regression models, 168
Regression, weighted least squares for, 169
Residuals
  for detecting outliers, 253
  standardized, 255
Response measures, validity in, 138

S

Sample sizes, 5
  effect of unequal, 276
  small, 128
Sampling bias, in effect size estimates, 286

SAS Proc GLM, 171, 173, 187, 240, 241
Scale of measurement, 6
Scaling, in multiplicative models, 321
Scheffé's method, for simultaneous comparisons, 161
Secondary analysis, definition of, 12
Senility study, 316
Sex differences
in cognitive abilities, 16
in conformity, 21, 153–155, 158, 160, 161, 164, 171, 173
in field articulation ability, 20
in quantitative ability, 17
in verbal ability, 18
in visual-spatial ability, 19, 260, 261
Shrinkage, of effect size estimates, 82, 303
Significance, of combined results, 27
Significance tests, combined, 2
Simulation studies, 175
Simultaneous comparisons, Scheffe's method for, 161
Specification error, 172, 240, 248
Specification statistics, to detect outliers, 261
Spline approximations, 296
Square-mean-root, for unequal sample sizes, 68, 276
Standardized mean differences
for effect size, 78
mean of, 79, 104
mean-squared error of, 104
variance of, 80, 104
Standardized residuals, 255
Statistical packages, for computations, 173, 174, 241, 261
Student achievement, 26

T

T-distribution, noncentral, 62, 89, 209, 307
Teacher indirectness, 26
Testing, for model specification, 172
Tests
of homogeneity, 122, 136, 142, 153, 234
one-tailed, 29
Tippett's method
for combined tests, 32
monotonicity of, 46
Transformations, of effect sizes, 184, 274
Transformed estimates, of effect size, 119
Trimming, of data, 250
Truncated distributions
effect sizes in, 290
random effects models in, 306
Twin study, 272
Type I error, 161, 286

U

Unbiased estimate
of correlation variance component, 244
of effect size, 81
Unequal sample sizes, square-mean-root for, 68, 276
Uniform distribution, 34

V

Validity
corrections for, 141
effect of, 138
Variances
heterogeneity of, 11
of effect size estimates, 193
of standardized mean differences, 80, 104
Variance-stabilizing transformation, 88, 317
of correlation coefficient, 120
of effect sizes, 119, 274
in test of homogeneity, 124, 235
Variance, systematic, 10
Vote-counting procedures
confidence intervals using, 53
description of, 47
inadequacy of, 48
limitations of, 67, 300
power of, 51
Vote-counts, confidence intervals using, 53

W

Wechsler Adult Intelligence Scale, 18
Weighted estimators
distribution of, 114
of effect sizes, 302
efficiency of, 113
Weighted least squares, for regression, 169
Wilkinson's method, for combined tests, 32

Z

z-transformation, of correlation coefficients, 120